Nonequilibrium Processes in Catalysis

Oleg V. Krylov, Prof., D.Sc.
Laboratory of Heterogeneous Catalysis
Institute of Chemical Physics
Moscow, Russia

Boris R. Shub, Prof., D.Sc.
Laboratory of Nonequilibrium Processes in Catalysis
Institute of Chemical Physics
Moscow, Russia

CRC Press
Boca Raton Ann Arbor London Tokyo

Library of Congress Cataloging-in-Publication Data

Krylov, O. V. (Oleg Valentinovich)
[Neravnovesnye protsessy v katalize. English]
Nonequilibrium processes in catalysis / Oleg V. Krylov, Boris R. Shub.
p. cm.
Includes bibliographical references and index.
ISBN 0-8493-4478-6
1. Catalysis. 2. Nonequilibrium thermodynamics. I. Shub, B. R. (Boris Ruvimovich) II. Title.
QD505.K7913 1993
541.3′95--dc20 93-9504
CIP

Direct all inquiries to CRC Press, Inc., 2000 Corporate Blvd., N.W., Boca Raton, Florida 33431.

International Standard Book Number 0-8493-4478-6
Library of Congress Card Number 93-9504
Printed in the United States of America 1 2 3 4 5 6 7 8 9 0
Printed on acid-free paper

FOREWORD

This book treats elementary surface processes and heterogeneous catalysis from the unique perspective of the authors. The Institute of Chemical Physics in Moscow has long been the center for research of gas-surface interactions and catalysis on oxide surfaces. The chapters reflect the knowledge gained by studies in both fields of research.

During the past twenty years, development of modern surface science techniques has permitted investigation of surfaces on the molecular level. The information gained through studies of energy transfer between incident gas molecules and the surface, using molecular beam-surface scattering, has provided a new view of elementary surface reaction processes that are also often important in heterogeneous catalytic reactions.

Surface science approaches to the understanding of catalysis have been gaining prominence worldwide. This book provides this important surface science view of catalysis from the perspective of nonequilibrium, irreversible processes.

G. A. Somorjai
Berkeley, California

PREFACE

The study of nonequilibrium processes in catalysis gives us an insight into the mechanism of catalytic transformations. For any catalytic reaction the following stages are necessary: the energy gain that is needed for the activation, the chemical conversion itself, and ultimately, the loss of excess energy by reaction products. The processes of activation and deactivation of solids are also of importance.

In the 1970s experimental and theoretical activity in this field began under our supervision at the Institute of Chemical Physics in Moscow. First results have been generalized in our collected volume, *Problems in Kinetics and Catalysis, Volume 17, Nonstationary and Nonequilibrium Processes in Heterogeneous Catalysis,* edited by O. V. Krylov and M. D. Shibanova, Nauka, Moscow, 1978 (in Russian).

Since that time, an increasing number of new methods have made their appearance in catalytic research. An extensive growth of publications in the last few years requires a new overview of this subject.

This book is devoted to experimental works and modern methods of investigating nonequilibrium processes in catalysis. The latter include the energy exchange between active species and a solid, and the appearance of nonequilibrium (''dissipative'') structures in the solid itself. The thermodynamics of irreversible processes is applied to analyze possible catalyst transformations.

The surface migration of adsorbed particles and its role in catalytic transformations, as well as chemoenergetic stimulation (i.e., the acceleration of a chemical reaction on account of excitation of intermediate reagents, being outputs of another reaction), are also considered in this book. We have supplemented the English edition with references up to 1992. The works performed at the Institute of Chemical Physics are covered in greatest detail.

Because the subject under consideration is, undoubtedly, at the origin of its development and there is no consensus on many questions, some results have not yet received an adequate explanation.

We hope that this book will be of interest to both academic and industrial scientists specializing in the field of adsorption and catalysis.

We are indebted for any critical remarks from our readers.

The authors are grateful to A. D. Berman, Ju. M. Gershenzon, V. F. Kiselev, S. A. Kovalevsky, M. A. Kozhushner, V. G. Kustarev, A. S. Prostnev, Ju. N. Rufov, M. E. Ryskin, and A. V. Skljarov for many stimulating discussions on different problems presented in this book.

O. V. Krylov
B. R. Shub
Moscow, Russia

THE AUTHORS

Oleg V. Krylov, D.Sc., is a professor and principal researcher at the Laboratory of Heterogeneous Catalysis, Institute of Chemical Physics, Russian Academy of Science, Moscow, Russia.

Professor Krylov graduated in 1947 from the Ivanovo Institute of Chemical Engineering, Ivanovo, Russia, and received his Ph.D. degree in 1951 from the Institute of Physical Chemistry, Moscow and his D.Sc. degree in 1964 from the Institute of Chemical Physics, Moscow.

Professor Krylov is a member of several scientific organizations: International Academy of Creative Endeavours, International Committee of the World Congresses on Catalytic Oxidation, Council of the International Congress on Catalysis, Presidium of the Council on Catalysis of Russia, and serves on the editorial boards of several scientific journals.

Professor Krylov has presented over 30 invited lectures at international meetings, over 50 invited lectures at national meetings, and many guest lectures at universities and institutes.

Professor Krylov has published more than 400 scientific papers and several books, many of them published in English. Among these titles are: O. V. Krylov, *Catalysis by Nonmetals. Rules for Catalyst Selection,* Academic Press, 1970; V. F. Kiselev and O. V. Krylov, *Adsorption Processes on Semiconductor and Dielectric Surfaces I,* Springer, 1984; V. F. Kiselev and O. V. Krylov, *Electronic Phenomena in Adsorption and Catalysis on Semiconductors and Dielectrics,* Springer, 1987; and V. F. Kiselev and O. V. Krylov, *Adsorption and Catalysis on Transition Metals and Their Oxides,* Springer, 1989. His current major scientific interests are heterogeneous catalysis, kinetics of heterogeneous processes, ecology, chemistry of natural gas, and petroleum refining.

Boris R. Shub, D.Sc., Professor, is Head of the Laboratory of Nonequilibrium Processes in Catalysis at the Institute of Chemical Physics, Russian Academy of Sciences, Moscow, Russia.

Professor Shub graduated from the Moscow Institute of Fine Chemical Technology, Moscow, Russia, in 1962 and received his Ph.D. degree from the Institute of Chemical Physics, Moscow, in 1969 and D.Sc. degree from the Institute of Chemical Physics, Moscow, in 1983.

Professor Shub is a member of several scientific bodies including the Scientific Council for Chemical Kinetics and Structure, and the Scientific Council for Catalysis.

Professor Shub has presented over 10 invited lectures at international meetings, over 20 invited lectures at national meetings, and some guest lectures at universities and institutes.

Professor Shub has won the S. Z. Roginsky Prize, and has published more than 100 scientific papers. His current scientific interests include physics and chemistry of heterogeneous processes and catalysis, dynamics of elementary processes and surface spectroscopy, ecology, and aircraft thermal protection.

TABLE OF CONTENTS

Nonequilibrium Processes in Catalysis

Chapter 1

INTRODUCTION

Any chemical reaction always leads to disturbance of the equilibrium energy distribution function. There are two main reasons for that. The first one is connected with the decrease of a concentration of more energetic reagent particles because the rate of their consumption far exceeds the rate of their restoration.

The second reason is that energy release in the course of an exothermic chemical reaction results in disturbance of equilibrium energy distribution. This energy gives rise to translational, vibrational, rotational, and electronic product excitation, along with solid excitation.

Particularly in catalysis, due to gas-surface interaction, the second type of nonequilibrium takes place mostly because adsorption is almost always an exothermic process (chemisorption heat for most molecules is of the order of tens and hundreds of kilojoules per mole).

Because energy exchange processes between excited molecules and solids of different physical and chemical natures are very complex and diverse, most progress in this area may be achieved only on the basis of fundamental study of some principal points, among which are the detailed kinetic analysis of energy exchange and identification of microscopic mechanisms of heterogeneous molecular energy deactivation. A knowledge of elementary stages of energy exchange mechanism on the surface is necessary to solve the questions connected with nonequilibrium processes in the gas phase and to create coverages with given characteristics with respect to the energy exchange with the gas phase.

These problems are of great importance for fundamental purposes and have practical significance in heterogeneous catalysis, laser chemistry, heterogeneous atmospheric chemistry, plasma chemistry, heat protection of aircrafts, and microelectronics, as well as in many other areas of science and technology.

Deactivation of excited molecules is the necessary elementary stage of most surface processes. The main characteristic of energy exchange between gas phase and solid surface is the average thermal accommodation coefficient ϵ, introduced by Knudsen

$$\epsilon = \frac{E_{inc} - E_{ref}}{E_{inc} - E_{surf}} \tag{1.1}$$

where E_{inc} and E_{ref} are average energies of incident and reflected particles, and E_{surf} is the average energy of a particle being in thermal equilibrium with the surface.

Generally, the thermal accommodation coefficient includes three partial coefficients which characterize translational, rotational, and vibrational energy exchange. For equilibrium gas we have

$$\epsilon = \frac{\epsilon_t(C_t + R/2) + \epsilon_r C_r + \epsilon_v C_v}{C_t + C_r + C_v + R/2} \tag{1.2}$$

where ϵ_t, ϵ_r, and ϵ_v are partial accommodation coefficients for translational, rotational, and vibrational energies, respectively; C_t, C_r, and C_v are partial mole heat capacities; and R is the universal gas constant.

The most up-to-date experiments have been carried out under conditions wherein essentially the whole of the excitation energy was located either in the first vibrational level or in the longest-lived electronically excited metastable one. In this case, the accommodation coefficient has especially clear physical meaning, namely, the probability to lose the excitation quantum during one collision with the surface. The probability of generating an excited molecule during the exothermic step of the reaction depends strongly on the mechanism of its deactivation.

It is of interest to trace the history of the problem of excited particles and nonequilibrium processes in catalysis. The idea of taking advantage of the energy released at the exothermic step of a catalytic process for the acceleration of the same reaction was often discussed in literature, beginning with the papers of Langmuir, who had suggested the concept of precursor or preadsorbed state, back in 1929.[1] Later this proposal was supported in experiments of Taylor and Langmuir,[2] who showed that the sticking coefficient of Cs atoms on W surface remained close to unity even for almost monolayer coverage $\theta = 0.98$, when Cs atoms collided mainly with occupied sites.

As early as 1934, Zeldovitsch and Roginsky[3] treated in detail the assumed role of the precursor in catalysis. To explain the peculiarity of the catalytic oxidation of CO on MnO_2 they suggested the existence of an intermediate state of the CO molecule with high energy from which it either transformed to the chemisorbed state (inactive for catalysis) or was oxidated by oxygen with the formation of CO_2

$$\mathrm{CO\ (gas)} \rightarrow \mathrm{CO\ (prec)} + \mathrm{O_2} \rightarrow \begin{cases} \mathrm{CO\ (chem)} \\ \mathrm{CO_2\ (gas)} \end{cases}$$

From time to time, these ideas were later discussed in catalysis. In doing so, it was postulated that the adsorption, being essentially an exothermic process, was a stage of the chemical activation similar to that in the gas phase.

In catalysis, especially in selective catalysis, the energy localization in one of the bonds of the reacting molecule is required to cause its break. Thus, the ideas of the resonance between one or another energy level of a catalyst and a reacting molecule were put forward by many authors in their theories of catalysis.

Scheve and Schulz[4] studied the adsorption, CO oxidation, and N_2O decomposition on transition metal oxides; it has been shown that the maximum of catalytic activity is observed when the thermal activation energy of the charge carriers in solids is equal to the energy level of the adsorbed molecule. Gardner[5] correlated infrared vibrational frequencies of adsorbed molecules with the number of valence electrons in the metal-adsorbate and concluded that ionic forms of adsorbate with a certain charge had greater bond rupture probability. Among more recent publications on the energy levels resonance between the reacting molecule and the catalyst, the papers by Gagarin and Kolbanovsky[6] should be mentioned. Johnson et al.[7] have found a correlation between the oxidation rate of toluene on oxide catalysts and the vibrational frequency of the metal-oxygen bond. A similar correlation between the hydrogenation rate of ethylene and propylene and the vibrations of metal-hydrogen bond was observed,[8] the maximum of absorption being at $\nu \simeq 2000$ cm^{-1}.

Kobozev[9] has formulated the ''aggravation'' principle according to which, for a particular active site structure, the greater is the molecular weight of a homogeneous catalyst, the higher is its catalytic activity. This fact was explained by the ability of multiatom molecules or solids to remain in the excited state for a long time.

The similarity between some characteristic features of the catalysis and the chain reactions theory was pointed out in some works.[10-14] In particular, it was shown that the stages of chain initiation, growth, and termination could be distinguished in some catalytic reactions. Roginsky was one of the first who paid attention to this fact.[14] In subsequent work, proposals were made that in catalysis it was possible to generate the active particles like free radicals which consecutively continued the chain, and to produce an overequilibrium concentration of active particles due to the energy of exothermic steps of the reaction. These views were most consistently followed by Semenov and Voevodsky[10] who had supposed that molecules on the surface could dissociate into free radicals by using an adsorption heat (with a possible involvement of free electrons of a solid). Later, these free radicals could take part in the chain reaction on the catalyst surface. For example, ethylene hydrogenation proceeds in accordance with the scheme

$$Z \ldots H + C_2H_4 \rightarrow Z \ldots CH_2CH_3$$

$$Z \ldots CH_2CH_3 + H_2 \rightarrow C_2H_6 + Z \ldots H$$

where Z is an active surface site.

Theoretical treatment of energy exchange between adsorbed molecules and solids as started in the 1930s, when Lennard-Jones and co-workers[15] estimated the probability of quantum mechanical transitions in the ground state for an impacting atom. Both phonon and electron excitations in a solid were taken into account. A critical review of the Lennard-Jones approach has been made by Bonch-Bruevich.[16] Zwanzig,[17] Cabrera,[18] McCarrol,[19,20] and

Mazhuga and Sokolov[21] solved the problem of the solid-state relaxation during the atomic adsorption and origin of nonequilibrium states on the surface in one-dimensional approximation. The calculations have demonstrated that long-lived excited particles could appear on the surface during adsorption and catalysis and take part in further transformations.

The publications of experimental results on measuring the partial accommodation coefficients stimulated the development of theoretical insights into the heterogeneous deactivation mechanisms of excited molecules. The papers in this research area can be divided into two different groups. The articles of the first group are devoted to macroscopic kinetic mechanisms where the energy transfer to the solid or to the adsorbed molecule is not taken into account; such mechanisms do not practically depend on the kind of excitation: vibrational, electronic, etc. In the second group, more emphasis is placed on so-called microscopic deactivation mechanisms of energy transfer to different degrees of freedom of the solid or the adsorbed molecule.

Another aspect of the problem of nonequilibrium catalytic process is the reconstruction of the surface and the catalyst bulk. It is evident that surface reconstruction and ''catalytic corrosion'' (fast catalyst reconstruction during a catalytic process) are completely determined by the energy liberated in various exothermic processes.

The existence of the surface itself (the lattice edge) causes the disturbance of the interatomic spacings and even the total surface reconstruction. This reconstruction is induced by the changes of surface electron energy levels and, as a consequence, by phonon spectrum modification. Such a reconstruction, changing the distance between the surface atoms, as well as the active site configurations, occurs in catalysis and adsorption and strongly influences catalytic activity and adsorptivity.

The reconstruction of the catalyst surface layer during the reaction has been well known for a long time. Here the fundamental works by Roginsky, Tretyakov, and Shekhter[22] on electron microscopy of platinum and palladium are worth mentioning. The catalytic reactions proceeding on these metals give rise to deep surface loosening. This phenomenon was called ''the catalytic corrosion''. Later it was examined by Turkevich et al.[23] The application of the low energy electron diffraction (LEED) method allowed the observance of these changes in the uppermost monoatomic layer during adsorption and catalysis.[24]

Rozovsky[25] had shown that the stationary surface composition of the catalyst might differ significantly from the equilibrium one during the process. In catalysis, the catalyst activation depends on the reaction rate taking place; the higher the reaction rate is, the greater the catalyst surface transformation is produced by the reaction.

If the situation is far from equilibrium, the ordered, so-called ''dissipative'' structures may appear. It is conventional to refer to the structures with not only spatial, but temporal ordering as well. The general rules of their formation are studied by new subjects: thermodynamics of irreversible processes[26] and synergetics.[27] The presence of dissipative structures in heterogeneous catalysis is attested to by the critical phenomena (abrupt changes in reaction rates at definite surface coverages, temperature, etc.), the multiplicity of stationary regimes, autooscillations, and autowaves.

The concept of the reaction affinity introduced in thermodynamics of irreversible processes permits the state of necessary conditions for the reaction to become irreversible. In heterogeneous catalysis, the continuous shift of equilibrium of intermediate stages may occur as a consequence of their spatial separation, so that different steps of the process may take place on different spatially separated active sites, or even on different phases. In addition to concentration gradients, temperature gradients can also occur in heterogeneous catalysis, resulting in the shift of intermediate equilibrium.

The question of interest is the chemoenergetic stimulation or, in other words, nonthermal application of the overequilibrium energy of a chemical reaction. The calculation indicates that at sufficiently large excitation energies this effect can be of prime significance. Its time is several orders greater than the deactivation time of the excitation.

This calculation results in important conclusions in catalysis. The generation of excited particles on the surface must lead to the increase in the reaction products yield. It is essential that the energy of excitation be transferred to the surface. As a result, the overequilibrium concentration of active sites can arise in conditions of catalysis. The relaxation time of these sites far exceeds the lifetime of excited molecules on the surface. That is why in heterogeneous catalysis the acceleration of the reaction through the elevated concentration of the overequilibrium active sites is the main channel for the chemoenergetic stimulation. In this case, the calculation demonstrates that the stationary regime of the catalytic reaction could be achieved in a very long time, this period approaching an order of magnitude to the lifetime of the catalyst.

Finally, let us discuss the definition or general features of catalysis. It is often thought that the essence of catalysis consists of a reaction proceeding via a new chemical path (Figure 1.1, Curve 2) which includes a set of stages with the lower activation energies in comparison with the noncatalytic process (Figure 1.1, Curve 1). Such a definition does not completely reflect all features of the catalytic mechanism. In particular, it does not involve the possible role of nonequilibrium active sites in catalysis, the production of particles with enhanced free energy, the shift of intermediate equilibrium through the concentration and other gradients. The affinity of the reaction and its stages are wider concepts taking into account not only the energy change, but the concentration as well.

According to Prigogine, one of the founders of thermodynamics of irreversible processes, the main function of a catalyst is to support the affinity between last intermediate and final products at a very high level.[26] The

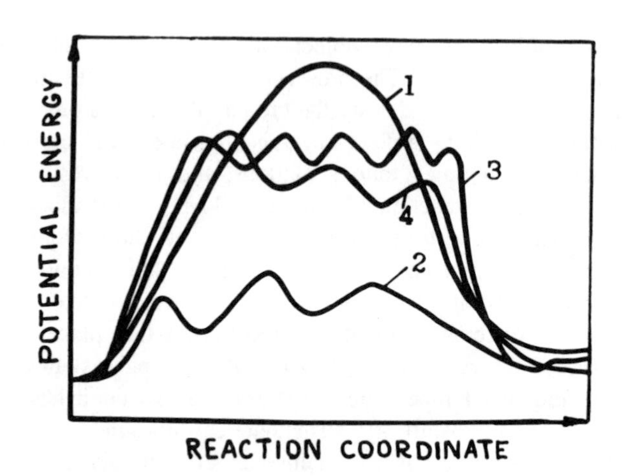

FIGURE 1.1. Energy profiles of noncatalytic (1) and catalytic (2–4) reactions.

efficiency of the catalytic system increases when some stages are far from the equilibrium. The entropy production rate in this case is extremely enhanced.

Curves 3 and 4 in Figure 1.1 show schematically a possible energy profile for a catalytic reaction proceeding through intermediate steps with increased energy. One can conceive of a case when the activation energy of the multistep catalytic reaction (as a matter of fact, all catalytic processes are multistep) is greater than that of the noncatalytic reaction. Nevertheless, the reaction will take place with a higher total rate and selectivity because of the shift of intermediate equilibrium.

Thus, the problem of measuring and calculating the rates of energy exchange between adsorbed molecules and solids is the main purpose for the theory of chemisorption and catalysis. The quantitative characteristics of these processes make it possible to determine the mechanisms of physical and chemical adsorption and, consequently, to understand the nature of elementary acts in catalysis. It is necessary to study how various excitation levels influence the adsorption and catalysis and to solve the reverse problem of the energy distribution in the products of desorption and catalytic reaction. That will allow us to take advantage of the "level" (state-to-state) description for the kinetics of the heterogeneous catalysis. It is also of great importance to investigate the "excitation" of catalyst (its reconstruction and activation under the influence of chemisorption and catalytic reaction) and to study the deviation from the equilibrium as well as the formation of dissipative structures.

It is impossible to cover all questions about nonequilibrium processes on the gas-solid interface in one book. We shall concentrate on such aspects of the problem as inner molecular energy deactivation and its effect on surface migration, energy distribution in catalytic products, and the ejection of excited particles from the surface to gas phase as a consequence of exothermic processes during gas-surface interaction. We shall dwell on the possibility of chemical activation, i.e., the acceleration of one reaction at the cost of another one generating the active particles. Some general features of chain processes, catalysis, and dissipative structures in catalysis will also be discussed. Finally, some examples of catalyst structure transformation and surface reconstruction due to the oxidation catalytic reactions will be presented.

Chapter 2

RESEARCH METHODS

The following concepts are used to characterize catalytic activity: (1) the rate per unit of mass, volume, or catalyst surface area (so-called specific catalytic activity); (2) the rate constant; and (3) the so-called turnover number $n = r/N$, where r is a reaction rate and N is a number of active sites on the surface. Boudart[28] points out that for many heterogeneous reactions the turnover number n ranges from 10^{-2} to 10^{2} s^{-1}. The upper of these limits is due to the diffusion restrictions when the reaction is performed at atmospheric pressure.

In general, the turnover number n is not constant for a given catalytic reaction. However, the statement[28,29] that the magnitude of n usually ranges from 10^{-2} to 10^{2} (and characteristic time τ from 10^{2} to 10^{-2} s, respectively) is of great importance for us. As with these values of characteristic times and turnover number, the amount of reaction products can be analytically measured in ordinary catalytic experiments. Therefore, when analyzing the times of various elementary stages of a process we will in the first place consider whether they belong to the interval 10^{-2} . . . 10^{2} s, i.e., and whether they may substantially affect the catalysis rate.

Figure 2.1 is a representation the table of published characteristic times of heterogeneous processes. It is evident that the times of many elementary processes are within the range 10^{-2} . . . 10^{2} s. These processes are adsorption and desorption, the majority of chemical stages, phase transformation, surface diffusion, and electron capture by so-called slow-surface states. At the same time, the energy exchange of atoms and molecules with the surface occurs as a rule, more rapidly; characteristic times for these processes are of the order 10^{-12} to 10^{-4} s. Nevertheless, it will be shown in the following chapters that these stages may strongly influence the rates of heterogeneous catalytic processes.

The up-to-date success in chemical kinetics proved the insufficiency of the concepts of rate and rate constant in characterizing reagents conversion dynamics. The principal purpose of chemical kinetics is to ascertain the relation between the dynamic characteristics and the topology of the potential energy surface in chemical transformation. From this viewpoint, the kinetic investigations, which measure integral rate constants and their temperature dependencies, are of little interest. Indeed, the thermal equilibrium initial conditions correspond to the Maxwell-Boltzmann energy distribution of translational and inner reagent states. Therefore, the nature of the individual collision process remains unknown due to averaging over the thermal distribution. The total rate constant k(T) of a chemical process is connected to the reaction cross-section α_{ij} at constant relative translational energy E_t in the following way

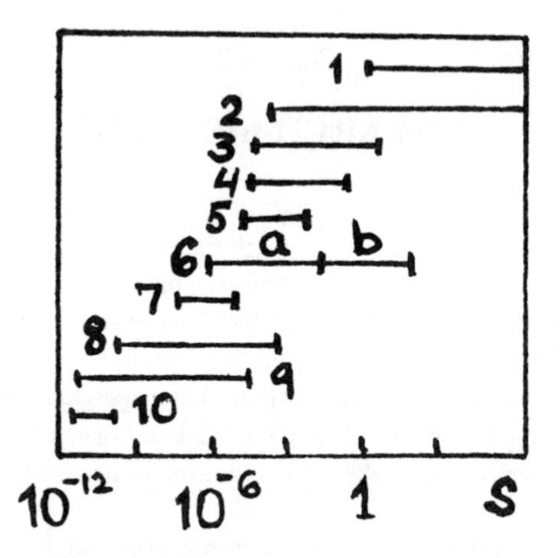

FIGURE 2.1. The characteristic times of heterogeneous processes.[29] (1) Volume diffusion; (2) monomolecular desorption; (3) bimolecular desorption; (4) surface molecular diffusion; (5) proton jumps; (6) electron capture in the surface state (a, short-lived, b, long-lived states); (7) spontaneous phase transition; (8) vibrational relaxation; (9) nonactivated adsorption; (10) translational and rotational relaxation.

$$k(T) = (n\mu)^{-1/2} (2/k_BT)^{3/2} \sum_{ij}\sum F_i(T) \int E_t\alpha_{ij}(E_t)\exp(-E_t/k_BT)dE_t \quad (2.1)$$

where i and j are initial and final states of the reagents, μ is the reduced mass, and $F_i(T)$ is the α part of molecules at the ith initial inner state.[30]

Consequently, it is clear that there is no way to extract information about α_{ij} if k(T) is the only known function; i.e., it is impossible to solve the main problem.

Usually, translational and inner degrees of freedom of reagents are in a state of thermodynamic equilibrium. Unfortunately, such conditions do not allow us to answer the question what energy forms are most effective to overcome the potential barrier and, consequently, to accelerate the reaction.

2.1. BEAM METHODS

2.1.1. MOLECULAR BEAMS

One of the commonly used methods in surface science is the molecular beam method.[31-35] It allows us to investigate the energy exchange between molecules and surface, as well as to study the elementary stages of chemisorption and catalysis. The flow of reagent molecules exits from the high pressure chamber through a small inlet into the vacuum chamber and strikes the target — catalyst. In the beam intermolecular collisions and those of molecules against the walls as well as molecular diffusion can be neglected.

Molecular energy in the beam is defined by the source design. The thermal sources, so-called Knudsen cells in which the gas is maintained under the temperature from 77 to 3000 K, are in common use. Nozzle size must be less than the molecular mean free path. In such a beam, molecules move almost parallel to each other. Molecular velocity distribution J(v) has the form

$$J(v) \sim \frac{v^3m^2}{(2k_BT)^2} \exp[-mv^2/(2k_BT)] \tag{2.2}$$

where m is a molecular mass, T is a source temperature, and k_B is the Boltzmann constant.

It should be mentioned that the distribution law J(v) for the molecules ejected from such a nozzle in the form of a parallel beam differs from the Maxwell distribution of bulk gas molecules. In the former case the average molecular energy is $2k_BT$; in the latter it is $3/2k_BT$.

The density of molecular flux (in s^{-1} cm^{-2}) is given by

$$J = 8.9\ 10^{19}p_iA_s/1^2(MT)^{1/2} \cos^2\varphi \tag{2.3}$$

where p_i is the source pressure (Pa), A_s is an area of source outlet (cm^2), 1 is a distance from the surface (cm), M is a molecular mass, φ is an angle between the beam axis and an angle normal to the surface.

From Equations 2.2 and 2.3 it is clear that Knudsen-type sources allow us to obtain the beams with flux densities of (2 to 4) $\times$ 10^{14} $s^{-1}cm^{-2}$ and velocities from $6 \cdot 10^4$ to $4 \cdot 10^5$ cm/s, that corresponds to molecular energies from 0.008 to 0.32 eV.

The term "beam temperature" is widely used in the literature. But it should be noted that the comparison of molecular energy in the beam with the gas temperature is correct only in the case of the Maxwell-Boltzmann distribution. As a rule, the average inner molecular energy $\langle E_r \rangle / k_B = T_r$ and $\langle E_v \rangle / k_B = T_v$. By analogy, for translational energy we must write $\langle E_t \rangle / 2k_B = T_t$ where E_t is the translational energy of one molecule in the beam. Another, but incorrect definition, $\langle E_t \rangle / k_B = T_t$, is also used sometimes.

The supersonic nozzle permits us to obtain molecular beams with much higher energies. The supersonic beam allows us to narrow down the velocity distribution 100-fold in comparison with the Maxwell distribution of thermal sources. Despite a high density, molecules in the supersonic beam have parallel trajectories and do not collide with each other. Using high pressure (>0.1 MPa) in the source allows us to reach the flux density of the order of 10^{16} to 10^{18} $s^{-1}cm^{-2}$ with the molecular energy of 1 to 2 eV. The kinetic energy may be increased up to 10 eV when applying the diluted mixtures of heavy gases with a fast light gas, for example, with He. Further energy increase in molecular beams is possible with the aid of the electric discharge, shock wave tubes, and other special methods.

Combining the heating and the supersonic ejection makes it possible to obtain molecular beams with different ratios between the translational and vibrational energies. The presence of a neutral gas at a moderate pressure

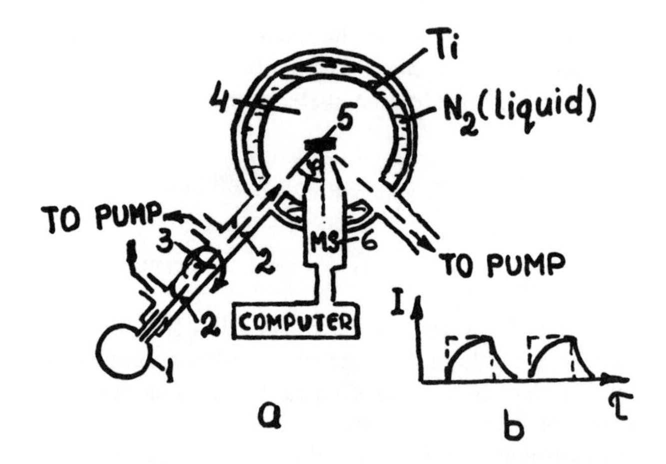

FIGURE 2.2. The scheme of the experimental setup for studies of molecular beam interactions with surface (a), and the current I of the modulated molecular beam interacting with surface as a function of time τ (b). (1) High-pressure chamber with the gas under investigation; (2) collimation inlets; (3) beam chopper; (4) vacuum chamber; (5) sample; (6) quadrupole mass spectrometer.

suppresses the translational energy, whereas the vibrational excitation of molecules in the beam remains practically unchanged.

The mechanism of elementary interaction of molecules with the surface as well as the mechanism of multistep catalytic reaction may be studied by the modulated molecular beams technique.

Figure 2.2. shows the installation used at the Institute of Chemical Physics. The reagent beam passes from the chamber (1), through the outlet (2), and the beam chopper (3) in the form of a rotatable breaker plate; then the beam enters into a vacuum chamber (4). The vacuum chamber is coated with titanium getter and is cooled by liquid nitrogen. The rotatable sample is placed in the center of the chamber that varies the angle of beam incidence. Molecules that are reflected by and desorbed from the catalyst are analyzed by the quadrupole mass spectrometer (6). One of the advantages of the molecular beam modulation technique is an increase in the signal-to-noise ratio by a factor of 3 orders of magnitude (residual gases in the chamber may be a source of noise origin). The beam time of flight is approximately 10^{-4} s. At a modulation frequency of 1000 Hz, the phase shift of 35° relative to the input beam pulse occurs. The phase shift may be measured with the accuracy of $\pm 1°$, hence, the method allows us to study the constants up to 10^4 s^{-1}.

The characteristic behavior of the measured ratio of output-to-input signal amplitudes and the phase shift enables us to draw a conclusion about the mechanism of surface processes.

One of the most important characteristics is the velocity of molecules desorbing or scattering at the surface. Velocity measurements can be performed using the time-of-flight mass spectrometer or the rotatable disk velocity selector. More often, the angular distribution of reflected, scattered, or desorbed molecules is measured to study the energy distribution of molecules. Maxwell had pointed out as early as the 19th century that in the absence of chemical transformations there are two extreme cases: molecules undergo either (1) the mirror reflection, or (2) adsorption followed by the thermalization with the surface and vaporization. When the complete thermal accommodation is reached, the reflected (or desorbed) beam intensity must be proportional to $\cos\varphi$, where φ is the angle between the reflected beam direction and the angle normal to the target surface

$$I = I_o \cos\varphi \tag{2.4}$$

This is the so-called Knudsen law.

In practice, Equation 2.4, as well as the mirror reflection, occurs very rarely. The intermediate situation of the partial thermal accommodation is actually observed. It means that during the multiphonon interaction the molecule desorbs without the complete energy exchange with the solid. Intermediate cases are usually described by the law

$$I = I_o \cos^n\varphi \tag{2.5}$$

This law is to be considered an empirical one, because the theoretical justification of Equation 2.5 exists only for a limited number of n values. The high n values in Equation 2.5 are usually attributed to the departure of the distribution of desorbed (scattered) molecules from the Maxwell-Boltzmann distribution. In the case of a rough surface, the measured n factor is less than that of a smooth surface. The dependence of n on surface roughness has been studied by Savkin et al.[36] In summary, let us list the possibilities of the molecular beam method. It allows one to determine

1. The minimum lifetime of a reagent molecule on the surface required for the reaction occurrence
2. The probability of the reaction or adsorption per one collision with the surface (the sticking probability)
3. The dependence of the adsorption and reaction probability on molecular energy in the beam
4. The energy dependence of desorbed molecules on the surface temperature
5. The deactivation probability of excited molecules
6. The energy (translational, vibrational, rotational) distribution of the reaction products or scattered molecules
7. The angular distribution of desorbed and scattered molecules

One may also study the dependence of the reaction rate on surface coverage with a controlled amount of an adsorbate.

2.1.2. ION BEAMS

It is clear from the previous discussion that the molecular beam technique is rather complicated. Moreover, the application of this method is restricted by velocity values of the order of $2 \cdot 10^5$ cm/s. For sufficiently light molecules, e.g., CO, NO, the maximum translational energy of particles in the molecular beam does not exceed 1 to 1.5 eV without application of special methods of excitation.

One of the main advantages of the ion beam technique is its wide accessibility. The interaction of ions with solid surfaces has been a subject of study in many laboratories throughout the world for a long time. The information accumulated with the aid of this method is reviewed in several papers and monographs.[37-38] The character of the problem defines the proper choice of experimental procedure and, consequently, the information. The following problems are the most important

1. Solid surface study, for example, ion scattering spectroscopy (ISS)
2. Plasma treatment of solid surfaces
3. Heterogeneous plasma chemistry
4. The development of novel methods for the kinetic control of heterogeneous processes
5. Study of the elementary stages of the atomic and molecular interaction with solid surfaces

Industrial and scientific necessity for these investigations is stimulated by the demand for novel technologies, materials, methods of protection against the destruction by corpuscular fluxes, etc.

Most of the works, devoted to the interaction of ions with solid surfaces, have an industrial trend, so the integral character of this information does not allow us to use these results for understanding elementary heterogeneous processes. Winters realized that the difficulties of molecular beam method in the study of these problems can be circumvented by using the simpler low energy ion technique.[39] It is well known that if the velocity $v << v_o$ ($v \sim \Delta/a$, $\Delta \simeq 0.5 \div 5$ [a.u.], $a \sim 2\sqrt{2\varphi}$, where φ is a work function of a metal), the collision between an atomic ion and a metal surface is preceded by the ion neutralization. It would appear natural that in the case of low-energy molecular ions the act of heterogeneous chemical transformation is preceded by the cascade of electron-vibrational transitions which result in the neutralization of incident ions and deactivation of the resulting molecules. Varying the parameters of the ion beam, i.e., the charge sign, the electron state, the distribution over vibrational levels, initial energy and the angle of incidence, alters the parameters of neutral molecules resulting from the neutralization; then these molecules undergo different chemical transformation such as the impact dissociation, dissociative adsorption, sputtering, etc.

In principle, one can formulate the conditions in which the neutral molecules will have dominant electronic, vibrational, or only translational excitation. A fundamentally new opportunity of studying the contribution of different degrees of freedom of gas reactants in the activation effect appears.[40,41] It is obvious that successive realization of this opportunity will depend on many circumstances, but one of the most important is the experimental possibility of wide variation of the parameters of the electron-vibrational molecular distribution near the surface. Also of importance is the question about the sensitivity of the process which will be chosen to detect the acts of the surface reactions and to study these distributions. To estimate the range of possible variations in the electron-vibrational distribution near the surface one must consider the evolution of the electron system of the ion approaching perpendicularly to the solid surface from a vacuum.

A separate task is the choice of a channel of surface transformation which could be convenient for the registration of activation effects. Its solution depends on the surface under investigation, the nature of the partner in the collision process, as well as on the notions about the dynamics of heterogeneous processes with the participation of electronically and vibrationally excited particles. Very recently the development of such notions has been started.

The assumption of applicability of the ion beam method for the study of collisions between molecules and the solid surfaces is based on the proposal of fast relaxation of the electron subsystem of the target. The irreversibility of electron transitions in the ion-surface system is a sufficient condition for the system of kinetic equations to be valid for describing the electron-vibrational transitions during the interaction of a slow ion with the surface. Some estimations and considerations concerning the relaxation mechanism of the electron excitation in metals and that of atomic particles near the solid surface have been presented by Bodrov et al.[40]

These authors have considered a simple model for the cascade of electron-vibrational transitions in which they neglected the perturbation of electron terms of the molecule near the surface, image forces, and rotational degrees of freedom. Equations for the calculation of electron-vibrational distributions of forming molecules, as well as some computational programs are suggested. Let us briefly consider some characteristic features of collision processes between ions and solid surfaces.

The energy diagrams in Figure 2.3 allow us to understand the evolution of the electron-vibrational subsystem of negative ions in the vicinity of the metal surface. Negative ions have one weakly bound electron with the binding energy of 0.1 to 1 eV. The values of electron-vibrational linkage parameters are small, and a small number of stationary vibrational states should be taken into account. If the energy of vibrational excitation exceeds the ionization energy, an electron leaves the negative ion via the mechanism of electron-vibrational interaction in a time of 10^{-10} to 10^{-12} s. The neutralization near the metal surface takes place in accordance with the mechanism of a weakly bound electron tunneling into free metal levels. The loss of a weakly bound electron results in the formation of a molecule in the electron ground state,

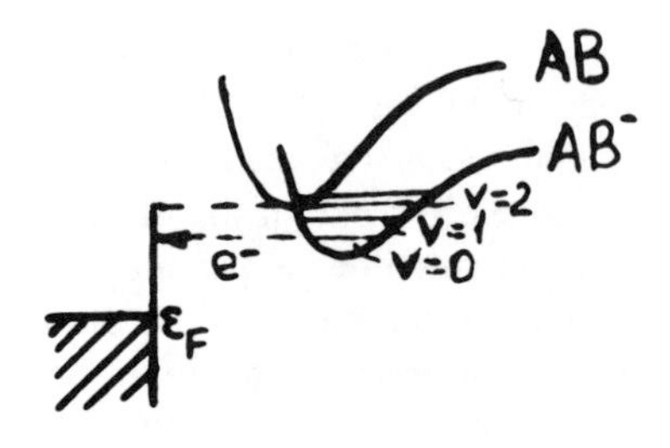

FIGURE 2.3. The diagram of the energy levels in a metal and potential energy of AB^- molecular ion and diatomic AB molecule interacting with the metal surface (the occupied levels are shown by hatching; ϵ_F is the Fermi level in the metal; unoccupied states are above ϵ_F; $v = 0, 1, 2, \ldots$ are vibrational levels of AB^- ion).

and the vibrational excitation energy of molecules formed near the metal surface is small.

Thus, the low-energy beam experiments with negative molecular ions may provide information about the interaction between the solid surface and neutral molecules in the ground or low-lying excited states.

The energy diagram shown in Figure 2.4 illustrates the main features of CO^+ neutralization in the vicinity of the metal surface. In this case, one can see that acts of neutralization may lead to the formation of the electronically excited molecule which deactivates partly or completely as the surface is approached. In contrast to the negative ions, parameters of electron-vibrational linkage may exceed unity. Hence, it is clear that the beam experiments with slow positive molecular ions may serve as a source of information on collision processes between metal surfaces and highly excited molecules with the excitation energy of the order of the ionization potential or electron work function.

The simplest two-level model system is studied to demonstrate the possibilities of such experiments.[40] Besides, the relation between the vibrational distributions of ions and forming molecules is considered. It was concluded that the beam experiments with slow positive molecular ions may be used to study heterogeneous processes which involve electronically and vibrationally excited molecules. When the normal velocity component of the beam is decreased, the electron excitation energy of forming molecules drops, but their vibrational excitation energy increases.

Thus, the ion beam experiments may serve as a source of unique information on reactivity of electronically and vibrationally excited molecules in the course of their transformations on the solid surface.

In collaboration with Bykov and Kovalevsky,[42] we suggested a very simple and effective method to investigate the interaction between ions and solid surfaces. It is based on the ion acceleration using the cyclotron resonance technique. The adsorption probability is measured by the pressure change of the gas under investigation.

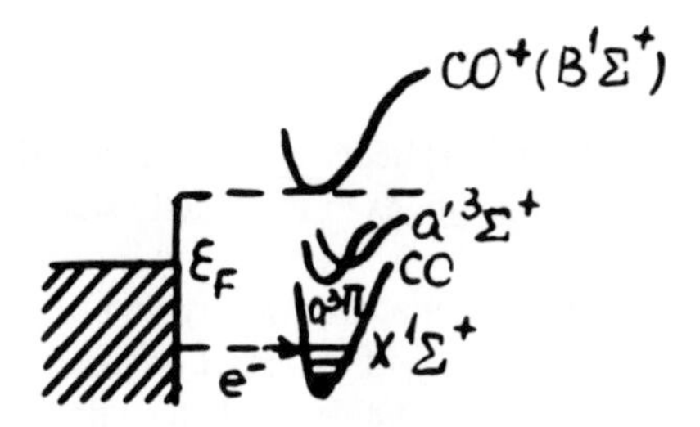

FIGURE 2.4. The diagram of the energy levels in a metal and potential energy curves of CO molecule and CO^+ ion interacting with the metal surface.

To generate the ions, a beam of collimated electrons 0.1 mm in diameter with the energy of 100eV is used. The beam is directed along the axis of the omegatron detector and is parallel to the magnetic field H. The radiofrequency electric field E ($E \perp H$) accelerates resonantly the ions which strike the collector made of the material to be investigated. The energy E_i of resonant ions is proportional to H^2L^2, where L is the distance between the collector and electron beam axis H. Varying the magnetic intensity enables the ions generated to strike the collector with a desirable energy.

In experiments the kinetics of the pressure change in the closed tube may be described with high accuracy by the following equation

$$\ln(c_o/c) = -\alpha_i\epsilon + \alpha_d \frac{ILe}{V} t \quad (2.6)$$

where c_o and c are initial and current concentrations of the gas particles; α_i is the ionization cross-section by the electron impact, α_d is the molecular ion dissociation cross-section, I is the electron current intensity, e is the electron charge, V is the volume of the reactor, ϵ is the ion adsorption probability, and t is the time.

The ion adsorption probability ϵ is derived from the measurements of the pressure in the cell in the presence and the absence of the high-frequency field resonance. It should be mentioned that the ion cyclotron resonance technique combined with the suggested method of measuring the adsorption probability has, in our opinion, some advantages over the beam experiments in which Auger-electron spectroscopy is used for adsorbed particles registration.[43] Varying the magnetic intensity changes the average ion energy at the target with the precision of several hundreds of electronvolts. The available energies range from thermal values to hundreds and thousands of electron volts. Moreover, the application of the omegatron gauge for measuring the partial pressures of the gases studied increases significantly the sensitivity of the method.

2.2. LASER METHODS

The fast development of laser technique has resulted in the appearance of a number of new research methods in the kinetic studies of the interaction of molecules with the surface.

Thanks to both their high monochromaticity and high power, lasers can (1) produce independent excitation of different degrees of freedom in order for the reaction to be directed into the desirable channel; (2) detect with high precision and identify intermediate products of the reaction; measure the energy state of the gas phase molecule over all degrees of freedom: translational, rotational, vibrational; and (3) measure the lifetime of excited states on the surface.[44]

2.2.1. IDENTIFICATION OF GAS-PHASE PARTICLES

Numerous reviews and books[45-48] have been devoted to laser spectroscopy. We shall briefly consider some of the methods and compare different methods widely used at the Institute of Chemical Physics in Moscow.

Laser absorption spectroscopy (LAS), one of the oldest spectroscopic methods, has become a state-of-the-art instrument due to laser development. The use of laser as a light source increases tremendously the sensitivity of the method and its spectral resolution. For example, a spectrometer with such parameters has been constructed on the basis of tunable semiconductor lasers;[49] its wavelength ranges from 4 to 20.5 μ and the spectral resolution is 10^{-3} cm^{-1}.

The method of laser magnetic resonance (LMR)[50] stems from the fact that the energy of the laser quantum is chosen approximately equal to the energy difference between two levels in the fine structure of a radical. Conditions for the resonant absorption may be obtained by supplying the external magnetic field. Different modifications of LMR as well as its combinations with other methods are described in the literature.[51-53] Note that the first LMR instrument, based on the CO_2-laser, was developed at the Institute of Chemical Physics in 1975.[54] This method is especially convenient for the highly sensitive identification of multiatom radicals in hetero- and homogeneous processes.

The method yielding considerable opportunities is the intercavity laser spectroscopy (ILS). It was developed at the Institute of Physics in Moscow in 1970, and was applied for the first time as a kinetic method at the Institute of Chemical Physics.[55,56] This method is based on the idea that particles with narrow absorption lines within the laser generation band are placed in the laser cavity with a wide bandwidth. The parameters of the laser media (neodymium glass, dyes, etc.) are selected so that the light amplification in the media compensates the energy loss at the mirrors but not in the substance under investigation. The method is similar to that of absorption spectroscopy, but has a giant optical length reaching up to 10^5 m. It is important that such an optical length is realized in a reactor of small sizes.[56] The heterogeneous

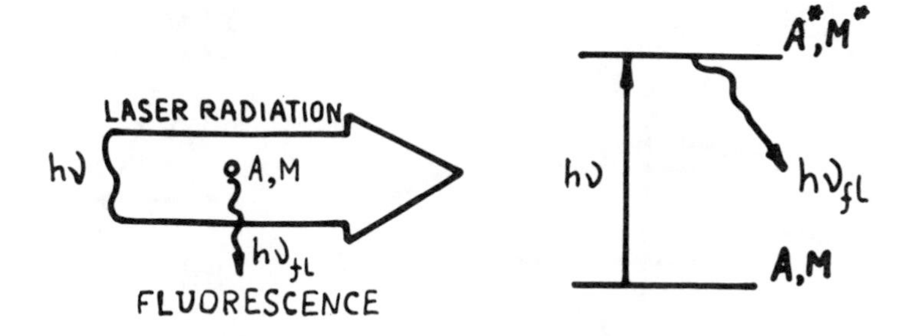

FIGURE 2.5. The principal scheme of the laser-induced fluorescence (A, atoms; M, molecules).

and homogeneous deactivation of vibrationally excited NH_2 radicals was studied for the first time by this method.[57]

Laser-induced fluorescence (LIF) is based on the laser excitation of molecular and atomic electron states which then fluoresce; this fluorescence is observed in the direction perpendicular to the laser beam axis (Figure 2.5). When the laser wavelength λ is tuned to the molecular transition $E_i \rightarrow E_k$, the number of absorbed quanta per second per path Δx equals

$$n_{abs} = N_i \alpha_{ik} n_l \Delta x \tag{2.7}$$

where N_i is the concentration of absorbing molecules, α_{ik} is the absorption cross-section, and n_l is a number of incident photons.

The number of photons per second emitted as a result of fluorescence from the E_k level is

$$N_{fl} = N_k A_k = n_{abs} \eta_k \tag{2.8}$$

where A_k is the total spontaneous transition probability, $\eta_k = A_k (A_k + R_k)$ is the quantum yield of the transition, and R_k is the nonradiative transition probability.

At $\eta_k = 1$ in stationary conditions, the number of fluorescent photons is equal to the number of absorbed photons. The sensitivity of the method is determined by the emission flux on the cathode of the photomultiplier.

When the laser wavelength λ_l is varied in the absorption spectra range of the molecule under investigation, the total fluorescence intensity $I_{fl}(\lambda_l) \sim n_l \alpha_{ik} N_i$ is detected as a function of λ. The spectrum obtained is similar to the absorption one and is called "the excitation spectrum". Due to extremely high sensitivity,[58] the excitation spectroscopy is successful in the detection of very small amounts of radicals and short-lived intermediate products of chemical reactions.[59] Apart from the measurement of low concentrations, LIF provides the detailed information about the distribution $N_i (\nu_i'', J_j'')$ of reaction products over states. Indeed, adjusting consecutively the laser wavelength

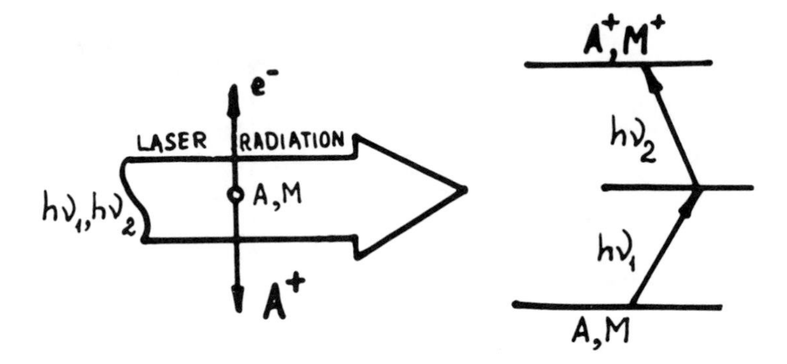

FIGURE 2.6. The principal scheme of the multiphoton ionization.

into two different absorbing transitions 1 → k and 2 → m, one can obtain the relative population N_1/N_2 directly from the ratio-measured signals

$$\frac{n_{1fl}}{n_{2fl}} = \frac{N_1\alpha_{1m}}{N_2\alpha_{2m}}$$

providing that the absorption cross-sections are known.[45]

The method is applicable for molecules with sufficiently high fluorescent efficiency and available spectroscopic parameters. To define relative quantum-state populations it is also necessary to know Franck-Condon factors for vibrational levels and Honl-London factors for rotational lines. That is why the number of objects suitable for LIF investigations is limited. Nevertheless, the method has a large sensitivity ($\sim 10^6$ particles per cm^{-3}),[47] and a high spatial and time resolution. In principle, the spatial resolution is limited by λ^3 and time resolution might reach 10^{-12} s.

LIF is often used in combination with other laser methods.[51] For example, the absorbing sample is placed in the laser cavity, or is induced by the second laser that can determine precisely the states of the molecule throughout the fluorescence period induced by the first laser.

The method of resonance enhanced multiphoton ionization (REMPI) has the highest sensitivity and flexibility.[48,60] In this method, the laser pulses stimulate the transitions which lead to resonant excitation of molecules or atoms to high-lying levels and further ionization (Figure 2.6). A molecular or atomic ion formed in this process is usually detected by the mass spectrometer. There are no limitations on the energy level structure, the decay rate, or nature of particles under study for REMPI applicability. The dependence of the multiphoton ionization signal on laser wavelength is determined by energy level population, so this dependence allows us to calculate the distribution of the inner states of the molecule under investigation. Zare et al.[61,62] applied the REMPI method to the study of the associative desorption of hydrogen from different faces of single copper crystals.

A number of laser research methods, such as coherent anti-Stokes Raman spectroscopy (CARS),[63] optical acoustic spectroscopy (OAS),[57] multiphoton excitation (MPE),[64] Doppler high resolution laser spectroscopy,[65] and others, might be applied in the near future to the study of the quantum kinetics of processes at the interface between gas and solid. The following expanded table of parameters taken from the review of Sarkisov and Cheskis[47] (Table 2.1) compares different laser methods.

In order to study such molecules as N_2, H_2, and O_2 using the highly sensitive LIF or REMPI, the excitation in a vacuum in the ultraviolet (UV) region is required; that presents a severe experimental problem. These difficulties can be overcome with the help of electron beam-induced fluorescence, (EBIF) where excitation and ionization are produced by the electron beam. The electronically excited molecule fluoresces, and its fluorescence spectrum provides the information about the electron-excited state. The lines of the excited molecule must be well known to apply EBIF, as well as MPI or MIF. Up to now, the EBIF method has been applied only to the study of the N_2 molecule.[66]

Fourier transform infrared spectroscopy (FTIRS) was used to identify vibrational and rotational levels of CO_2 which are a product of CO oxidation on Pt.[67]

Further improvement of the above methods of particle identification makes it possible to determine not only the vibrational and rotational states of a molecule, but also to measure the "translational" temperature (the velocity distribution) of every state on account of broadening of the spectral lines due to the Doppler effect. The Doppler effect means that the frequency ν of the molecular transition depends on the velocity v of the moving molecule

$$\nu = \nu_o(1 + v/c) \tag{2.9}$$

where c is the velocity of light. For the isotropic molecular distribution in the sample the spectral line has the Gaussian shape

$$S(\nu) = S(\nu_o) \exp\left\{ -\ln 2 \left[\frac{\nu - \nu_o}{(\Delta\nu)_d} \right]^2 \right\} \tag{2.10}$$

where $2(\Delta\nu)_d$ is the Doppler width which depends only on translational temperature T_t

$$\frac{(\Delta\nu)_d}{\nu_o} = \frac{2\sqrt{2R \ln 2}}{c} \left(\frac{T_t}{M}\right)^{1/2} \tag{2.11}$$

In the visible range of spectrum, the Doppler broadening $(\Delta\nu)_d$ is of the order of 0.1 cm^{-1} ($300 < T_t < 1000$ K). To resolve the line shape at this value of $(\Delta\nu)_d$, the trial laser must have the line width about 0.01 cm^{-1} or less.

TABLE 2.1
Comparative Characteristics of Laser Methods

Parameter	Methods					
	LAS	ILS	LMR	LIF	CARS	REMPI
Time resolution(s)	10^{-6}	10^{-6}	10^{-2}:10^{-3}	10^{-8}:10^{-9}	10^{-8}	10^{-8}:10^{-9}
Spatial resolution	1 mm	1 mm	1 mm	10^{-6} cm^{-3}	$10^{-5} \div 10^{-6}$	10^{-6} cm^{-3}
Sensitivity (cm^{-3})	$10^{12} \div 10^{13}$	$10^{8} \div 10^{10}$	$10^{8} \div 10^{10}$	$10^{6} \div 10^{10}$	$10^{13} \div 10^{14}$	10^{3}
Spectral resolution (cm^{-1})	0.03	0.03	$10^{-2} \div 10^{-4}$	10^{-3}	$10^{-2} \div 10^{-3}$	10^{-3}
Spectral range (μ)	$0.2 \div 10$	$0.4 \div 1.2$ $2.7 \div 3.0$	$4.0 \div 2000$	$0.2 \div 1.0$	IR range	$0.2 \div 1.0$

2.2.2. SURFACE INVESTIGATIONS

Laser tools are very useful for selective excitation of a certain bond of adsorbed particles. The application of lasers in experiments on selective excitation of particles at the surfaces faces the problem of significant thermal effects. Fast heating usually results in the thermal desorption or thermally induced surface reactions instead of selective excitation. Laser-induced desorption is used to determine the kinetic rate constants of surface processes.[68] The pulsed lasers or catalysts with effective heat removal (metals) are used to avoid the thermal effects.

The second harmonic generation (SHG) method is a very convenient one in studies of fast processes. The SHG effect is forbidden under the electric-dipole approximation in a medium with inversion symmetry. At a surface the inversion symmetry is necessarily broken. For this reason, the process is highly surface specific. An estimation[69] shows that using 10-ns pulse at a power of 10 MW/cm^2 enables one to reach SHG sensitivity of 10^4 photons per pulse. This is the evidence in favor of a submonolayer character of SHG sensitivity in surface studies.

The response time in the SHG method is due only to electron relaxation. Very rapid processes have been successfully examined in some cases. For example, the dynamics of [111] Si surface melting has been studied using 75-fs laser pulses at 610 nm.[70] The crystal order disappears already in a time of 100 fs, i.e., in such a short time the surface links "move apart" and lose their regular order.

Now only electron transitions are studied with the aid of SHG. This is an apparent disadvantage of SHG because, for example, the vibrational transitions are more important in the studies of the energy relaxation processes. Unfortunately, IR detectors have insufficient sensitivity. The sum frequency generation (SFG) method has been proposed to measure the vibrational transitions using the optical photomultiplyer.[71] A tunable IR laser has been used to study the vibrational adsorbate transitions. The resulting spectrum was transformed into visible range by summing with the frequency of the second laser.

Now consider the question about measuring the lifetimes of excited states at the surface. The possibility of determining the lifetime $\Delta\tau$ using IR spectral line width $\Delta\nu$ is often discussed. According to the Heisenberg uncertainty principle, the intrinsic line width $\Delta\nu_{rad} \sim h\Delta\tau$, where h is the Planck constant. But the width of a spectral line is determined commonly by external reasons, e.g., by the Doppler broadening in the gas phase. In general $\Delta\nu >> \Delta\nu_{rad}$. Dephasing plays an important role in the line broadening even in the gas phase.

On the surface, the width of the IR spectrum line is significantly greater than that of the gas phase. For example, the vibrational line width $\Delta\nu$ of C=O bond in the absorbed layer is about 10 cm^{-1}, and it is 10^2 times greater than that of the gas value. If the vibrational line broadening due to the surface nonuniformity is negligible, the "pure" individual vibrational line width may

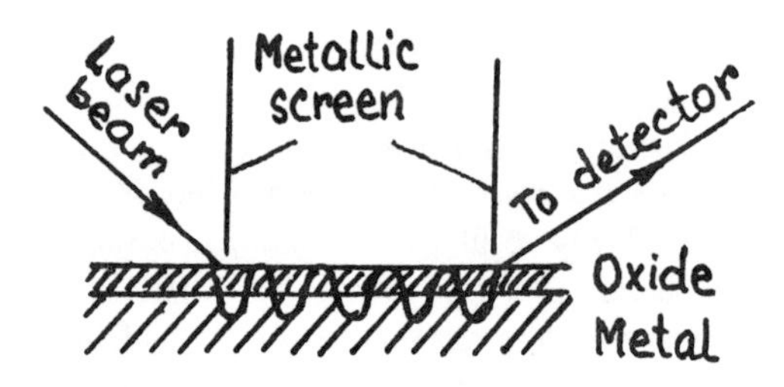

FIGURE 2.7. The principal scheme of SEW experiment.

serve as a measure for the rate of the energy transfer from an adsorbed molecule to the solid. The broadening at the surface includes the local interactions between neighbors; the resonance between the local vibrations and lattice phonons; the excitation of the electron-hole pair in the solid, vibrational dephasing due to solid vibrations; scattering by defects, etc.[72]

The vibrational line width may be determined by IR-spectroscopic tools. The resolution of ordinary IR devices is $\Delta\nu \approx 1 \div 5\ cm^{-1}$. The HREELS (high resolution electron energy loss spectroscopy) method has an even worse resolution, $\Delta\nu = 30 \div 50\ cm^{-1}$. A new method for studying the vibrational spectra of molecules adsorbed on surfaces of small areas has been developed at the Institute of Chemical Physics and the Institute of Spectroscopy, surface electromagnetic waves (SEW).[73] To excite the adsorbate, the laser beam was directed onto the surface at a definite angle (Figure 2.7). The adsorbed particles absorb the light which propagates along the surface. The method combines the high sensitivity with the high resolution of IRS. Using SEW, the spectrum of hydrogen adsorbed on a single W crystal[74] and the spectrum of methane on the smooth Al_2O_9 surface of 1 cm^2 have been obtained.[75]

Recently, the picosecond pulse laser technique has been used in order for the actual lifetimes of short-lived excited surface states to be determined.[76] In a typical experiment, the sample is exposed to laser radiation (e.g., using a Nd laser) with a few picoseconds duration. Then it is probed with the aid of the second laser for some time after the flash.

2.3. EXCITED MOLECULAR DEACTIVATION STUDIED BY FLOW METHODS

The flow methods can be used for studying the kinetics of heterogeneous deactivation of excited molecules when the molecular beam method is inapplicable; i.e., when the accommodation coefficient is small.

A number of flow methods have been developed at the Institute of Chemical Physics to study deactivation of excited molecules on solid surfaces.

The typical experimental arrangement in the flow studies includes the source of excited molecules, the reactor which can vary the time of gas contact with the surface under investigation, and the cell for measuring the concentration of excited molecules. The subject of studies is the dependence of the concentration of the excited molecules on the time of their contact with the reactor surface covered by the substrates under investigation. Also of interest are the temperature and the pressure dependencies. Using this method, the time dependence of the concentration of the reaction is determined by measuring the concentration at different points along the axis of the typically cylindrical reactor. The reacting gases are continuously pumping through the reactor-tube to maintain the steady conditions.

Thus, the flow methods make it possible to carry out the spatially resolved measurements of the reactant concentrations and to determine the rate constants for heterogeneous deactivation of excited particles. Glass or quartz cylinder tubes 15 to 60 mm in diameter and 20 to 200 cm in length were used in these experiments. A typical value of the flow velocity is 10^9 cm/s, and the time of the process in the tube is about 1 ms. The tube dimensions given above are limited by the requirement to maintain the unidimensional flow of the gas. That is also the reason for limiting the pressure by the range from $0.5 \cdot 10^2$ to $1 \cdot 10^3$ Pa. To avoid the systematic errors in deactivation measurements a high precision in measuring the rate of reagent feed is required.

The perturbation of the unidimensional flow may occur for various reasons; in particular, it may be disturbed by the Poiseuille gradient of pressure or concentration gradients along the tube. These effects may result in significant backflow diffusion. Unfortunately, the attempts to eliminate these effects impose the opposing requirements on the flow velocity u, so the optimal value of u exists. This optimal velocity cannot sometimes be used because of some additional limitations imposed on the studied system. Backflow diffusion is negligible when $Dk/u^{-2} >> 1$, where D is the diffusion coefficient of excited molecules, and k is an effective rate constant. The deactivation process of the first order is given by

$$n = n_o \cdot \exp(-kt) \tag{2.12}$$

where n_o is a concentration of excited particles at the reaction outlet in the absence of sample and n is a current concentration. If the bulk processes can be neglected, the deactivation rate constant (i.e., reverse lifetime of an excited molecule in the reactor) depends on reactor sizes, on the probability of a molecule to be deactivated per one collision with the surface (the accommodation coefficient, ϵ), and on the diffusion coefficient of excited molecules.

The aim of these investigations is to obtain the accommodation coefficient ϵ on the basis of information about the effective deactivation rate constant, so it is necessary to discuss the transfer processes of excited molecules in the gas phase, as well as its behavior near the surface (the boundary conditions). In the limit of fast diffusion, when heterogeneous deactivation of excited molecules is the limiting step of the process (the kinetic regime), the calculation of ϵ is rather simple. If this is the case, diffusion equalizes the concentration of excited molecules over the reactor cross-section, so the deactivation rate is connected with a number of molecule-surface collisions by the known relation of the kinetic theory

$$V \frac{dn}{dt} = -\epsilon S \frac{n\overline{v}}{4} \quad (2.13)$$

where V is the reactor volume, S is its surface area, and $\bar{v}$ is the mean thermal velocity of excited molecules.

Integration of this equation gives

$$n = n_o \exp\left[-\frac{S}{V}\frac{\epsilon\overline{v}}{4}t\right] \quad (2.14)$$

and we have

$$k = \frac{S}{V}\frac{\epsilon\overline{v}}{4} \quad (2.15)$$

The case of comparable rates of diffusion and heterogeneous deactivation (the diffusion regime) was earlier considered in for the cylindrical reactor.[77,78] Cylindrical reactors are sometimes inconvenient especially for studying the heterogeneous deactivation on metals.

The coefficients of accommodation on the surfaces of metal wires were measured in the following manner. The sample under investigation had the form of a cylinder. The cylindrical reactor was made of, or covered by, a substrate that weakly deactivates the excited molecules. The sample was inserted into the reactor so they were coaxial. The inlet concentration of excited particles was maintained constant by the microwave discharge or the thermal source. At the outlet of the reactor, the relative intensity of the radiation of the excited molecules was detected as a function of sample length.[78]

The dependence of the relative signal intensity on the sample length allows one to calculate the value of the accommodation coefficient. To perform the calculations, we should take into account that the reactor consists of two parts. The first consists of two coaxial cylinders, and the second is represented by the rest of the cylindrical reactor. The solution of the task for the second part of the reactor is well known;[77,78] the dependence of the concentration of excited molecules along the cylinder axis is exponential. As we have already shown the exponential factor depends on the reactor substrate. Deactivation in the first part of the reactor also has an exponential character. Thus, at the outlet of the reactor, the dependence of a relative signal on the length of the inserted part of the sample is given by

$$I_1/I_2 = \exp[-(k_1 - k_2)(l_1 - l_2)] \tag{2.16}$$

where k_1 is the rate constant for deactivation in the first part of the reactor, k_2 is that for the second part, and $(l_1 - l_2)$ is the variation of length of the cylindrical sample inserted along the reactor axis. Substituting the length l for the time $t = l/u$ (u is the flow velocity), we have

$$I_1/I_2 = \exp[-(k_{i,eff} - k_{2,eff})(t_1 - t_2)] \tag{2.17}$$

One can determine $k_{1,eff}$ and the accommodation coefficient ϵ, for the surface under investigation by measuring the relative signal intensity, providing $k_{2,eff}$ is known. The relationship between k_{eff} and ϵ can be derived from the diffusion equation for vibrationally excited molecules between two coaxial cylinders

$$D\left[\frac{\partial^2 n}{\partial r^2} + \frac{1}{r}\frac{\partial n}{\partial r} + \frac{\partial^2 n}{\partial x^2}\right] = u(r)\frac{\partial n}{\partial x} \tag{2.18}$$

with the boundary conditions at the inner and outer cylinders, respectively

$$D_o\left[\frac{\partial n}{\partial r}\right]_{r=r_o} = \frac{\epsilon u_o}{4} n \tag{2.19}$$

$$D_1\left[\frac{\partial n}{\partial r}\right]_{r=r_1} = \frac{\epsilon u_1}{4} n \tag{2.20}$$

where r and x are the radial and the axial cylindrical coordinates, r_o and r_1 are the radii of the inner and the outer cylinders, respectively; u is the flow velocity; and D_o and D_1 are diffusion coefficients for excited molecules at the temperature of the inner and the outer cylinders, respectively.

The dependence of the diffusion coefficient on the coordinate across flow direction is conditioned by the temperature profile due to the difference between the temperatures of inner and outer cylinders of the reactor. In collaboration with Vasiljev, Kovalevsky, and Ryskin[79] we have developed the method of solution of the task for this reactor, which takes into account the velocity and temperature profiles, longitudinal and transverse diffusion, as well as correct boundary conditions.

It should be mentioned that the reactors of this type can measure the accommodation coefficients in a wider ϵ range in comparison with ordinary cylindrical reactors. Besides this, the surface cleanliness in the experiments can be improved due to the possibility of electric heating of the metal under study.

Thus, measuring the rate constant for deactivation of excited molecules, one can determine the accommodation coefficient ϵ with a given accuracy, so the precision of ϵ depends on that of the rate constant. Because the relationship between ϵ and k is nonlinear, the error ϵ is influenced by both Δk and k. The approximate formula for the overall resistance[80] gives the dependence of $\Delta\epsilon$ on Δk

$$1/k = 1/k_d + a/\epsilon \tag{2.21}$$

where k_d is the diffusion rate constant for deactivation and a is a constant.

On differentiation of this equation, we obtain the relation between relative errors in the accommodation coefficient and the rate constant

$$\frac{\Delta\epsilon}{\epsilon} = \frac{k_d}{k_d - k} \cdot \frac{\Delta k}{k} \tag{2.22}$$

This formula allows us to conclude that $\Delta\epsilon/\epsilon \approx \Delta k/k$ in the kinetic regime ($k << k_d$). In the diffusion regime, when $k \approx k_d$, the value of $k_d/(k_d - k)$ might become much greater than unity, so the precision of the measured accommodation coefficient decreases significantly. The accuracy of determination of $\Delta k/k$ has been estimated by Bray et al.[64] and is approximately equal to 10% at low deactivation rates (the kinetic regime) and 20% at high rates of deactivation (the diffusion regime).

The applicability of the flow methods for studying the heterogeneous deactivation of excited molecules is defined by the presence of effective sources of excited particles as well as the development of sensitive tools for measuring their concentration. The microwave discharge is a common source of excited particles. The detection is made with the aid of optical titration, ESR, LMR, and optical and mass spectroscopes. Concrete examples are given in Chapter 3.

Chapter 3

HETEROGENEOUS DEACTIVATION OF EXCITED MOLECULES

3.1. HETEROGENEOUS DEACTIVATION OF VIBRATIONALLY EXCITED MOLECULES

Previous to our studies, several researchers developed the methods for studying the heterogeneous deactivation* of vibrationally excited molecules and obtained the values of accommodation coefficients for different surfaces. In this section we review the results of work carried out mainly at the Institute of Chemical Physics.

3.1.1. HISTORICAL REVIEW

The first experiments on the deactivation of vibrationally excited molecules were done by Shafer and Klingenberg[81] in 1954. They studied systems in complete vibrational-translational equilibrium. The work had been initiated in connection with the authors' suggestion that some heterogeneous catalytic reactions, such as dissociation, can govern the rate of vibrational energy exchange between an adsorbate and a solid.

The rate constant of ethane decomposition and the accommodation of its vibrational energy turned out to change in a similar manner depending on the surface composition. When calculating the accommodation coefficient of the vibrational energy, these authors assumed that the accommodation coefficients of translational and rotational energy are equal to unity. In this case, even small errors in Knudsen's accommodation coefficient ϵ give rise to significant errors in the accommodation coefficient of the vibrational energy.

The problem of heterogeneous deactivation of vibrationally excited molecules has become of special interest in connection with the creation of the low-pressure gas laser when the vibrational energy transfer from the gas bulk to the reactor wall should be taken into account. At the same time, the progress in laser equipment can substantially extend the studies in heterogeneous deactivation using direct methods of generation and registration of vibrationally excited molecules.

The research done so far can be subdivided into two groups relative to research methods; there are direct methods based on kinetic studies of the luminescence decay of the adsorbed molecules and indirect ones in which the information on the accommodation coefficients is obtained from the

* We use the term "deactivation" for the transition from the excited to the ground state. The terms "quenching", "relaxation", and "decay" are also widely employed.

measurements of the heat release pressure changes in gas irradiated by the infrared (IR) laser, the shifts of the flammability limit, etc.

In 1962, Millikan[82] developed a pulse method to study the deactivation of radiating dipole molecules. The principle of the method is the following. The gas under investigation enclosed in a reactor made from the studied material is irradiated by an intense pulse generating the vibrationally excited molecules. As a radiation source, the high-powered lasers with modulated Q-factor are usually employed. On cutting off the radiation, the fluorescence kinetics of excited molecules is measured. Usually it is described by the exponential dependence

$$I = I_o \exp(-\beta t) \tag{3.1}$$

where β is the effective constant of the deactivation rate of excited molecules. In general it includes three processes: (1) the collisional or nonradiational gas deactivation in the bulk, in this case its rate depends only on the gas pressure and temperature; (2) the heterogeneous deactivation depending on the rate of diffusion to the surface, i.e., on the gas pressure and temperature and on the geometry of the reactor; and (3) the spontaneous radiation transitions.

Thus, the effective deactivation rate constant β is a sum of three rate constants in the general case

$$\beta = k + \delta + \pi \tag{3.2}$$

Here k is the rate constant for homogeneous deactivation and depends linearly on the pressure; δ is the heterogeneous deactivation rate constant which takes into account the diffusion

$$\delta = k_{dif}k_{kin}/(k_{dif} + k_{kin}) \tag{3.3}$$

and π is the rate constant for radiative deactivation.

Pressure dependence of different rate constants for deactivation of vibrationally excited CO_2 molecules is shown in Figure 3.1. As it is seen π is practically constant at a high pressure, but increases with pressure decrease approaching the Einstein coefficient for spontaneous radiation. At a high pressure, the main contribution to the effective rate constant is made by the homogeneous deactivation, but at low pressure by radiative and heterogeneous processes. It should be emphasized that in order to determine the constant δ, very low pressure ($\sim 10^{-1}$ Pa) conditions are required, and consequently, the accumulation of detected signal must be applied.

The accommodation coefficients (Table 3.1) depend slightly on the nature of the surface, which favors the proposal, suggested by Knudsen, that deactivation takes place on the surface completely covered with the layer of adsorbed molecules. The data on deactivation of vibrationally excited CO_2 molecules, obtained by Kovacz et al.,[84] also support this suggestion for some

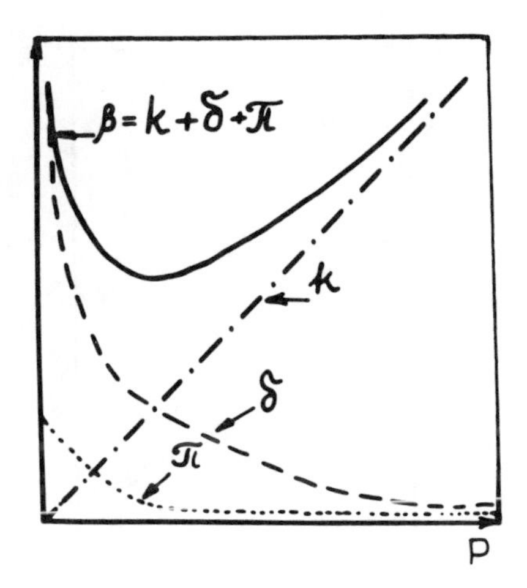

FIGURE 3.1. Pressure dependence of deactivation rate constants of vibrationally excited CO_2 and N_2O molecules.[83]

TABLE 3.1
Accommodation Coefficients of Vibrationally Excited Molecules[83]

Molecule	Surface	T (K)	ε
CO_2	Pyrex®	300	0.4
	Pyrex®	300	0.22
	Brass	300	0.22
	Teflon®	300	0.22
	Quartz	300–1000	0.45–0.05
N_2O	Pyrex®	300	0.29
	Quartz	300–1000	0.33–0.05
HCl	Pyrex®	300	0.45
CO	Pyrex®	300	0.2

surfaces: Pyrex®, brass, Teflon®, etc. These authors arrived at the following mechanism of deactivation. Because all measurements were carried out at room temperature the investigated surfaces are covered with a layer of adsorbed CO_2 molecules. Therefore, deactivation occurs via resonant exchange between the adsorbed molecules and vibrationally excited gas-phase molecules.

The results obtained by Rozenshtein[85] also confirm the fact that excited molecules deactivate in the collision with the surface, covered with adsorbed particles. The temperature dependence of the accommodation coefficient for CO_2 and N_2O on quartz surface, studied in this paper, is shown in Figure 3.2. It may be concluded that with the temperature increase from 300 to 500 K, the accommodation coefficient drops rapidly, probably due to desorption of adsorbed molecules. Further temperature increase results in significant

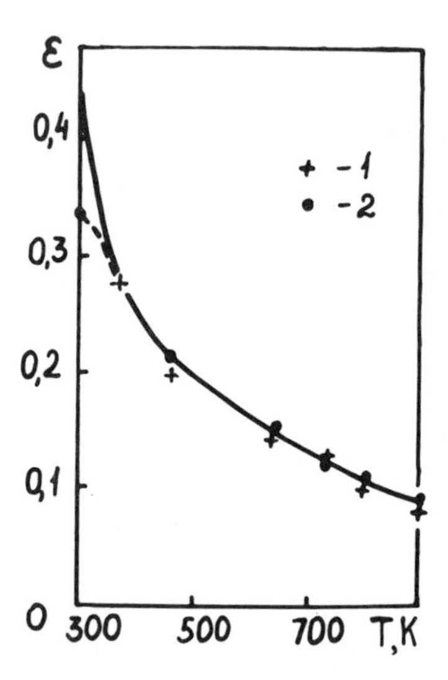

FIGURE 3.2. Temperature dependence of CO_2(1) and N_2O(2) accommodation coefficients on the quartz surface.[83]

decrease in accommodation coefficient which depends, in general, on the nature of the solid. This interpretation is in accordance with the results of CO_2 adsorption on quartz obtained by Villermaux,[86] who has shown that at 500 K the CO_2 desorption from the surface is almost complete.

Shäfer with co-workers[81,87-89] used indirect methods, based on the measurement of the heat removal from thin metal wires at different temperatures to the strip made of the same material which is placed near the wire. Using the equation for thermal accommodation coefficient and supposing that $\epsilon_t = \epsilon_r$ (i.e., the accommodation coefficient for translational energy is equal to that of rotational energy), or $\epsilon_r = k\epsilon_t$, or $\epsilon_t = 1$, these authors have calculated the accommodation coefficients for vibrational energy of CS_2, CO_2, NO_2 molecular adsorption on platinum in the temperature range 273 to 373 K. Deactivation of vibrationally excited N_2O molecules on Pd and Ag has been studied in the same temperature range. In all cases, the accommodation coefficient decreases with increasing temperature. Similar results were obtained for more complex molecules, such as ethane, methane, freon.

Villermaux[86] has suggested measuring the pressure change of hydrogen doped with mercury irradiated by a powerful mercury-discharge lamp. Electronically excited mercury atoms in 6^3P state, generated by irradiation, collide with hydrogen molecules, resulting in their dissociation. The gas-phase recombination of hydrogen atoms leads to the formation of vibrationally excited molecules. It is clear that the rate of heat removal from the bulk depends on the process of heterogeneous deactivation of excited molecules at the reactor walls. The kinetic measurements of pressure change, when the lamp had been turned off, allowed them to calculate the accommodation coefficient of vibrationally excited hydrogen molecules, which was found to equal $\epsilon = 5.10^{-4}$ for quartz at room temperature.

Hunter[90] also determined the accommodation coefficients by measuring the pressure change. If the gas in the cell is exposed to IR radiation, its pressure is increased. A new stationary pressure value is related to the rate of energy removal from the bulk to the wall. Using these relationships, one can calculate the accommodation coefficient. The coefficients for vibrational energy accommodation of CO_2, N_2O, CH_4 on NaCl surface were obtained in this way.

Fedotov, Sarkisov, and Vedeneev[91] have determined the accommodation coefficients of vibrationally excited hydrogen and deuterium on the surface of molybdenum glass, which, in their opinion, was covered with a layer of adsorbed DF molecules. The method is based on the measurement of the upper ignition limit of fluorine mixture with hydrogen or deuterium. Being highly exothermic, the reaction generates the vibrationally excited HF* or DF* molecules which transfer their energy to hydrogen or deuterium. Interacting with fluorine, the vibrationally excited hydrogen initiates new chains. At a sufficiently low pressure the upper ignition limit is strongly influenced by the heterogeneous deactivation of vibrationally excited hydrogen, so it becomes possible to determine the accommodation coefficient using the value of the ignition limit. The experiments were carried out for hydrogen at 300 K and for deuterium in the temperature range 77 to 275 K. The accommodation coefficient grows with the temperature increase. These authors explain their result by the resonant energy exchange between hydrogen and hydrogen fluoride molecules.

Most of the work on heterogeneous deactivation of vibrationally excited molecules has been carried out using flow methods.[92-98] In this case, the subject of studies was the dependence of the concentration of vibrationally excited molecules on the contact time with the reactor surface or with the studied material deposited on the reactor walls. To obtain the accommodation coefficient, one should analyze the processes occurring in a flow reactor. Nalbandjan and Voevodsky[99] have shown that the kinetic equation for deactivation of active particles in the reactor

$$dn/dt = w_o - kn + D\Delta n \tag{3.4}$$

is equivalent to the equation

$$dn/dt = w_o - k\bar{n} + k_t\bar{n} \tag{3.5}$$

where n is the local concentration of active particles, w_o is the rate of generation of active particles, k is the rate constant for homogeneous processes, D is the diffusion coefficient, Δ is the Laplacian operator, k_t is the effective constant for heterogeneous deactivation, and $\bar{n}$ is the average bulk concentration of active particles.

If the rate of the process is limited by the surface processes, but not by the transport of active particles to the surface, then at kinetic regime, the constant k_t is given by

$$k_t = \epsilon v/d \tag{3.6}$$

where ϵ is the probability for an active particle to be deactivated at one collision with the surface, v is the effective thermal velocity of the particles, and d is the reactor diameter.

Neglecting the diffusion along the tube and the gas viscosity, Equations 3.4 and 3.5 are applicable for flow conditions, but in this case $t = x/u_{\alpha v}$, where x is the axial coordinate and $u_{\alpha v}$ is the flux velocity averaged over the flow cross-section. At high values of ϵ, it is necessary to take into account both the diffusion and the velocity profile perpendicular to the flow; that was done by Gershenzon et al.[77,78]

A typical flow experimental arrangement for studying the heterogeneous deactivation of excited particles consists of a source of active particles, the reactor with variable time of gas-surface contact, and the cell for the measurement of the concentration of active particles.

Electric discharge is common and one of the most powerful sources of generation of both vibrationally and electronically excited particles. Unfortunately, discharge is a non-selective source; therefore, in each individual case special efforts are taken to eliminate the unneeded particles. Moreover, reaction of the discharge products with the reactor walls often results in the time dependencies of experimental results.[98] The use of thermal source allows us to avoid these difficulties, but generates significantly lower concentrations of excited particles and requires special tools for gas "cooling", i.e., for thermalization of translational degrees of freedom.

The method of detection for vibrationally excited particles is chosen starting from the structure of a molecule under study. Raman spectroscopy is used for nonpolar molecules, in particular, for excited nitrogen molecules.[98] Vibrationally excited hydrogen molecules were detected with the aid of vacuum UV absorption.[96] To detect the nonpolar molecules, the optical titration is often employed.[78,92,100] This method is based on the registration of light emitted by a small amount of an added dipole reagent which has resonant levels with the particle under investigation. For example, in order to detect the vibrationally excited molecules of deuterium and nitrogen, CO_2 and N_2O additions are employed

$$D_2^*(N_2^*) + CO_2 \rightarrow D_2(N_2) + CO_2^* \rightarrow CO_2 + h\nu$$

Analysis of the processes when the titrated CO_2 addition is mixing with nitrogen shows that the intensity of the light emitted by CO_2 is proportional to the concentration of vibrationally excited nitrogen. This process is used as

TABLE 3.2
The Accommodation Coefficients of Vibrationally Excited Molecules Measured by Flow Methods

Molecule	Surface	T (K)	ϵ	Ref.
N_2	Pyrex®	350	4.5 10^{-4}	93
			4.6 10^{-4}	94
		300	6 10^{-4}	98
	Molybdenum glass	282–603	1–3 10^{-4}	97
	Quartz	300	7 10^{-4}	98
		300–700	(2.3–3.1) 10^{-3}	85
	Stainless steel	300	1.2 10^{-3}	98
	(type 304)			
	(type 321)	300	1 10^{-3}	98
	Steel	300	2.6 10^{-3}	98
	(type 4130)			
	Aluminum alloy			
	(type 5052)	300	1.8 10^{-3}	98
	(type 6061)	300	1.3 10^{-3}	98
	Teflon®	300	6 10^{-4}	98
	Copper	300	1.1 10^{-3}	98
	Silver	295	1.4 10^{-2}	78
	Aluminum oxide	300	1.4 10^{-4}	98
	Boric acid	300–700	3 10^{-3}–5 10^{-4}	85
CO	Pyrex®	300	1.86 10^{-2}	94
CO_2	Pyrex®	300	0.18	94
	Molybdenum glass	300–500	0.45–0.18	85
N_2O	Pyrex®	300	0.14	94
	Molybdenum glass	300–530	0.03–0.01	85
HF	Glass	300	9 10^{-3}	95
H_2	Pyrex®	300	1 10^{-4}	96
	Teflon®	300	0	96
D_2	Quartz	300	9.5 10^{-5}	100
	Teflon®	300	0	100
	NaCl	293–403	(6.3–4.5) 10^{-4}	101

the basis for the method of measuring the accommodation coefficient for fast deactivation of molecules diluted by slowly deactivated gas which is an energy buffer for rapidly relaxing molecules.[65] This method allows us to measure even the higher values of accommodation coefficients (Table 3.2).

It should be mentioned that the applicability of the molecular beam technique in examining the heterogeneous deactivation of vibrationally excited molecules is restricted by high values of accommodation coefficients, at least, greater than 0.1.[35] So, the flow method is the only available one for molecules with small ϵ values. Apparent disadvantages of the flow apparatus include, in particular, the absence of the strict control over the surface cleanliness and roughness. Indeed, at a pressure of 50 to 500 Pa, the gas contamination of even 10^{-6}% covers the surface by a monolayer in 1 min.

3.1.2. DEACTIVATION ON METAL SURFACE

We have studied the deactivation of nitrogen on silver, as well as deuterium and nitrogen on copper and tungsten.[78,79] These materials have been chosen for the following reasons: silver and copper do not chemisorb nitrogen, but tungsten adsorbs nitrogen dissociatively. Copper does not adsorb deuterium, but tungsten adsorbs deuterium dissociatively. In addition, our study of heterogeneous deactivation of vibrationally excited molecules on metals is practically the first investigation of such a kind.

Silver

The reactor was presented by the ensemble of six glass tubes 1.4 cm in diameter and of different lengths (from 21 to 49 cm), that allowed us to obtain the kinetic dependency by varying the contact time at a constant flow velocity. "Active" nitrogen passed through each of these tubes via a multiway cock. The inner surfaces of the tubes were covered by silver. By measuring the relative signal intensity in the optical cell downstream from the tube reactor, the kinetic order and the effective rate constant for deactivation of vibrationally excited molecules were determined. The TiO_2 insert was interposed between the discharge and the reactor. The presence of TiO_2 significantly decreases the concentration of atomic nitrogen and metastable molecules;[78] this excludes the influence of the homogeneous deactivation of molecular nitrogen on atoms. Morgan and Schiff[93] have measured the probability of vibrationally excited nitrogen molecules deactivation in collisions with nonexcited molecules. The value of this probability turned out to be of the order of 10^{-10}, so, under the conditions of our experiments, the deactivation process takes place only on the reactor surface.

Measurements of accommodation coefficients were carried out at a pressure of 500 Pa at 195 K. It was found that deactivation is of the first order. In this case the effective rate constant is given by:

$$k_{eff} = -\frac{u}{l_i - l_j} \ln \frac{I_i}{I_j} \tag{3.7}$$

where u is the mean gas velocity in the tube section, l_i is a length of ith tube, and I_i is the signal of "active" nitrogen luminescence on passing through the ith tube.

The obtained value of effective rate constant for deactivation, k_{eff} = $(160 \pm 20)\ s^{-1}$, is close to that of the diffusion process ($k_{dif} = 230\ s^{-1}$); therefore, it is necessary to take into account the diffusion and Poiseuille profile of the flow velocity in order to determine the vibrational energy accommodation coefficient. Doing so and using the measured value of k_{eff}, we have determined the coefficient of vibrational energy accommodation of molecular nitrogen on silver, $\epsilon = (1.4 \pm 0.4) \cdot 10^{-2}$. Making some assumptions about the mechanism of the process, the obtained value of the accommodation coefficient may be used to evaluate the lifetime of vibration excitation τ_u at the surface. The most probable microscopic mechanism of deactivation for physisorbed molecules, e.g., nitrogen on silver, is the adsorption mechanism.[85,93,94] It is described by the following scheme

$$N_2^* + Z \underset{k_{-1}}{\overset{k_1}{\rightleftarrows}} N_2^*Z; \qquad N_2^*Z \overset{k_2}{\rightarrow} N_2Z; \qquad N_2Z \underset{k_{-3}}{\overset{k_3}{\rightleftarrows}} N_2 + Z$$

where N_2^* is a vibrationally excited molecule of nitrogen and Z is an adsorption site.

Let the concentration of adsorbed vibrationally excited nitrogen be quasi-stationary. Taking into account that the accommodation coefficient is the ratio of a number of deactivated molecules to a number of collisions of these molecules with the surface, we have

$$\epsilon = -\frac{V}{S}\frac{d[N_2^*]/dt}{[N_2^*]\bar{v}/4} = \frac{4k_1k_2(1-\theta)N_o}{(k_2+k_1)\bar{v}} \tag{3.8}$$

where $\bar{v}$ is the mean molecule thermal velocity, V/S is the ratio of volume to surface, θ is the relative surface coverage, N_o is the number of adsorption sites, and $[N_2^*]$ is the concentration of excited gas-phase nitrogen molecules.

If the rate of deactivation far exceeds that of desorption, i.e., $k_2 >> k_{-1}$, then

$$\epsilon = 4k_1(1-\theta)N_o/\bar{v} \tag{3.9}$$

and the accommodation coefficient is equal to the sticking coefficient.

As a rule, the sticking coefficients of diatomic homonuclear molecules exceed the accommodation coefficients of vibrational energy. So, the assumption that the process is limited by vibration quantum transfer to the solid is preferable, i.e.,

$$k_2 << k_1; \quad \epsilon = \frac{4k_1k_2}{k_{-1}\bar{\bar{V}}}(1-\theta)\sigma_o \tag{3.10}$$

Provided that the surface coverage θ obeys the Langmuir isotherm, one may derive the equilibrium constant for adsorption

$$K = \frac{k_1}{k_{-1}} = \frac{\theta}{(1-\theta)[N_2]} \tag{3.11}$$

then

$$\epsilon = 4k_2\theta\sigma_o/\bar{v}[N_2] \tag{3.12}$$

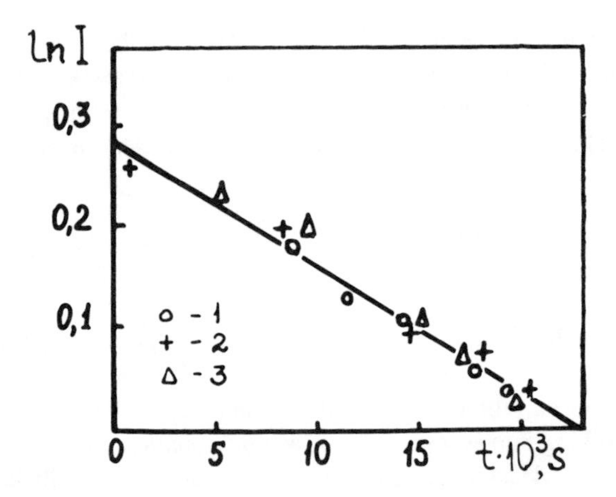

FIGURE 3.3. The logarithm of the signal intensity vs. the contact time of excited nitrogen molecules with copper at 293 K (1–3: different samples) at a pressure of 360 Pa.

The surface coverage can be evaluated using De-Boer equation: $N_o = n\tau$, where N_o is a number of molecules adsorbed on the surface, n is a number of collisions of molecules with the surface per unit time, and τ is a characteristic lifetime of a molecule at the surface, which can be estimated from the Frenkel formula

$$\tau = \tau_o \exp(Q/RT) \tag{3.13}$$

where Q is the adsorption heat. Then, the accommodation coefficient is given by

$$\epsilon = k\tau_o \exp(Q/RT) \tag{3.14}$$

Now the lifetime of vibrational excitation on the surface $\tau_x = 1/k_2$ can be estimated using the measured accommodation coefficient ϵ and the value of adsorption heat of nitrogen 13 kJ/mol at Ag surface.[102] The time τ_x is equal to 10^{-8} s. Similar estimations have been made for molybdenum glass[92]: the deactivation time for vibrationally excited nitrogen has been found to be $\tau_x = 10^{-8}$ s. This time is close to the calculated time of a two-phonon vibrational decay mechanism.[92]

Thus, the study of the kinetics of excited molecule deactivation (irrespective of the excitation type) allows us to evaluate the excitation lifetime without application of special laser methods. Unfortunately, the results obtained are strongly affected by the kinetic model involved.

Tungsten and Copper

We have studied the deactivation of vibrationally excited D_2 and N_2 molecules on tungsten and copper.[79] The results obtained correspond to the first-order equation (Figure 3.3) and demonstrate that the effective rate constant for vibrationally excited nitrogen deactivation at the copper surface depends weakly on pressure. At a lower pressure, deactivation takes place in the kinetic regime because the diffusion coefficient increases. At a pressure of 60 Pa, the effective rate constant proved to equal $k_{eff} = 26 \pm 3\ s^{-1}$. The accommodation coefficient calculated from Equation 3.8 is $(5.0 \pm 0.6)\ 10^{-3}$. If the diffusion is negligible, the accommodation coefficient can be estimated using Equation 3.9. Its value is $(3.9 \pm 0.5)\ 10^{-3}$. Therefore, to calculate the accommodation coefficient using the data obtained at a pressure of 150 Pa, we applied Equation 3.9. The ϵ value obtained $(4.7 \pm 0.5)\ 10^{-3}$ is close to one obtained at a pressure of 360 Pa with allowance for diffusion.

In the case of vibrational deactivation of excited molecular nitrogen on the tungsten surface, the experimental dependence of the logarithm of relative signal intensity on the contact time of gas with the sample is linear irrespective of the gas pressure. The resulting value of the effective rate constant has proved to be $14 \pm 2\ s^{-1}$. It is much less than the value obtained for nitrogen deactivation on copper that is evidence of the kinetic regime. The value of the N_2^* accommodation coefficient, calculated using Equation 3.8, is $(2.2 \pm 0.3)\ 10^{-3}$.

Measurements, made at different pressures, demonstrated the independence of the effective deactivation rate constant on pressure. Because the diffusion coefficient of deuterium is significantly less than that of molecular nitrogen, the diffusion can be neglected. Equation 3.8 gives the values of the accommodation coefficients for molecular deuterium deactivation on tungsten surface $(1.0 \pm 0.2)\ 10^{-3}$ and on copper $(1.2 \pm 0.2)\ 10^{-3}$. Thus, the accommodation coefficient of deuterium is less than that of nitrogen on the same surface. This is likely the consequence of differences in both the adsorptivities and relaxation rates of these gases. It is known that the latter depend strongly on a number of phonons involved in the process of vibrational relaxation.

For both gases the accommodation coefficient on copper exceeds that on tungsten, which does adsorb nitrogen and deuterium. Because the measurements were carried out at a pressure when the tungsten surface was covered completely, we suppose that deactivation occurs preferably during the physical adsorption of an excited molecule on the chemisorbed layer on tungsten. Deactivation is more effective on the clean surface (e.g., silver, copper, which do not chemisorb nitrogen) because there is a possibility to transfer the vibrational energy of physisorbed molecule to free electrons of the solid.

Microscopic Mechanism of Deactivation

Consider the excitation of electron-hole pairs as a possible microscopic mechanism for deactivation of a diatomic vibrationally excited molecule near the metal surface.[103,104] Let us calculate the probability of one-quantum vibrational transition due to energy transfer to one electron of the metal. Suppose that an excited molecule is physisorbed as it follows from our experiment, so the metal does not perturb strongly electron wave functions of the molecule, and its vibrational frequency is not sufficiently changed. To describe the metal, we use the model of free electron gas in $z > 0$ half-space with the electron potential in the form of a rectangular step. The molecule is modeled by a harmonic oscillator.

In one-electron approximation, the Hamiltonian of the system is given by

$$H = H_e + H_M + V \tag{3.15}$$

where H_e and H_M are electron and molecular Hamiltonians, respectively, and V is the potential of electron-molecular interaction.

The total potential φ for a metal electron is the sum of the molecular and induced potentials. The latter appear due to the influence of the rest of metal electrons disturbed from their equilibrium positions by the electric field of the molecule. The expressions for the total potential φ have been obtained for the model of "mirror reflection of electrons" from the vacuum-metal boundary for the dipole and quadrupole oriented perpendicular and parallel to the surface. The deactivation probability has been calculated to the first order of the perturbation theory in the operator V. It was assumed that $k_B T << h\omega << \epsilon_F$, where k_B is the Boltzmann constant, T is the temperature, $h\omega$ is the energy of vibrational quantum, and ϵ_F is the Fermi level of the metal.

For molecules with the dipole transition, the values of the deactivation probability for the perpendicular and parallel molecular orientation with respect to the metal surface are given by

$$w_D^{\perp} = 8\ 10^{11} \frac{1}{M} \left(\frac{d\mu}{dR}\right)^2 J_1^{\perp} s^{-1} \tag{3.16}$$

$$w_D^{\prime\prime} = 4\ 10^{11} \frac{1}{M} \left(\frac{d\mu}{dR}\right)^2 J_1^{\prime\prime}\ s^{-1} \tag{3.17}$$

where M is the reduced molecular mass, $d\mu/dR$ is the derivative of the dipole momentum with respect to internuclear distance taken at the equilibrium point, and $J_1^{\perp}$, $J_1^{\prime\prime}$ are integrals depending on the distance to the metal surface.

For molecules with the quadrupole transition

$$w_Q^{\perp} = 6.6\ 10^{11} \frac{1}{M} \left(\frac{dQ}{dR}\right)^2 \frac{J_2^{\perp} d}{r_s^2}\ s^{-1} \tag{3.18}$$

$$w_Q^{\prime\prime} = 2.5\ 10^{11} \frac{1}{M} \left(\frac{dQ}{dR}\right)^2 \frac{J_2^{\prime\prime} d}{r_s^2}\ s^{-1} \tag{3.19}$$

Here dQ/dR is the derivative of the quadrupole momentum, and r_e is the mean distance between electrons in the metal in atomic units. It is connected with the electron density n by the following relation

$$n^{-1} = \frac{4}{3} \pi (r_s r_B)^3 \tag{3.20}$$

where r_B is the Bohr radius. For typical metals $r_s = 2 \div 6$. The calculations have demonstrated that for $\hbar\omega/\epsilon_F << 0.3$, the magnitudes of all above integrals in Equations 3.16 to 3.19 depend weakly on the energy of the vibrational quantum.

The contribution of exchange interaction to the deactivation probability has also been estimated in the first order of the perturbation theory. Calculations have shown that in the case of physisorbed molecules the multipole contribution is dominant for polar molecules. For nonpolar molecules the values of exchange and quadrupole contributions may be of the same order.

Consider Equations 3.18 and 3.19 to discuss the results and to compare them with the experiment. At room temperature the molecular lifetime in a physisorbed state is 1 to 2 orders longer than the time of flight d/v in the vicinity of the surface in the collision process without adsorption (here d is a characteristic length of the surface potential). So deactivation occurs predominantly from the adsorbed state. The vibrational accommodation coefficient in this case is given by

$$\epsilon = s\tau_a w(d) \tag{3.21}$$

where s is the sticking probability, τ_a is the lifetime in the adsorbed state estimated using the Frenkel formula in Equation 3.13. Numerical estimations of the deactivation probability and the accommodation coefficient for nitrogen on silver ($r_s = 2.9$) give in $\omega \approx 10^8$ s^{-1} and $s = 5\ 10^{-3}$. In this case, Q_α value was taken to be 15 kJ/mol,[102] $d \approx 0.2$ nm, dQ/dR = 2.8 D/nm, $s \approx 1$ at 300 K. The obtained ϵ value coincides with that measured experimentally $1.4 \cdot 10^{-2}$. Similar estimations for CO molecule deactivation on silver ($d\mu/dR = 3.14$ D/nm) give $w \approx 7 \cdot 10^9$ s^{-1}. Note that w(d) does not practically depend on temperature; hence, the temperature dependence of the accommodation coefficient is due only to τ_α, i.e., in fact, to the adsorption heat.

At 600 to 1000 K, the exponential dependence of ϵ on T is changed by a smoother one because deactivation in the adsorbed state is no longer. Below the room temperature this law ceases to be true when the relative coverage becomes of the order of unity. Note that at 300 K, $p \approx 10^2$ Pa, and $Q_\alpha \approx 8 \div 20$ kJ/mol, the relative coverage is $10^{-5} \div 10^{-6}$.

Rather weak dependence of the deactivation probability on the distance between the molecule and the metal surface leads to independence of the accommodation coefficient on the initial surface cleanliness. For example, in the case of a quadrupole molecule an increase in the distance to the surface of 0.1 to 0.15 nm, that corresponds approximately to additional monolayer coverage, results in the decrease of w magnitude tenfold. However, if the molecules from monolayer coverage have vibration frequencies resonant to those of studied molecules, the mechanism of excitation transfer may change (exchange mechanism). Because the probability of resonant v-v exchange, as a rule, is greater than the w value from Equations 3.16 to 3.19, the excitation is transferred firstly to an adsorbed molecule and then to conduction electrons. If this is the case, the accommodation coefficient is defined by the probability of v-v exchange. According to Equations 3.16 to 3.19, the probability of electron-hole excitation depends weakly on the nature of a metal but is affected by the type of the molecule transition (dipole or quadrupole). That is the reason, in our opinion, why the accommodation coefficient obtained for N_2 and D_2 on different metals varies within one order of magnitude, $10^{-2} \div 10^{-3}$ at 300 K, despite considerable difference in vibrational quanta of these molecules. This fact can be explained neither by the multiphonon mechanism[105] nor by the mechanism of vibrational energy transfer to the translational motion of a molecule.[106] In both cases, w must depend exponentially on the vibrational quantum energy. Moreover, the temperature dependence of ϵ for these mechanisms differs substantially from that observed in experiments.[106]

In conclusion, it should be noted that we have considered deactivation mechanisms for diatomic molecules only; however, Equations 3.16 to 3.19 are valid for more complex molecules provided that $d\mu/dR$ and dQ/dR are derivatives with respect to the normal coordinate of proper modes. It should also be realized that since vibrational quanta of complex molecules are small, the mechanism of the energy transfer may change. In this case, the relaxation of a vibrational quantum into a few lattice phonons might become an alternative to the electron deactivation mechanism. Taking into account that physisorption heats for complex molecules are substantially higher than those for diatomic molecules ($Q_{NH_3} \approx 25$ to 36 kJ/mol), it becomes clear why the values of accommodation coefficients for complex molecules on metals are on the order of unity.

All facts mentioned above allow us to draw the conclusion that in the case of diatomic molecules (N_2 and D_2) on metals, the electron mechanism is the main channel of deactivation.

3.1.3. DEACTIVATION ON SEMICONDUCTOR SURFACE

We have obtained the experimental values of vibrational accommodation coefficients for molecular nitrogen and deuterium on the silicon surface; they are $1 \cdot 10^{-2}$ and $4 \cdot 10^{-3}$, respectively.[107] These values far exceed those for dielectrics. It seems difficult to understand because the concentration of conduction electrons in our sample is less than $2 \cdot 10^{15}$ cm^{-3} (the n-type conductivity). If the vibrational spectrum of a solid does not contain the phonons with energies comparable to the molecular vibration frequency ω, the multiphoton exchange mechanism[101] also gives very low deactivation rates. Note that the frequency of optical phonons in silicon is 500 cm^{-1},[108] compared with 3120 cm^{-1} for D_2 and 2360 cm^{-1} for N_2.

Consider the possible contribution of bulk semiconductor electrons.[109] If the energy $h\omega$ of a molecular vibrational quantum far exceeds the semiconductor energy gap E_g, deactivation can occur through the electron mechanism described previously for metals. For intrinsic semiconductor, such as germanium with $E_g = 0.74$ eV or silicon with $E_g = 1.1$ eV, the interband transitions for one-quantum molecular ones are forbidden. An estimation of the contribution of bulk impurity electron states reveals that this mechanism is also negligible.[109]

Consider one more possible channel of excitation energy loss. It is well known that electron states corresponding to surface "dangling" bonds arise on a clean semiconductor surface. These states are localized in the vicinity of the surface since their wave functions rapidly decay into the bulk with a characteristic length of a few angstroms. Their density may be as much as 10^{15} cm^{-2}. Let us estimate the probability of transition from the first to the ground vibrational level in a diatomic molecule localized at a distance d from the semiconductor surface, when the vibrational energy is transferred to one of surface electrons. The Hamiltonian of the system is again taken in the form Equation 3.15 with the same molecular Hamiltonian H_M. As previously, we suppose that excited molecules cannot chemisorb on the semiconductor surface, and a characteristic molecular size is $a << d$.

The deactivation probability w of a dipole molecule with perpendicular and parallel orientation relative to the surface is given by[109]

$$w_D^{\perp} = 10^{12} \frac{m^*}{\epsilon_s + 1} \frac{1}{M} \left(\frac{d\mu}{dR}\right)^2 \frac{1}{k_F d} \quad s^{-1} \tag{3.22}$$

$$w_D'' = \frac{1}{2} w_D^{\perp} \quad s^{-1} \tag{3.23}$$

where m* is the effective mass of surface electrons in units of the free electron mass m_e, ϵ is the dielectric constant of semiconductor, and k_F is the Fermi wave vector for 2-D electrons.

For molecules with the quadrupole transition we have

$$w_Q^{\perp} = 1.2\ 10^{11} \frac{m^*}{\epsilon_s + 1} \frac{1}{M} \left(\frac{dQ}{dR}\right)^2 \frac{1}{k_F d^3} \quad s^{-1} \tag{3.24}$$

$$w_Q'' = \frac{3}{8} w_Q^{\perp} \quad s^{-1} \tag{3.25}$$

where d is in angstroms, and k_F in $Å^{-1}$. Equations 3.22 to 3.25 are derived in suggestion that $(h\omega/\epsilon_F) << 1$.

Let us apply these formulas for N_2 and D_2 deactivation at the silicon surface. For deuterium molecules the reduced mass M = 1 a.u.m., dQ/dR = 2.9. Let $d + z^* = 3Å$, $\epsilon_s = 12$. At the frequency corresponding to D_2 vibrational transition $(v = 1) \rightarrow (v = 0)$, we have $k_F = 1.15\ Å^{-1}$, $m^* = 5m_e$, $\Delta k = 0.2k_F$ that is consistent with values of the electron density n = $2.1 10^{15}\ cm^{-2}$ and a characteristic width of the surface band ~1 eV at Si(111) surface.[110] From Equation 3.24 for perpendicular orientation we have $w^{\perp} \sim 5 \cdot 10^9\ s^{-1}$ by assuming that $\Delta kd\ I << 1$ and $k_F d >> 1$. Taking the adsorption heat $Q_\alpha = 8.5$ kJ/mol, we obtain $\epsilon = 1.4\ 10^{-2}$. This value slightly exceeds the experimental one $4 \cdot 10^{-3}$. The above estimation is correct when the surface band is not completely filled. Actually, it is known that the silicon surface is reconstructed depending on the mode of preparation. For example, two structures (2 × 1) and (7 × 7) are observed on Si(111) surface. The reconstruction causes the single surface band to split into two subbands, the lower being completely filled and the upper empty. In the case of (7 × 7) structure, these subbands are in contact and even overlap each other, whereas a gap of ~0.15 eV exists in the case of (2 × 1) structure. The gap existence does not impose limitations on the transitions for D_2 and N_2 molecules, but may decrease the probability value because the electron matrix elements and the density of states may be less than it was assumed in our model. That would be the reason why the theoretical ϵ value is slightly overestimated.

Because the values dQ/dR for nitrogen and deuterium are almost equal (2.8 and 2.9 D, respectively), the accommodation coefficient of D_2 must be 7 times greater than that of N_2 in view of the difference in mass values. However, it does not agree with the experimental results. This discrepancy is explained by different adsorption heats, for nitrogen $Q_\alpha = 19.3$ kJ/mol, and for deuterium $Q_\alpha = 8.4$ kJ/mol, that leads to substantially different lifetimes of these molecules in the adsorbed state. At a room temperature N_2 lifetime is 20 times greater than that of D_2.

In experimental conditions the silicon surface is permanently exposed to the nitrogen or deuterium flow containing atomic species. The surface oxide film seems to be removed, which is indirectly confirmed by the fact that the measured values of accommodation coefficients differ substantially from those for the quartz surface.[101] The presence of atoms in the flow results in their adsorption and partial coverage of the silicon surface. The coverage degree is governed by the two competitive mechanisms: atomic sticking and their recombination through the Rideal or Langmuir-Hinshelwood scheme. The recombination coefficient of H atoms on Si(111) is $(2 \div 3)\ 10^{-2}$.[111] In suggesting that the atomic sticking coefficient is sufficiently large, for example, 0.1 to 0.3, approximately 10 to 20% of the surface area remains unoccupied. Note that the recombination coefficient was obtained with the supposition of negligible excitation of molecules created in the recombination process. The correction for this effect may increase substantially the part of unoccupied surface.

Because the surface is still covered partly by atoms, there is one more possibility to explain the deactivation mechanism. It is well known that the SiD_4 molecule has vibrational modes with frequencies of 1545 and 1597 cm^{-1}, whereas $h\omega = 3118\ cm^{-1}$ for deuterium. A comparison between these values shows that a conversion of one deuterium quantum into two quanta of SiD_4 vibrational is possible. In the framework of this mechanism it is still difficult to explain such a good correlation between accommodation coefficients for D_2 and N_2 molecules.

The mechanism of vibrational energy transfer to electrons of surface states is applicable to semiconductors without an energy gap (or with a small one) between the conduction and valence surface bands. There are electron surface states at the metal surfaces as well. They are usually located a few electronvolts lower than the Fermi level and are of little importance in vibrational energy deactivation. If deactivation of electronically excited molecules is considered, this channel should be taken into account both for semiconductors and metals.

The above scheme of vibrational energy transfer suggests a new method for the study of energy structure of electron surface states. It is founded on the investigation of heterogeneous deactivation of different vibrationally excited molecules regarded as energy probes. The main advantage of this method is a small contribution from bulk semiconductor bands.

3.1.4. DEACTIVATION ON NaCl AND TEFLON® SURFACES

The choice of these solids is due to the following reasons. First, sodium chloride and Teflon® are dielectrics. Second, NaCl is an ionic crystal with a well-known phonon spectrum. Teflon® surface is of special interest, since it is shown[85] that heterogeneous deactivation of vibrationally excited deuterium, as well as its atomic recombination, is extremely weak on the Teflon® surface. Therefore, Teflon® seems to be the most suitable coverage for the study of net homogeneous processes with hydrogen participation. Interaction of nitrogen with the Teflon® surface has not been investigated.

Deactivation of deuterium molecules on NaCl surface at 193 K is first order in $[D_2]$ with $k_{eff} = (44\pm)\ s^{-1}$. This k_{eff}, value is indicative of almost absolute heterogeneous nature of the process occurring in the kinetic regime. In this case the accommodation coefficient of the vibrational energy turned out to be $\epsilon = kd/v = (6.3 \pm 0.7)\ 10^{-4}$, where d is the reactor diameter and v is the thermal velocity of D_2^* molecules at 293 K. For other temperatures $\epsilon = (5.6\pm)\ 10^{-4}$ at 330 K, $(5.4 \pm 0.7)10^{-4}$ at 358 K, and (4.5 ± 0.8) 10^{-4} at 403 K.

The descending temperature dependence may be explained by the adsorption-phonon mechanism of deactivation, according to which it occurs from the adsorbed molecular state, and the vibrational energy is transferred to phonons in the solid. The decrease of the accommodation coefficient with the temperature growth results from the reduced molecular lifetime in the adsorbed state. When a substantial part of the surface is covered with adsorbed molecules, an increase of the accommodation coefficient with the temperature growth is possible provided that the adsorbed molecules have vibrational frequencies resonant with those of deactivating molecules. A similar effect has been observed in studies of deactivation of vibrationally excited deuterium molecules on the reactor walls covered with a layer of adsorbed molecules.[91] Clearly, the accommodation coefficient in this case far exceeds the value obtained in our experiments.

Yellow luminescence has been observed in studies of the atomic recombination of deuterium on NaCl films deposited on the quartz tube. Its radiation spectrum corresponds to the sodium D-line. The radiation intensity increases with increasing temperature. Curiously, it was found that the radiation intensity depends on the film thickness. The greatest effect was observed for thin films, but heating even up to 673 K failed to induce the luminescence when the film thickness was 0.2 to 0.3 mm. The recombination coefficient of D atoms has been measured on sufficiently thick films in the absence of radiation and found to be $\gamma = 4\ 10^{-4}$. It was greatly increased on thin films in the presence of radiation. The latter effect can be attributed to the appearance of Na atoms in electronically excited states.

Deactivation of vibrationally excited nitrogen molecules on Teflon® has been studied at a room temperature and nitrogen pressure in the range 100 to 300 Pa.[101] The value of the accommodation coefficient varied in the course of experiment that is probably due to the treatment of the Teflon® surface with "active" nitrogen. The greatest value of the accommodation coefficient obtained in a set of seven experiments is $4.5\ 10^{-4}$, so the actual value $\epsilon < 4.5\ 10^{-4}$. Black et al.[98] have obtained $\epsilon = 6 \cdot 10^{-4}$ that is very close to our value.

3.1.5. D_2 DEACTIVATION ON MOLECULAR CO_2 CRYSTAL SURFACE

Energy exchange processes between D_2 and CO_2 molecules are well understood in the gas phase;[112] that is why the molecular CO_2 crystal has been chosen for our purposes. It may be suggested that energy exchange between the D_2^* molecule and the molecular CO_2 crystal will have the same characteristic features as in the gas phase. The measurements were carried out using a flow vacuum setup which is schematically shown in Figure 3.4. Gaseous CO_2 was employed for the titration.

It has been shown previously[85] that in similar conditions the vibrational temperature of deuterium is approximately 1100 K, which corresponds to the occupancy of the first vibrational level, ~1% of all molecules. This is also confirmed by measurements of vibrational distribution of hydrogen molecules downstream from the microwave discharge.[113] It was found that the concentration of H_2 (v = 1) is approximately two orders lower than the concentration of unexcited molecules, and H_2 (v = 2) concentration is two orders lower

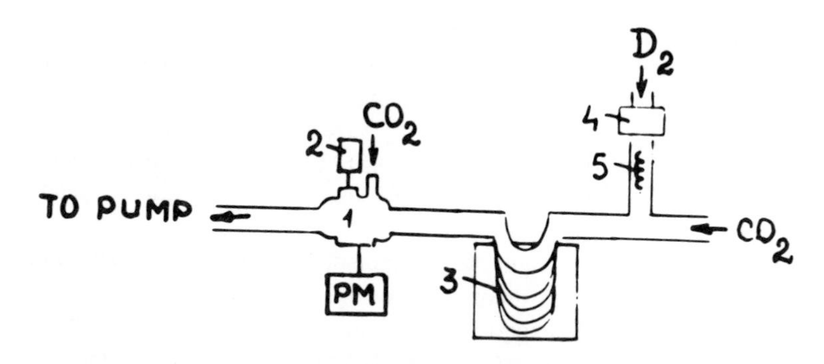

FIGURE 3.4. The scheme of the flow vacuum setup for studies in deactivation of excited D_2 molecules on solid CO_2: (1) titration cell; (2) detector; (3) reactor made of tubes with different lengths; (4) discharge tube; (5) catalyst of atomic recombination.

than that of H_2 (v = 1). So, we have dealt in our experiments mainly with D_2 molecules in the first vibrational level.

All deactivation processes in the gas phase should be excluded to obtain the correct results about the surface under investigation. In our conditions the most effective deactivation partners of D_2^* molecules are deuterium atoms and resonant molecules similar to CO_2. Nevertheless, it was shown that deactivation of vibrationally excited D_2^* molecules interacting with D atoms in the gas phase does not lead to appreciable errors in measurements.[101]

The measured value of the D atom concentration in our experiments is $2.6 \cdot 10^{14}$ cm^{-9}. To diminish the atomic concentration, a thin platinum wire was placed in the flow downstream of the discharge. This results in the decrease of the atomic concentration by a factor of 15 to 20. In this case, the effective deactivation rate constant for D atoms is $k_{eff} = k[D] = 1.5$ to 2 s^{-1} at room temperature and 0.4 s^{-1} at temperatures lower than 150 K, which is beyond the accuracy of our measurements. The freezing out of the CO_2 layer on the reactor walls was performed from CO_2 flow at 163 to 169 K. A thin solid CO_2 layer appeared on the walls. In the first place the accommodation coefficient was determined at a temperature of liquid nitrogen. Later this value was used as the reference point in the temperature dependence. The reactor for measurements at constant temperature consists of three U-form glass tubes of the same diameter, but different lengths placed in the Dewar vessel. The reaction order and the effective rate constant of vibrationally excited molecules were determined from the relative signal intensity. Because a signal variation is defined only by the difference in the tube lengths, the effective rate constant due to D_2^* interaction with the CO_2 layer is measured. The results obtained have shown that deactivation is of the first order in $[D_2^*]$ with the effective deactivation rate constant $k_{eff.} = (34.2 \pm 5)$ s^{-1}.

Consider the diffusion processes involved in our case. At 273 K and at atmospheric pressure, the deuterium coefficient of self-diffusion D for non-excited molecules is equal to 0.84 cm^2/s. For homonuclear diatomic molecules the diffusion coefficient of excited molecules does not differ from that of nonexcited ones.[114] In our conditions the diffusion rate constant $k_{eff.} = 14.63\ D/d^2 = 1500\ s^{-1}$ is many times higher than the effective deactivation rate constant, and, therefore, the diffusion may be neglected. The accommodation coefficient equals $\epsilon = k_{eff.}\ d/v = (4.4 \pm 0.5) \cdot 10^{-4}$. Check tests for different pressures have shown that the accommodation coefficient did not depend on pressure.

Temperature dependence of the accommodation coefficient is determined from the concentration measurements of excited molecules, when the reactor is being defrozen. In the experimental temperature range, deactivation is of the first order. The effective rate constant and the accommodation coefficient are defined by equations

$$\ln \frac{I(T)}{I(T_o)} = \left[- \frac{k(T)}{u(T)} - \frac{k(T_o)}{u(T_o)} \right] 1 \tag{3.26}$$

$$\epsilon(T) = \left[k(T_o) \frac{u(T)}{u(T_o)} - \ln \frac{I(T)}{I(T_o)} \frac{u(T)}{1} \right] \frac{2R}{u(T)} \tag{3.27}$$

where $k(T_o)$ is the effective rate constant at 77 K, u(T) is a flow velocity at a given temperature, v(T) is a thermal velocity of molecules, and l is the reaction length.

On increasing the reaction temperature up to 140 K, the signal increases that is indicative of decreasing the accommodation coefficient. The temperature dependence of the accommodation coefficient at 230 Pa, $v(T_o) = 6.9$ m/s, and in the vicinity of 77 K is well described by the following empirical equation

$$\epsilon = 1.45 \cdot 10^{-4} \exp(200/RT) \tag{3.28}$$

The number of different possible microscopic deactivation mechanisms for the molecular crystal is not large, as in the case of metals or semiconductors. There are no free electrons in the CO_2 molecular crystal, and excitation energies of bound electrons are much greater than the deuterium vibrational quantum. So, there is no need to consider the question of energy transfer to electrons in the CO_2 crystal.

In principle, heterogeneous deactivation of any excited molecule can occur in two ways. The first is the so-called impact deactivation, i.e., energy loss in inelastic molecular scattering at the surface. The second proceeds via preliminary adsorption of the deactivated particle. Then the deactivation probability is given by

$$\epsilon = (1 - s)w_{imp} + snw_{ad} \tag{3.29}$$

where s is the sticking coefficient, w_{imp} and w_{ad} are deactivation probabilities via the impact and adsorption mechanisms, respectively, and n is the number of molecular vibrations in the adsorption well over a period of the adsorption lifetime.

Consider the question of what channel is dominant. Experimental data on the sticking coefficient of deuterium on CO_2 surface are absent. Therefore, we are forced to use a simple mechanical model which is in fair agreement with the experimental data.[115,116] The calculations have shown that the molecular sticking probability is equal to unity, when its kinetic energy is less than the critical sticking energy E_{or}, and it is equal to zero in the opposite case. The critical sticking energy is expressed as follows

$$E_{cr} = 4\pi^2(m/M)^2 Q_e \exp[-(mQ_e/MQ_\alpha)^{1/2}] \tag{3.30}$$

where m and M are molecular and adsorbent masses, respectively, Q is the sublimation heat of an adsorbent (CO_2 in our case), and Q_α is the adsorption heat.

As it follows from Equation 3.30, E_{cr} strongly depends on the mass ratio of an adsorbed molecule to that of adsorbent. Karlov and Shaitan[117] estimated the critical sticking energy for deuterium on the CO_2 surface on the supposition that the deuterium physisorption heat is an average energy between the CO_2 and deuterium sublimation heats. They found E_{cr} = 2.6 kJ/mol that gives rise to values of the sticking coefficient in the range from 0.99 to 0.92 in the temperature interval 77 to 125 K.

Thus, the impact deactivation channel can be neglected on account of high values of the accommodation coefficients as well as the number of molecular collisions with the surface in the adsorption well is much greater than unity. Hence, Equation 3.29 reduces to

$$\epsilon = snw_\alpha \exp(Q/RT) \tag{3.31}$$

where Q is the depth of the adsorption well.

When deactivation of an excited molecule occurs through the adsorption state, there are three possible microscopic mechanisms of energy losses: (1) the transfer of molecular vibrational energy into its own translational degrees of freedom, (2) energy exchange with the lattice phonons, and (3) the transfer to the inner molecular vibrations of CO_2 molecules.

Consider each of these possibilities as applied to our task, but note that there are, undoubtedly, systems in which one or the other of these channels predominates.

To estimate the probability of vibrational energy transfer to its own translational motion, we take advantage of the theory which has been developed for collision of vibrationally excited molecules with inert ones in the gas phase. The Landau-Teller approach is inapplicable in this case, because the deuterium vibrational quantum far exceeds the molecular kinetic energy. Therefore, we invoke the quantum transition probability calculated by Jackson and Mott with the aid of the distorted wave method. Taking into account long-range as well as short-range interaction at low temperature, the averaged transition probability can be expressed in a form

$$w = \frac{8\pi^2}{\hbar^2}\frac{m^2}{M}\frac{\hbar\omega}{\alpha^2}\exp\left(-\frac{2\pi m}{\alpha\hbar}\sqrt{\frac{2\hbar\omega}{m}} + \frac{2\pi^2 mkT}{\alpha^2\hbar^2} + \frac{Q}{RT}\right) \quad (3.32)$$

where m is the reduced mass of colliding particles, M is the reduced oscillator mass, $\hbar\varphi$ is the energy of vibrational quantum, and α is the depth of the adsorption well.

At 100 K and for Q ~ 4 kJ/mol, which is typical for physisorption, the transition probability is $\sim 10^{-8}$, which is 4 orders less than the experimental value of the accommodation coefficient. Thus, this mechanism is of little importance in D_2^* heterogeneous deactivation on the CO_2 surface.

The second mechanism is the vibrational energy exchange with CO_2 lattice phonons. The phonon spectrum of CO_2 crystal is well known. The frequencies of libration and translation vibrations are shown to be less than 150 cm^{-1} using IR and Raman spectroscopy. Thus, the transition under consideration is a multiphonon process with a very low probability. To estimate the probability of such a process we use experimental data on vibrationally excited molecular lifetimes in intergas solid matrices. Dibatt et al.[118] have measured the lifetimes of excited CO molecules in the argon matrix and that of ammonia in the argon and nitrogen matrices; they are 508 and 2 ms for CO and NH_3, respectively. The lifetime of vibrationally excited nitrogen in the argon matrix is ~1 s.[119] This is explained by the fact that deactivation of one vibrational nitrogen quantum requires excitation of more than 23 lattice phonons. So, the molecular deactivation probability per one vibration is of the order of 10^{-12}, which is also absolutely inconsistent with experimental results.

Finally, consider the transfer of D_2 vibrational energy to inner molecular CO_2 vibrations. The CO_2 molecule has the following modes: symmetric stretching vibration with ν_1 = 1388 cm^{-1}, asymmetric stretching vibration with ν_9 = 2349 cm^{-1}, and twice-degenerated bending mode with ν_2 = 667 cm^{-1}. Therefore, deuterium deactivation can occur via two channels

$$D_2(v = 1) + CO_2(000) \rightarrow D_2(v = 0) + CO_2(101) - 48\ cm^{-1}$$

$$D_2(v = 1) + CO_2(000) \rightarrow D_2(v = 0) + CO_2(001) + 620\ cm^{-1}$$

where 48 cm^{-1} and 620 cm^{-1} are resonance defects.

The first channel has been considered by Moore et al.[112] It is the authors' opinion that D_2^* vibrational energy transfers into energy of CO_2 asymmetric stretching and bending modes, but such a transition is forbidden by selection rules for dipole-quadrupole interaction. In order for the probability of this process to be estimated, we shall use the theory developed previously for the gas-phase processes. Deactivation via the first channel has a negative resonance defect. We assume that the energy required for this transition is taken either from acoustic phonons of the solid, or is due to the transition to a lower level in the adsorption well for vibrations of the deuterium molecule as a whole.

The latter can be treated as local vibrations of a light mass. Because the spectrum of such vibrations has a discrete character, it is difficult to expect that there is a proper level located 48 cm^{-1} lower than the initial one. Therefore, the transitions of the deuterium molecule in the adsorption well as a whole were neglected for this channel.

Now, suppose that the deficient energy is taken from the acoustic phonons in the solid.[101] Assume that CO_2 molecular vibrations as a whole can be modeled by a harmonic oscillator with a frequency of 48 cm.$^{-1}$ To describe these vibrations we take advantage of the model of radially vibrating shells. Suppose that the adsorbed D_2^* molecule moves in the repulsive potential perpendicular to the surface in a straight line between D_2^* and CO_2 molecular centers of mass. The Hamiltonian of the system can be written in a form

$$H = H_{mol} + H_{ads} + V \tag{3.33}$$

where H_{mol} is the Hamiltonian of the adsorbed molecule, H_{ads} is the adsorbent Hamiltonian, and V is the interaction potential.

The transition probability is defined in the first order of the perturbation theory as V. The repulsive potential between colliding molecules is given by

$$V = V \exp[\alpha(r + \chi_{kl} + x)] \tag{3.34}$$

where r is the distance between D_2 and CO_2 centers of mass, χ_{kl} is the sum of displacement projections of the kth deuterium atom and l^{th} CO_2 atom onto the surface normal, x is the projection of CO_2 displacement as a whole onto the surface normal, and α is a parameter. CO_2 motion as a whole is described by the harmonic wave functions, and D_2 molecular motion by quasiclassic wave functions of translational motion.

It has been shown[101] that for the parallel orientation of the CO_2 axis with respect to the surface normal, the collision does not give rise to the desired transition since in this case the excitation of CO_2 deformation modes is forbidden. In a similar manner, the perpendicular orientation is also ineffective because the valence mode of excitation is forbidden.

Taking into account that CO_2 molecules can occupy only the first vibrational level owing to low temperatures, the transition probability can be written in a form

$$w = \frac{2\pi}{\hbar^2}\frac{m\bar{E}}{4\pi\hbar\alpha^2}\left|10^{-2}\alpha^3\zeta\zeta_2\zeta_3\right|^2 \frac{\alpha^2\hbar^2}{2\mu\hbar\omega(CO_2)}\frac{\exp[-\hbar\omega(CO_2)/k_BT]}{1-\exp[-\hbar\omega(CO_2)/k_BT]} \quad (3.35)$$

where $\zeta_i = \hbar(2\mu_i h\, \nu_i)^{-1/2}$, μ_i is the reduced mass of the ith mode, ν_i is the frequency of the mode, ζ is connected with the reduced mass of D_2 molecule, ζ_2 is connected with the reduced mass of CO_2 molecule for bending vibrations, ζ_3 is the same parameter for asymmetric stretching vibrations, μ is the CO_2 molecular mass, $\omega(CO_2)$ is the vibrational frequency of CO_2 molecule as a whole, and E is the mean translational energy of the deuterium molecule which is assumed to be equal to k_BT in estimations.

Because $\hbar\omega(CO_2) < k_BT$, we can expand the exponential terms in Equation 3.35 into a series. On substitution of numerical values of parameters for the conditions in our experiment, we have

$$w \approx 10^{-3}\, T \left(\frac{3\alpha}{10^{10}}\right)^6 \quad (3.36)$$

Substituting Equation 3.36 into Equation 3.31, we obtain

$$\epsilon = 10^{-3}\, T \left(\frac{3\alpha}{10^{10}}\right)^6 e^{Q/RT}$$

Let us fit the α value and the depth of the adsorption well in such a way as to give the closest approach to the observed ϵ value and to its temperature dependence. The Q value of 1.5 kJ/mol is necessary to agree with the experimental temperature dependence, and $\alpha \approx 10^9$ cm^{-1} is required to coincide with the observed ϵ value. Such an α value far exceeds the usual magnitudes of repulsion parameters. For reasonable α values the accommodation coefficients are much lower than those obtained experimentally, so this mechanism is unlikely.

The second deactivation channel includes the transfer of deuterium energy to the asymmetric stretching mode of CO_2, the resonance deficient energy being transferred to rotational and translational degrees of freedom of the D_2 molecule. When the resonance deficient energy is transferred to D_2 translational motion, the transition probability can be defined in the framework of the above model. For a single collision with the CO_2 molecule we may write

$$w = \frac{2\pi}{h^2} S^2_{i\to f} < U(r) >^2 \sigma(E_i - E_t)$$

where $S_{i\to f}$ is the transition matrix element between the inner D_2 molecular vibrations and CO_2 asymmetric stretching mode, and $< U(r) >$ is a nondiagonal transition matrix element between wave functions of the translational motion. The transition matrix between wave functions of translational motion has been found even in the works of Jackson and Mott.[120] Using the above expression for the transition probability, the accommodation coefficient is found[101] to be on the order of 10^{-6}, which is also far less than the experimental value. So this mechanism is unlikely as well.

Winter[121] has measured experimentally the probability of vibrational energy exchange in gaseous mixtures of CO_2 and D_2. It turned out that below 500 K the exchange probability drops with increasing temperature. This dependence cannot be understood in the framework of only short-range exponential repulsion interaction. Sharma and Brau[122] have shown that only long-range multipole interaction enables us to obtain the correct explanation of the experimental temperature dependence.

Temperature decrease results in increasing importance of long-range forces and cannot, therefore, be ignored in our case, when the temperature is close to that of liquid nitrogen. The theory of vibrational-rotational transitions for long-range interactions has been developed by Sharma and Brau[122] for the dipole-dipole interactions and by Cross and Gordon[123] for dipole-quadrupole ones. In particular, atomic resonant exchange processes have been regarded. For example, the resonance deficiency for vibrational quantum transfer from the N_2 to the CO_2 molecule is only 18 cm^{-1}, which is much less than the translational energies of the molecules involved. This enables us to treat the translational motion classically. The probability of vibrational-rotational transition has been calculated to the first order of the perturbation theory. An excellent agreement with the experimental results has been found. Rozenshtein[85] revealed the quasiresonant vibrational energy exchange between D_2 and CO_2 in the gas phase; it was accompanied by simultaneous D_2 excitation from the 4th to the 6th rotational level.

If the adsorption potential for the deuterium molecule has the form of a wide potential well in which transitions occur due to the dipole-quadrupole interaction without overlapping the electron wave functions, we may think of the deuterium molecule as being freely rotating. It is well known that rotational-translational (R-T) energy exchange in molecular collisions occurs with a probability close to unity. This enables us to believe that the distribution over rotational levels is in equilibrium. At a low temperature corresponding to the conditions of our experiments, the ground and the first vibrational levels are occupied in the main. Therefore, the vibrational energy transition from the deuterium to the CO_2 molecule will occur with a very large resonance deficiency on the order of hundreds of inverse centimeters. Because deuterium molecules are physisorbed on the CO_2 surface, the binding energy is substantially less than that released in the exchange process. Thus, it will be transferred into the translational motion of the adsorbed molecule rather than into lattice phonons because their characteristic frequencies are substantially lower than the resonance deficiency.

The transition probability has been calculated to the first order of the perturbation theory. Quasiclassic wave functions were used for description of the relative deuterium motion as a whole. It was also assumed that molecular vibrations and rotations are not related to each other energetically. The interaction potential has been expanded in a series of multipole terms, since the long-range forces manifest themselves beyond the overlapping region of molecular electron clouds. A comprehensive description is given elsewhere.[125] On calculating the transitions matrix elements and summing over initial rotational states, we obtain the transition probability in the following form

$$
\begin{aligned}
w &= \frac{8\ 10^{-3} c^2 m < Q_1^{(1)} >< Q_2^{(2)} >}{\hbar^2 \alpha^2 r_f^8 k_B T \sqrt{\gamma}} \times \\
&\sum_{j=o}^{s} n(\Delta j) \sqrt{2\left[\frac{(E_o + \Delta j)^2}{E_o \sqrt{E_o}} - (E_o + \Delta j)^{1/2}\right]^{-1}} \times \\
&\exp\left[-\gamma(\sqrt{E_o + \Delta j} - \sqrt{E_o}) - \beta E_o\right] \\
\gamma &= \frac{2\pi\sqrt{2m}}{\alpha \hbar^2} \\
\beta &= \frac{1}{k_B T} \\
E_o &= \frac{0.14\ \gamma \sqrt{\Delta j}}{\beta}
\end{aligned}
\tag{3.37}
$$

Here c is a constant of the order of unity, m is the deuterium molecular mass, α is the parameter of the Morse potential, r_f is the closest distance between the molecules, $< Q >$ are multipole matrix elements, $n(\Delta j)$ is the statistical weight of initial rotational states, and Δj is the energy transferred motion.

Substituting the value of the total transition probability into Equation 3.31, we find that the absolute value and the temperature dependence of the accommodation coefficient are defined by the parameters α, Q, and r_1. Fitting these parameters on the basis of the literature data, we can obtain a reasonable agreement with the experimental accommodation coefficient.

The equilibrium distance r_f is unknown in this case, but it would be reasonable to suggest that it is of the order of equilibrium spacing in molecular crystals, i.e., 0.2 to 0.3 nm. If the α value of $4 \cdot 10^8\ cm^{-1}$ is chosen, the adsorption well depth is about 3.7 kJ/mol, which is close to the value of Karlov and Shaitan.[117] The r_f value is equal to 0.15 nm, and the adsorption well depth is ~6.3 kJ/mol which is in a good agreement with the experiment. Clearly, it is this mechanism that is responsible for the deactivation of deuterium vibrational energy on the CO_2 molecular crystal.

It is interesting to note that the realization of this mechanism enables us to separate vibrationally excited molecules from nonexcited ones, since at low temperatures when only the lowest rotational levels of the adsorbed molecule are excited, the energy released in the deactivation process is transferred mainly to the translational motion of the adsorbate. Hence, the dominant desorption of vibrationally excited molecules must be observed at low temperatures, and we get a novel method of isotope separation.

3.2. HETEROGENEOUS DEACTIVATION OF ELECTRONICALLY EXCITED PARTICLES

Interest in questions related to the reactivity of electronically excited molecules is not accidental. It is induced by studies in selective photochemical reactions, processes in the discharge, in gas lasers, and investigations in the chemistry of the atmosphere and others. In particular, it is known that the reactivity of electronically excited atoms of carbon, oxygen, sulfur, etc. and radicals differ strongly from their reactivity in the ground state.

The causes of the enhanced reactivity of excited particles will not be discussed here. Commonly, the excited electron and the electron remaining in the ground orbital are not already paired, so an excited molecule can participate in radical reactions. This excited electron has a weak energy with the nuclei and occupies a larger spatial region; as a consequence, it can be readily broken loose. The energy of electronically excited molecules can transfer the kinetic energy of heavy nuclei in the course of reaction (the nonadiabatic transition), which is why the activation energy of the process can be significantly lowered. The Woodword-Hoffmann rules for the orbital symmetry are the ones most used in the theory of molecular reactivity. For "allowed" reactions the ground states of reagents and reaction products must correlate with each other in symmetry. If the reagent ground state correlates with the excited state of the product, such a reaction is said to be "forbidden", i.e., it has a large energetic barrier. It is clear that electronic excitation of the reagent can remove this restriction, and the reaction can easily occur. A large number of papers have been devoted to the chemiluminescent phenomena arising in exothermic reactions.[126] At present, chemiluminiscence is a powerful investigational tool to study kinetics of different processes as well as for the reactivity and structure of reagent components. Radiation has been observed in a number of catalytic reactions such as hydrogen and methanol oxidation, N_2O decomposition, etc.[127] The appearance of electronically excited molecules was found in heterogeneous atomic recombination,[128-130] adsorboluminiscence,[131] nonequilibrium surface conductivity,[132] electron emission,[133] electron-hole pair injection,[132] and others. Studies of the reactivity of electronically excited molecules in adsorption and catalysis are still in the early stages.

There is an urgent need to understand the energy-exchange mechanism between electronically excited particles and the solid surface, since it plays an important part partially in all previously mentioned processes. At the present time, the data on the accommodation coefficients of electronic energy are few in number and extremely scattered. Also, the studies in heterogeneous relaxation of electronically excited particles are very individual as to the generation and registration methods. This, in turn, requires a separate approach for each case.

Heterogeneous deactivation of electronically excited particles is also of practical use in the problems of thermal protection of space aircrafts, thermonuclear reactions, etc. Such a situation, when the necessary data are practically absent, is partially the result of severe experimental difficulties. It is well known that the radiation lifetimes of electronically excited molecules for dipole transitions are of the order of 10^{-7} to 10^{-9} s. Therefore, none of the methods described in Section 3.1 for vibrational energy exchange are applicable in the case of electronically excited molecules. In the case of forbidden transitions, the lifetimes of electronically excited states are much longer; for example, the lifetime of singlet oxygen $O_2(^1\Delta_g)$ is of the order of 1 h. Long lifetimes of the metastable states enable us to use simple and accurate flow methods for studies in their heterogeneous deactivation, but a special registration technique is required in each individual case because the application of direct spectroscopic methods often present problems.

We have studied the heterogeneous deactivation of $Ar(^3P_{0,2})$, $N_2(A^3\Sigma_u^+)$, and $O_2(^1\Delta_g)$.[134-138] The choice of these particles was governed by the following reasons. First, all of them are metastable with long radiation lifetimes $r_r(r_r^{Ar} = 1.3$ s, $r_r^{N_2}r \simeq 10$ s, $r_r^{O_2} \simeq 4000$ s). Second, electronically excited argon, nitrogen, and oxygen all together cover a rather wide energy range from 1 eV for oxygen to 11.5 eV for argon which may by of importance in further investigations of different microscopic mechanisms. Finally, electronically excited nitrogen and oxygen are of significant interest for chemists in the area of catalysis, chemistry of the atmosphere, and laser engineering.

3.2.1. $Ar(^3P_{0,2})$ QUENCHING ON QUARTZ

Heterogeneous deactivation of metastable $Ar(^3P_{0,2})$ with the excitation energy 11.5 eV has received the most study on glass. Futcho and Craut[139] and Wieme and Wieme-Lenaerts[140] have investigated $Ar(^3P_{0,2})$ heterogeneous quenching on the walls of a discharge tube. This method is the most used, but has an important disadvantage connected with the possible treatment of walls by discharge products. For instance, Slovetsky and Todesaite[141] have shown that the discharge treatment results in increasing heterogeneous recombination rate of nitrogen atoms.

Kolts and Setser[142] have applied the technique without this disadvantage because excited argon atoms are removed from the discharge region by the flow. The method of photon resonance absorption was used to detect excited argon $^3P_{0,2}$ atoms. For this purpose a set of optical windows was disposed along the reactor tube, so the temperature range of measured accommodation coefficient was rather narrow.

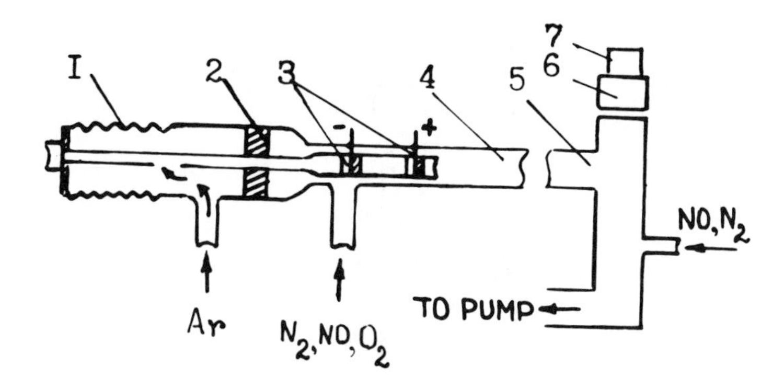

FIGURE 3.5. The scheme of the flow setup for studies of the deactivation of electronically excited Ar and N_2: (1) the system of sylphones; (2) sealing ring; (3) discharge electrodes; (4) reactor; (5) registration cell; (6) spectrometer; (7) photoelectron multiplier.

We suggest a new research method that differs from ones developed earlier. It enables us to lower a gas pressure significantly and to separate the process of heterogeneous deactivation in the pure state.

The flow setup for studies of deactivation of electronically excited molecules is shown in Figure 3.5. The detection of metastable argon atoms has been carried out using optical titration by nitrogen molecules, a small amount of which (10^{-1} to 1 Pa) was introduced into the registration cell. The direct registration of argon atoms is hindered because of measurement difficulties in the vacuum UV region.

The process of optical titration was described by Kolts et al.[143] They included two stages (see also Section 3.2.2)

$$Ar(^3P_{0,2}) + N_2 \longrightarrow Ar(^1S_o) + N_2(C^3\Pi_u)$$

$$N_2(C^3\Pi_u) \longrightarrow N_2(B^3\Pi_g) + h\nu\ (2^+ - \text{band system})$$

Under our conditions the radiation intensity of the second positive band system of molecular nitrogen is proportional to the concentration of metastable argon atoms $^3P_{0,2}$ in the registration cell. The radiation of the second positive band system $N_2(C \rightarrow B)$ in the interval 300 to 400 nm is shown in Figure 3.6. The spectrum shape was found to be independent of $Ar(^3P_{0,2})$ concentration; therefore, any line may serve for registration. We have chosen the strongest line in the band $N_2(C \rightarrow B)$.

Deactivation kinetics of metastable argon on quartz at 300 K has been in a very wide concentration range, which spanned approximately 5 orders of magnitude. The logarithm of relative intensity of the N_2 radiation band (λ = 337 nm) vs. the contact time (i.e., the time it takes for a given volume of the gas to be driven through the reactor). It is seen from Figure 3.7 that $Ar(^3P_{0,2})$ deactivation is the first-order process.

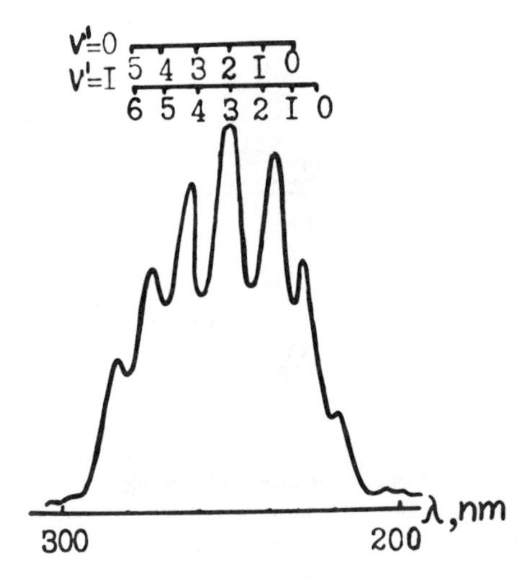

FIGURE 3.6. Radiation spectrum of the second positive N_2 band system that is observed in the registration cell.

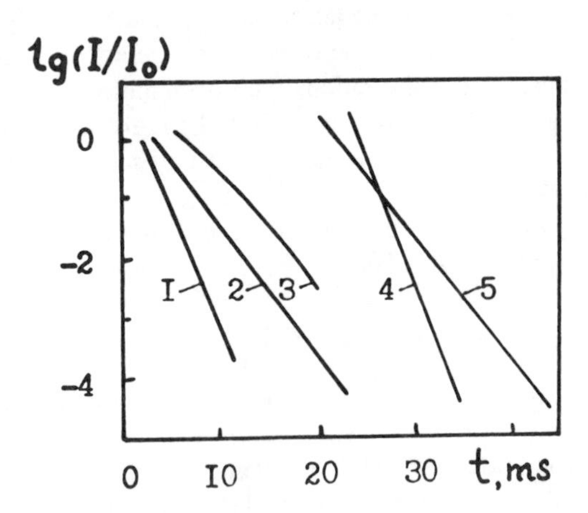

FIGURE 3.7. Kinetic curves of $Ar(P_2)$ deactivation at different pressures (Pa): (1) 500; (2) 90; (3) 150; (4) 500; (5) 280.

All experiments were carried out in such a way that the minimal distance from the source of metastable argon to the registration cell were longer than the characteristic length L required for the Poiseuille velocity profile to reach the steady state. The L value has been calculated according to the formula $L = 0.24\ rRe$, where r is the reactor radius and Re the Reynolds number.

Bray et al.[64] have analyzed the processes in a flow cylinder reactor in terms of heterogeneous deactivation of excited particles on the reactor walls in the diffusion region and first-order homogeneous deactivation. If the

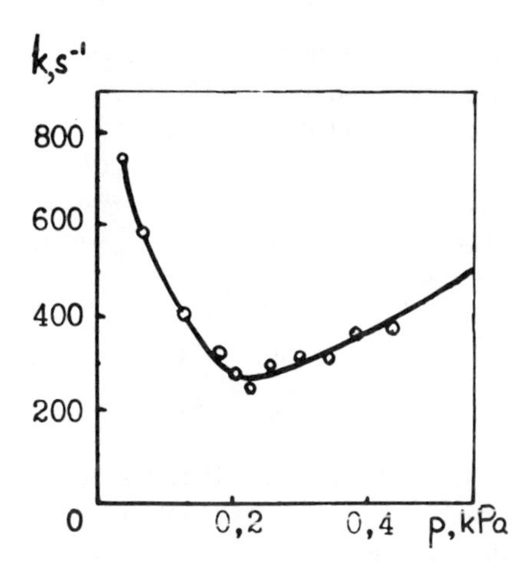

FIGURE 3.8. The rate constant of $Ar(P_2)$ deactivation as a function of argon pressure (points refer to experimental values; solid line is the calculation from Equation 3.39 with the obtained rate constants).

longitudinal diffusion is disregarded, the dependence of relative concentration of excited particles vs. the contact time may by written in the form[85]

$$\frac{14.63D^*}{4r^2} + (0.6 \div 0.63)\, k_{hom} = t^{-1} \ln\left(\frac{n}{n_o}\right) \tag{3.38}$$

where r is the reactor radius, D* is the diffusion coefficient of $Ar(^3P_{0,2})$ atoms in argon, k_{hom} is the effective rate constant of homogeneous deactivation, and n is the concentration of excited particles.

Taking into account collisions of $Ar(^3P_{0,2})$ with one and two ground-state (^1S_o) argon atoms, the effective rate constant of the homogeneous deactivation can be written in the form[134]

$$k_{hom} = k_{1hom}P + k_{2hom}P^2 + \tau_{rad}^{-1} \tag{3.39}$$

where k_{1hom} and k_{2hom} are the rate constant for binary and triple collision, respectively; P is the argon gas pressure; and τ_{rad} is the radiation lifetime of metastable argon.

The effective deactivation rate constant as a function of argon pressure in the reactor is shown in Figure 3.8. Taking into account Equations 3.38 and 3.39, we obtain the following values of rate constants: $D^* = (45 \pm 3)$ cm^2/s at 130 Pa, $k_{1hom} = (2.8 \pm 0.4)\ 10^{-15}$ cm^3/s, $k_{2hom} = (1.9 \pm 0.4)$ 10^{-32} cm^6/s. These results agree well with those of other authors. Koltz and Setser[142] showed that the output concentration of $Ar(^3P_0)$ comes to only $10 \div 15\%$ of the $Ar(^3P_2)$ concentration. Since the rate constants of $Ar(^3P_0)$ and $Ar(^3P_2)$ are of the same order on the quartz surface, it would be reasonable to suppose that the main contribution to the signal intensity is made by $Ar(^3P_2)$. Therefore, the obtained binary and triple rate constants as well as the diffusion coefficient D^*_o and the accommodation coefficient ϵ of argon on the quartz surface refer to the metastable atoms in the $^3P_2$ state.

For the purpose of further calculating the accommodation coefficient of electronically excited $Ar(^3P_{0,2})$, let us use the approximate formula of Frank-Kamenetsky for "series-resistors":

$$\frac{1}{k_{het}} = \frac{1}{k_{kin}} + \frac{1}{k_{diff}} \tag{3.40}$$

The value of the accommodation coefficient ϵ calculated from Equation 3.40 taking into account the possible experimental error is $\epsilon > 0.4$. This result agrees with the experimental data of Allision et al.[144] on molecular beam scattering. These authors have found that the reflection probability of $Ar(^3P_{0,2})$ from the glass surface is ~0.03. For such ϵ values the process proceeds in the diffusion region and does not depend on the surface by discharge products.

Note that the value of the diffusion coefficient D^* obtained in our experiments and in other works (see Table 3.2) differs from the self-diffusion coefficient of argon. The usual way to determine the diffusion coefficient of excited particles is as follows. It is suggested that the accommodation coefficient $\epsilon = 1$ should be used allowing us to calculate the diffusion coefficient. What actually happens is that the ϵ value may significantly differ from unity, while the deactivation rate constant remains diffusive. This may result in a large error in ϵ and D^* values. Therefore, the theoretical estimation of D^*_o is of interest.

A small D^*_o value seems to be connected with the resonance excitation transfer between $Ar(^3P_2)$ and $Ar(^1S_o)$. It counts in favor of this suggestion as well that the diffusion coefficient estimated using reasonable gas-kinetic parameters[142] (with exchange repulsion taken into account) far exceeds the experimentally measured value. Palkina et al.[145] in their theoretical estimations of D^*_o also arrived at a value corresponding to the upper bound of the experimental spread. We calculated the diffusion coefficient in the framework of the asymptotic method.[146] Account has been taken of the Δ term splitting the even and odd states which is due to the excited electron in $Ar(^3P_2)$; hence, no differentiation has been made between the terms with $\Omega = 0$, 1, and 2, where Ω is the projection of the total angular momentum onto the molecular axis. The following expression has been derived for the Δ-term electron wave functions using $Ar(^1S_o)$ and $Ar(^3P_2)$ parameters

$$\Delta = 0.44\, R^{3.884}\, (1 - 2.24\, R^{-1}) \exp(-1.625R) \tag{3.41}$$

where R is the distance between Ar atoms in atomic units.

It was shown previously that if double the cross-section of the excitation transfer exceeds the cross-section of diffusive elastic scattering, the diffusion coefficient should be substituted by twice the value of the excitation-transfer cross-section, i.e., $2\sigma_{e.t.}$. Because the distance of the excitation transfer is greater than the gas-kinetic molecular size, the trajectory distortions due to atomic interactions can be neglected. As a result, we obtain[147]

$$2\sigma_{e.t.} = \pi R_o^2 \tag{3.42}$$

where R_o is determined by the equation

$$\frac{1}{v_{rel}} \sqrt{\frac{\pi R_o}{2\gamma}} \Delta R_o = 2.8 \tag{3.43}$$

where v_{rel} is the relative velocity of the atoms and $\gamma = 1.625$.

Since R_o has a weak logarithmic dependence on the relative velocity v_{rel}, it is a common practice to substitute the value $v_{rel} = \sqrt{8k_BT/\pi\mu}$ in $\sigma_{e.t.}$ instead of averaging over the Maxwell distribution. As a result, we obtain $2\sigma_{e.t.} = 1.2\ 10^{-14}$ cm^2 at 300 K. The value of the diffusion coefficient is 43 cm^2/s at 133 Pa. The exchange repulsion gives rise to approximately 10% decrease in $\sigma_{e.t.}$ and increases D_o^* value corresponding to 48 cm^2/s. The calculated value of the diffusion coefficient of $^9P_2$ electronically excited argon atoms in gaseous argon is in good agreement with the experimental results.[134] We have calculated[147] the diffusion coefficient of all inert atoms in $(np^5)^9P_{9/2}$ and $S_2^{<9/2}$ states in their own gases. The experimental and theoretical results are compiled in Table 3.3.

As was previously intimated, the heterogeneous microscopic deactivation mechanisms of electronically excited particles are presently unknown except for the heterogeneous deactivation of metastable atoms on metal surfaces. It is known that their energy can result in the electron emission from the metal. Note that the excitation energy in this case must be greater than the metal work function. It is a common rule for metastable atoms of inert gases, but it is not necessary for other electronically excited particles, for instance, for singlet oxygen. Deactivation mechanisms on nonmetals are absent from all.

We suppose three possible channels for argon $^3P_2$ deactivation on quartz. The first is connected with the energy transfer from argon to excitons. It was shown[148] that excitons in quartz are responsible for the absorption in the region ~11.6 eV that is very close to the $^3P_2$ argon energy. Excitons appear due to the oxygen excitation in SiO_2 from the ground p_y-orbital localized in Si-O-Si plane and directed normally to the Si-Si axis, into the 3s orbital.

The second possible channel of $Ar(^3P_2)$ deactivation on quartz is connected with a small energy splitting between $^3P_2$ and $^3P_1$ states, because the transition allowed from the last to the ground state is possible. It should be emphasized that for physisorption an effective $^3P_2$ and $^3P_1$ mixing is possible; hence, this deactivation channel is one of the most probable.

TABLE 3.3
The Diffusion Coefficients (in cm^2/s) of Excited Inert Atoms X in Their Own Gases at 133 Pa

T (K)	Ne	Ar	Kr	Xe
77	18.2	4.0	2.5	1.5
77	37.5	9.4	7.1	2.7
77	15.4			
77	14 ± 3	2.1 ± 4		
77	16			
300	146.1	43.4	24.5	16.2
300	430.0	143.5	77.6	44.3
300	120 ± 10	67.5	30 ± 1.5	13 ± 1
300	200 ± 20	45 ± 4	20.3	16
300	150	54		
300	165 ± 30	54 ± 6		
300	170 ± 10	48 ± 6		
300	171	52 ± 5		
300		48 ± 5		
300		56 ± 3		
300		45 ± 3		

The third deactivation channel is connected with a high excitation energy of the Ar atom. This energy is sufficient for the rupture of practically any chemical bound; thus, a ''chemical'' way for $Ar(^3P_2)$ deactivation becomes possible. Recently, the dissociative methane chemisorption on Ni(111) upon bombardment by translationally excited argon, has been found[149] (for details see Section 4.2).

In general, all three deactivation channels can probably occur. Unfortunately, at present there are no detailed theoretical estimations for different mechanisms nor experimental data sufficient for unambiguous identification of a certain microscopic deactivation mechanism.

3.2.2. $N_2(A^3\Sigma_u^+)$ QUENCHING ON QUARTZ AND NICKEL

The processes involving atomic and molecular nitrogen in different electronic states are of partial interest for the study of nonequilibrium processes and especially for the chemistry of the upper atmosphere. Lin and Kaufman[150] have found that the reaction $N(^2D) + O_2 \rightarrow NO + O$ is the source of NO molecules in the upper atmospheric layers. A lot of work is devoted to investigation of the so-called ''active'' nitrogen, i.e., one passed through a discharge.[151] Atomic nitrogen recombination is accompanied by a yellow radiation and is due mainly to the selective population of vibrational levels with $0 \leqslant v \leqslant 12$ of the first positive band system: $N_2(B^3\Pi_g \rightarrow A^3\Sigma_u^+)$. Nitrogen potential terms are given in Figure 3.9. Campbell and Trush[152] proposed the mechanism of nitrogen fluorescence on the basis of a thorough study of $B^3\Pi_g \rightarrow A^3\Sigma_u^+$ radiation transition and literature analysis. They found

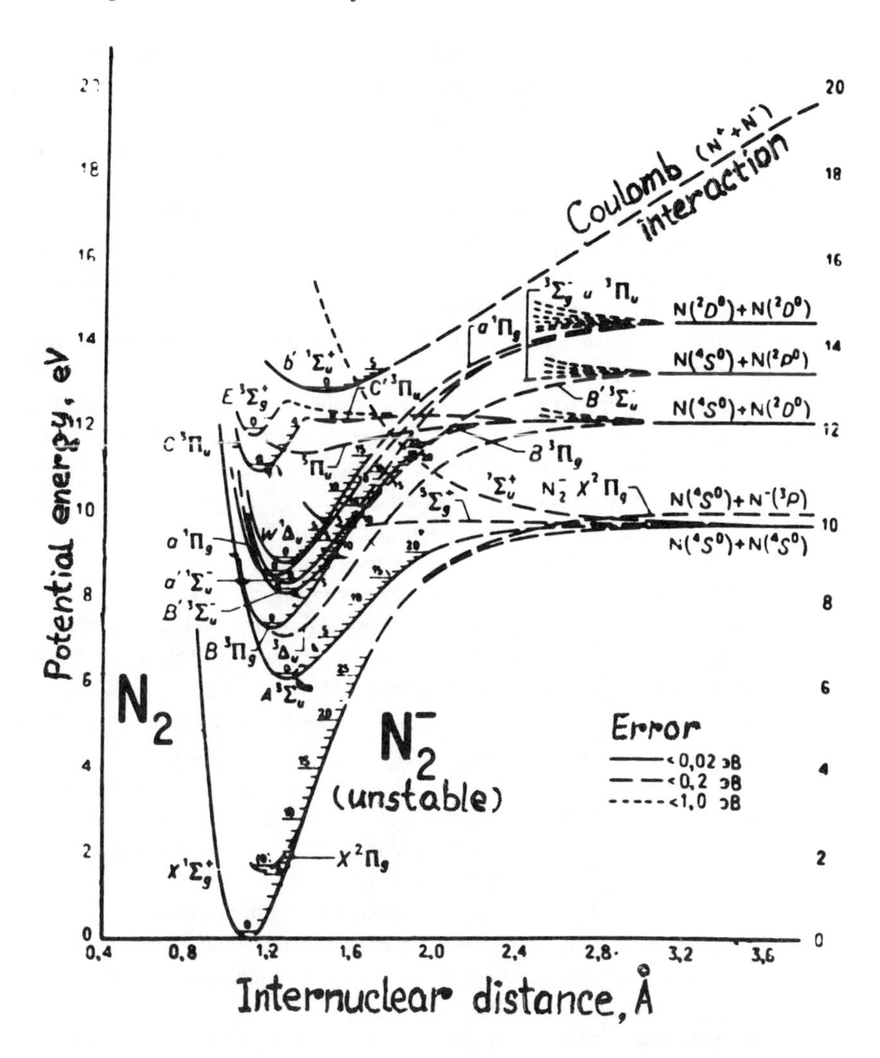

FIGURE 3.9. Potential energy terms of N_2 molecule.

that the excited state is populated as a result of the third-order recombination reaction.

The study of $N_2(A^3\Sigma_u^+)$ reactivity is of considerable importance, since this state has a sufficiently long lifetime of the order of 2 ÷ 3 s.[153,154] As early as 1960, Harteck et al.[128,129] revealed the formation of electronically excited nitrogen molecules in heterogeneous atomic recombination on a number of metal surfaces: Ni, Ag, Co, and Cu. These authors suggested also the possible radiation mechanisms.

We have repeatedly examined these reactions for a wider set of objects. In particular, in collaboration with Tabachnik[155] it was found that $N_2(A^3\Sigma_u^+)$ formation occurs on Ni, Cu, and Ag surfaces but not on Mo and W. This most likely is connected with easy production of tightly bound surface nitride layer on Mo and W. Note that, in principle, the formation of electronically excited nitrogen is very sensitive to the experimental conditions and that the reproducibility of the results is poor even on Ni surface, which is the most powerful $N_2(A^3\Sigma_u^+)$ source.

At present it is quite unclear why electronically excited molecules appear on certain surfaces. It is assumed that if the heat released in the exothermic process of heterogeneous atomic recombination is enough to create a molecule in the electronically excited state, the molecular yield is determined by the atomic recombination rate and the molecular lifetime on the surface or by their deactivation rate.

There are two research groups working in the field of $N_2(A^3\Sigma_u^+)$ heterogeneous deactivation. Vidaud et al.[156,157] have measured the accommodation coefficients of $N_2(A^3\Sigma_u^+)$ on Pyrex® and platinum surfaces: $\epsilon_{pyr} = 3\ 10^{-5}$ and $\epsilon_{Pt} = 3\ 10^{-3}$. Setser et al.,[158,159] in contrast to these results, have drawn a conclusion that the $N_2(A^3\Sigma_u^+)$ accommodation coefficient is always close to unity, as was evident from their studies on a number of surfaces.

Such a severe disagreement between the two groups of results stimulated our activity in reexamining this problem using a new method. Experiments were performed using the flow setup shown in Figure 3.5. The argon flow passes through the discharge of direct current that serves as a source of argon atoms excited in the $^3P_{0,2}$ states. Nitrogen is mixed to argon at the output of the discharge tube. Argon $^3P_{0,2}$ atoms transfer the energy to nitrogen molecules and excite them into the $C^3\Pi_u$ state. The metastable nitrogen molecules appear as a result of a set of allowed radiation transitions: $C^3\Pi_u \rightarrow B^3\Pi_g \rightarrow A^3\Sigma_u^+$. Since the cross-section of the argon deactivation by nitrogen molecules is almost gas kinetic,[158,160] and the radiation lifetimes are short, it would be reasonable to suppose that $N_2(A^3\Sigma_u^+)$ molecules are formed in the vicinity of the discharge tube. A thin capillary 2.5 mm in diameter and 3 cm in length at the end of the discharge tube keeps the nitrogen molecules from entering the discharge region due to the back diffusion.

Experiments were carried out in the pressure interval 100 to 800 Pa, and the ratio between nitrogen and argon rates of flow was varied from 0.1 to 0.3. The direct voltage supplied to the discharge electrodes was equal to 250 V, and the discharge current in argon was 0.6 mA. Contrary to the methods of Setser and Vidaud, a moving discharge source was applied to enhance the detection sensitivity. The metastable nitrogen concentration was measured using the exchange-luminescence connected with the excitation energy transfer from $N_2(A^3\Sigma_u^+)$ to NO molecules. The latter were introduced in small proportion (with a partial pressure ~1 Pa) immediately in the registration cell. The processes occurring in the optical titration are described by the following scheme

$$N_2(A^3\textstyle\sum_u^+)_{v=0.1} + NO(X^2\Pi) \longrightarrow N_2(X^1\textstyle\sum_g^+) + NO(A^2\textstyle\sum^+)$$

$$NO(A^2\textstyle\sum^+) \longrightarrow NO(X^2\Pi) + h\nu\ (\gamma\text{-transition, 200 to 300 nm})$$

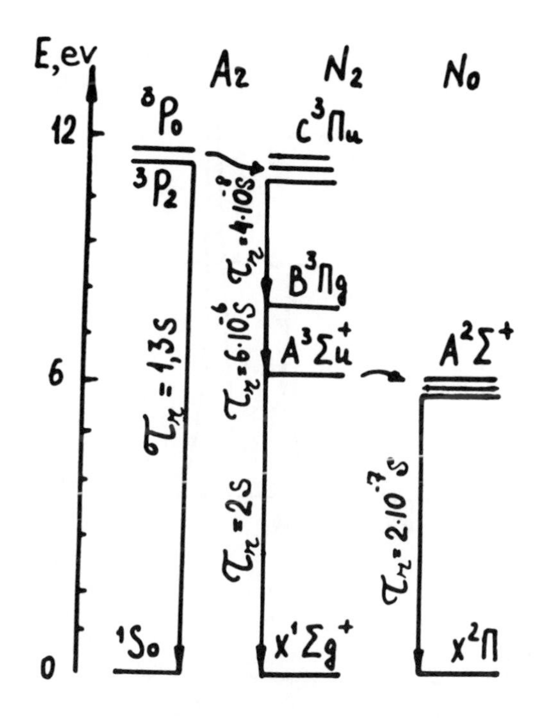

FIGURE 3.10. Energy levels of Ar, N_2, and NO.

The scheme of argon, nitrogen, and NO levels is demonstrated in Figure 3.10.

Because the cross-section of $N_2(A^3\Sigma_u^+)$ deactivation by NO molecules is close to gas kinetic,[143] and the radiation time of γ-transition is short, the γ-transition intensity may be thought of as being proportional to the concentration of metastable nitrogen molecules. The radiation spectrum of NO γ-transitions is represented in Figure 3.11. In the range 200 to 300 nm it was registered using spectrometer MDR-3 and photomultiplier FEU-39. It was found experimentally that the radiation spectrum of γ-transition did not depend on the $N_2(A^3\Sigma_u^+)$ concentration, so the most intense transition was chosen for the registration of metastable nitrogen

$$NO(A^3\Sigma_u^+)_{v'=o} \longrightarrow NO(X^2\pi)_{v''=2}(\lambda = 247\text{ nm})$$

Figure 3.12 represents the logarithm of the absolute signal intensity at λ = 247 nm, characterizing the concentration of metastable nitrogen as a fun^ction of distance between the discharge tube and the registration cell for different pressures and flow velocities. There are two distinct regions with the different $N_2(A^3\Sigma_u^+)$ deactivation kinetics. Consider the region of signal intensities located above the n* value in Figure 3.12. The intensity of the registered signal (λ = 247 nm) vs. the contact time of $N_2(A^3\Sigma_u^+)$ molecules in the reactor is represented in Figure 3.13. The contact time is given by

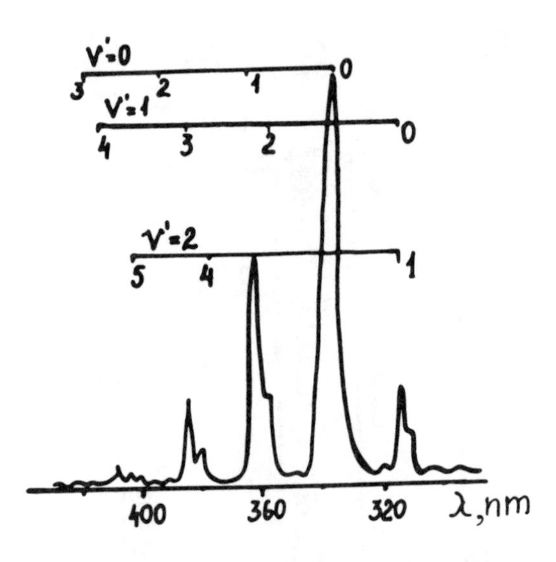

FIGURE 3.11. Radiation spectrum of NO γ-transition.

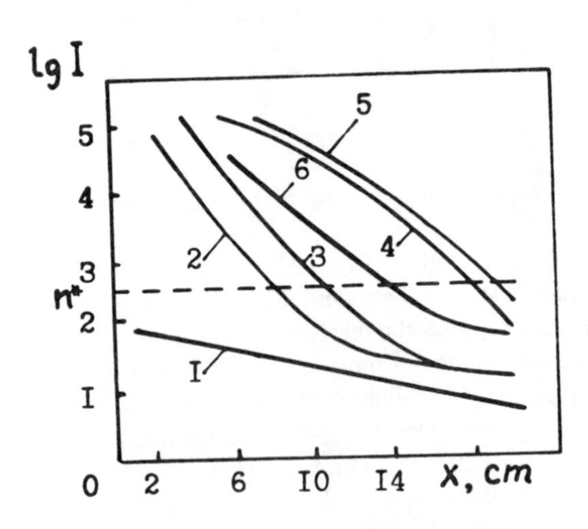

FIGURE 3.12. Kinetic curves of $N_2(A^3\Sigma_u^+)$ deactivation in the quartz reactor: (1) 150 Pa, u = 2.1 m/s; (2) 140 Pa, u = 8.4 m/s; (3) 150 Pa, u = 9.4 m/s; (4) 140 Pa, u = 17 m/s; (5) 550 Pa, u = 4.9 m/s; (6) 1100 Pa, u = 2.2 m/s.

t = $\bar{x}$/v, where x is the distance from the source of metastable molecules to the registration cell, and v is the gas velocity averaged over the flow cross-section. It is seen that in a wide concentration region ranging over three orders, $N_2(A^3\Sigma_u^+)$ deactivation obeys the first-order law. The effective rate constant k_{eff} vs. the total argon and nitrogen pressure P is represented in Figure 3.14. The heterogeneous deactivation of metastable nitrogen occurs in the diffusion regime, because the effective deactivation rate constant is independent of the flow velocity. The value D* = (154 ± 6) cm^2/s at 133 Pa has

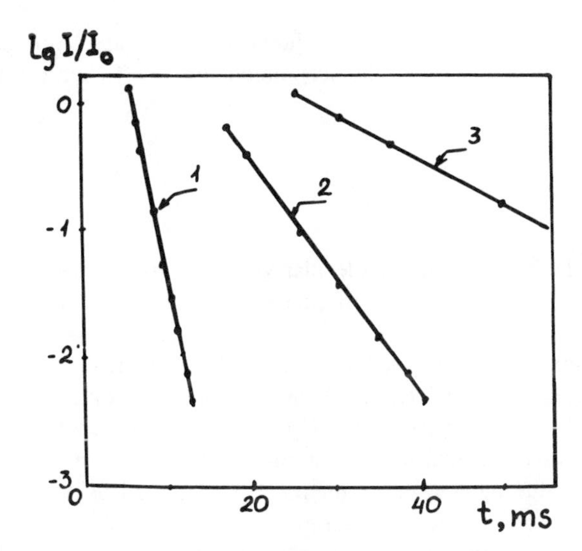

FIGURE 3.13. Logarithmic dependence of the relative signal (λ = 247nm) intensity on the time of $N_2(A^3\Sigma_u^+)$ residence in the reactor for 1g $\geq$ n*: (1) 1.4 10^2 Pa; (2) 5.5 10^2 Pa; (3) 11 10^2 Pa.

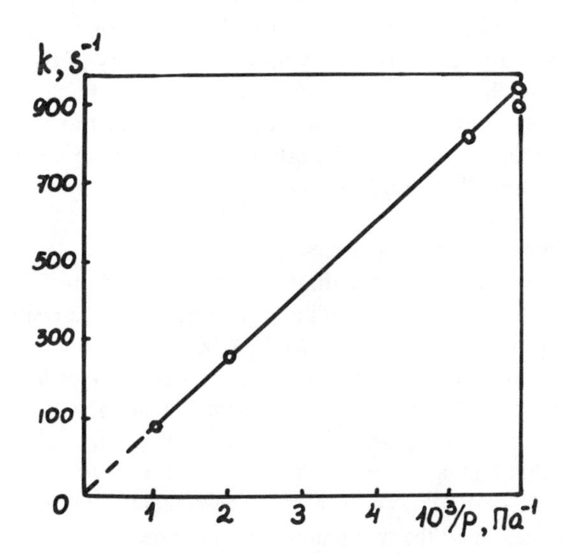

FIGURE 3.14. Effective rate constant of $N_2(A^3\Sigma_u^+)$ as a function of P^{-1}.

been derived from the dependence k_{eff} on P^{-1}. Estimations[101] show that the longitudinal diffusion can be neglected under our conditions. This fact is supported by the absence of k_{eff} dependence on the flow velocity at P = 1.4$10^2$ Pa within the limits of experimental error.

In the concentration region below n* (see Figure 3.12), the value of the effective deactivation rate constant turned out to be ~21 s^{-1} at P = 150 Pa and v = 2.1 m/s. Suppose that the homogeneous deactivation can be disregarded in this region. Then the accommodation coefficient can be derived from

$$\frac{\epsilon \bar{v}}{2r} = t\,'\ln\left(\frac{n}{n_o}\right) \tag{3.44}$$

where $\bar{v}$ is the mean thermal molecular velocity, r is the reactor radius, and n and n_o are the concentrations of excited particles. The ϵ value was found to be $7\ 10^{-4}$.

Let us discuss the obtained ϵ value. In the first region, where $n > n^*$, in a wide concentration interval of the metastable nitrogen, is linearly dependent on P^{-1} and our calculated $N_2(A^3\Sigma_u^+)$ diffusion coefficient k_{eff}, coincides with that obtained by Wright and Winkler.[151] As expected, the calculated D_o^* value agrees well with the $N_2(X'\Sigma_g^+)$ diffusion coefficient in nitrogen and argon in the temperature interval from 77 to 500 K. The D_o^* value can be calculated from the results of Wright and Winkler[151] as well. The value obtained coincides with ours provided that $k_{eff} \sim r^{-2}$, the r^2 values in two works differ by a factor of 2. The gaseous $N_2(A^3\Sigma_u^+)$ deactivation rate constants for collisions with O_2 and NO obtained in our work in collaboration with Tabachnik are in a good agreement with the literature data.[154]

All above-mentioned facts bear witness to the reality of the first region; hence, for $n > n^*$ the heterogeneous deactivation of metastable nitrogen may be considered as occurring with the diffusion rate constant in agreement with the results by Setser et al.[158,159] The estimation of the accommodation coefficient from the formula (Equation 3.44) within the limits of the experimental error gives a value $\epsilon > 0.5$.

In the second region, where $n < n^*$, if the effective deactivation rate constant remained the same, a tenfold decrease of the flow velocity would result in a drop of 19 orders of magnitude of the radiation intensity. Actually, it decreases approximately 100-fold. At $N_2(A^3\Sigma_u^+)$ concentrations lower than n^*, the deactivation rate constant drops with decreasing the concentration and reaches a value of 21 s^{-1} under our experimental conditions. The latter value corresponds to $\epsilon = 7\ 10^{-4}$, which agrees well with the results by Vidaud et al.[156,157] Note that homogeneous processes may also contribute to the value of k_{eff}; then $\epsilon \leqslant; 7 \cdot 10^{-4}$, and even better agreement is achieved.

Let us estimate now the maximum concentration of metastable nitrogen and the critical value of n^* in our conditions. With a knowledge of the deactivation rate constant for gaseous interaction between two $N_2(A^3\Sigma_u^+)$ molecules [$k = 2.6 \cdot 10^{-10}$ cm^3/(molecules · s)],[161] and assuming the contribution of this process to be 5% below the total value of k_{eff} at P = 100 Pa, which is in the limits of experimental error, we may evaluate the maximum concentration n_{max} in these conditions. It was found that $n_{max} \leqslant 10^{10}$ cm^{-3} in agreement with the experimental value of Setser et al. under similar conditions. From Figure 3.12 at P = 100 Pa we find that $n^* \lesssim 10^8$ cm^{-3}.

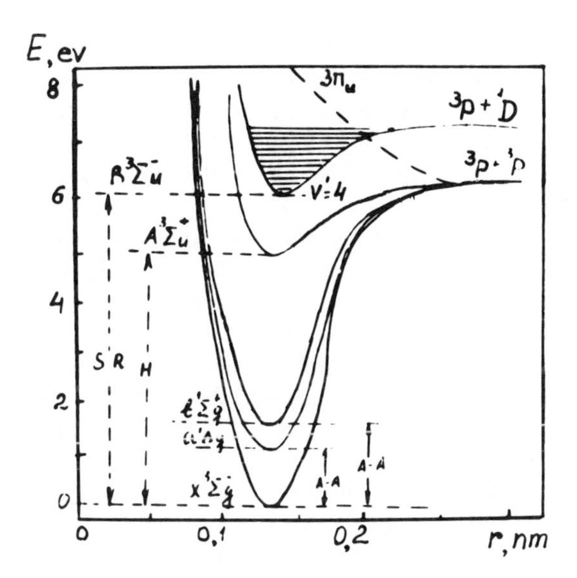

FIGURE 3.15. O_2 terms (S-R, Schumann-Runge bands; H, Hertzberg bands; A-A, atmospheric bands).

The results obtained on the quartz surface appear to be similar to those of nickel. The accommodation coefficient on Ni is $\epsilon \approx 5 \cdot 10^{-4}$. This result is not surprising because the nickel surface is oxidized and behaves in a similar way.

We considered a number of possible deactivation mechanisms leading to the $N_2(A^3\Sigma_u^+)$ concentration dependence.[155] All of them are to a large extent hypothetical. Therefore, we will not discuss them here. We can only state that at present the surface deactivation mechanism of electronically excited nitrogen molecules remains unclear, and further research activity is needed in this field.

3.2.3. QUENCHING OF SINGLET OXYGEN

The term "singlet oxygen" usually refers to two metastable forms: $a^1\Delta_g$ and $b^1\Sigma_g^+$. The ground state is $^3\Sigma_g^-$. The $O_2(^1\Delta_g)$ electron energy is 0.977 eV, and its lifetime is 64.4 min.[162] The second state $^1\Sigma_g^+$ with the energy 1.62 eV has a lifetime of 12 s.[163] Oxygen potential terms are presented in Figure 3.15. The band system responsible for the transition $a^1\Delta_g \rightarrow X^3\Sigma_g^-$ is called the atmospheric band.

Extremely long lifetimes of these states make possible the detailed investigation of the singlet oxygen reactivity. A lot of work was devoted to its reactivity in gas, liquid, and solid phases.[164,165] The deactivation of singlet oxygen in the gas phase along with the exchange processes with molecules capable of light irradiation[166] have been much studied. In particular, energy exchange between singlet oxygen and iodine provides the basis for creation of the so-called iodine laser.

Heterogeneous catalytic processes involving singlet oxygen are less well understood. There is experimental evidence of its reactions with multiatomic organic molecules adsorbed on silica gel.[167] The generation of singlet oxygen has been observed in the processes involving complex catalysts.[168] Vladimirova et al.[169] observed the enhanced adsorption effect for singlet oxygen on MgO.

In 1973, we suggested that electronically excited oxygen plays, possibly, an important role in catalytic oxidation reactions.[170] Oxygen has sufficiently high reactivity, but little is known about reactions of singlet oxygen in catalytic and adsorption processes. In connection with this fact there is a need to learn how to distinguish between reaction and deactivation interaction channels on the surface. A knowledge of heterogeneous deactivation of singlet oxygen is urgently required, because these processes are poorly understood. Nevertheless, some information is available in connection with investigations of gas-phase reactions involving singlet oxygen. The accommodation coefficients of $O_2(^1\Delta_g)$ deactivation on glass, Pyrex®,[171-174] stainless steel,[175] clean and with adsorbed HCl layer gold[176] have been measured. There are also qualitative data on singlet oxygen deactivation at the surfaces of platinum, iron, nickel, silver, and copper.[177]

We have carried out a detailed study of $O_2(^1\Delta_g)$ accommodation coefficients on a number of surfaces including studies on temperature and pressure dependence, influence of different adsorbed layers, as well as the study of singlet oxygen deactivation in the course of catalytic reaction.[178]

The main problem in investigations of heterogeneous deactivation of singlet oxygen is connected with the measurement of its concentration. Because the lifetime of the allowed electron transitions is typically of the order $10^{-6} \div 10^{-8}$ s, whereas the lifetime of singlet oxygen is of the order of 4000 s, the intensity of the $^1\Delta_g \rightarrow x^3\Sigma_g^-$ transition is 10 to 12 orders weaker than usually.

$O_2(^1\Delta_g)$ registration has been carried out using different methods such as thermal probing,[179] ESR spectroscopy,[180] photoelectron spectroscopy,[181] and photoionization methods.[181] It should be mentioned that all methods, apart from their weak sensitivity, are extremely complex, as applied to the study of heterogeneous deactivation of singlet oxygen.

Photometry of the transition $^1\Delta_g(v = 0) \rightarrow X^3\Sigma_g^-$ $(v = 0)$ is the most direct, but not the most sensitive, method of singlet oxygen detection. It is this method that has been used in our studies.

As has already been mentioned, both $^1\Delta_g$ and $^1\Sigma_g^+$ metastable states are called "singlet oxygen". Thanks to sufficient intensity of the radiation transition $^1\Sigma_g^+ \rightarrow X^3\Sigma_g^-$ there are no severe difficulties in detecting this state. Besides, there is little probability, especially at high pressures, of binary collisions between two molecules in $^1\Delta_g$ states that result in light irradiation in the regions 634 and 703 nm according to the scheme:

$$O_2(^1\Delta_g) + O_2(^1\Delta_g) \longrightarrow 2O_2\,(X^3\,\Sigma_g^-) + h\nu$$

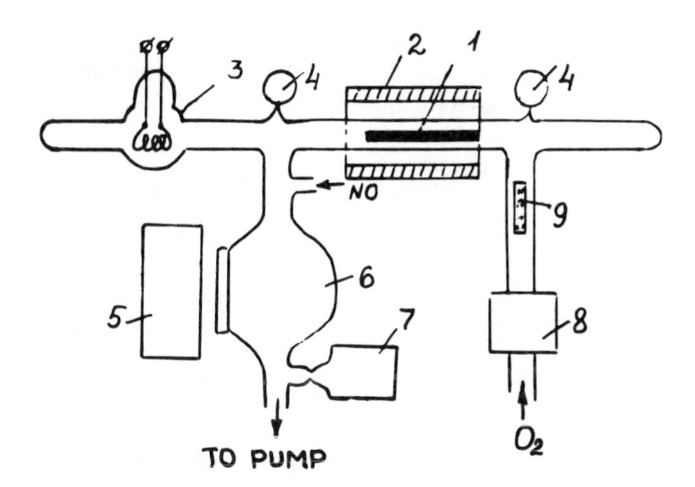

FIGURE 3.16. The scheme of the experimental setup for studies of singlet oxygen deactivation: (1) sample under study; (2) oven; (3) equipment for deposition of Pt film; (4) manometers; (5) registration block; (6) registration cell; (7) mass spectrometer; (8) microwave discharge; (9) tube made of molybdenum foil.

This method is also used for registration of singlet oxygen. The microwave discharge is a common source of singlet oxygen. Discharge products contain approximately 10% $O_2(^1\Delta_g)$ and 0.1% $O_2(^1\Sigma_g^+)$. From this point onward we take the term "singlet oxygen" to mean only the $^1\Delta_g$ state.

The flow setup used in our experiments is sketched in Figure 3.16. The flow velocity was varied from 1 to 5 m/s. High-purity oxygen at 66 to 930 Pa was drawn through the discharge tube (8) which served as a source of $O_2(^1\Delta_g)$. A considerable number of oxygen atoms were also generated by discharge. To decrease the concentration of atoms without appreciably decreasing singlet oxygen, inserts in a form of cylindric tubes from different metals were tested. It was found that the molybdenum-foil insert (9) lowers the concentration of oxygen atoms approximately tenfold, whereas the singlet oxygen concentration remains practically unchanged. Atomic concentration was measured by NO optical titration in the registration cell (6). Yellow-green radiation was used to detect oxygen atoms.[182]

The reactor had a form of glass tube 1.6 cm in diameter. Samples under investigation were introduced into the reactor with the aid of a magnet. The singlet oxygen concentration was determined when detecting the radiation in the region 1.27 μm. The radiation was collected by an optical system which consisted of a glass ball 20 cm in diameter covered by the aluminum mirror on the inside,[176] a fast diffraction spectrometer, a light filter, and a registration photodiode. The singlet intensity is proportional to the singlet oxygen concentration. An analysis of stable discharge products was carried out using a monopolar mass spectrometer. It was found that NO was the main impurity in the discharge products, its concentration being about 1%.

TABLE 3.4
Accommodation Coefficient of Singlet Oxygen on Different Surfaces at 298 K

Material	Accommodation coefficient	Material	Accommodation coefficient
Glass	$4.4 \cdot 10^{-5}$	Silicon	$7.3 \cdot 10^{-4}$
Aluminum	$5.9 \cdot 10^{-5}$	Copper	$8.5 \cdot 10^{-4}$
Titanium	$6.5 \cdot 10^{-5}$	Nickel	$2.7 \cdot 10^{-3}$
Molybdenum	$8.0 \cdot 10^{-5}$	Iron	$4.4 \cdot 10^{-3}$
Niobium	$1.2 \cdot 10^{-4}$	Silver	$1.1 \cdot 10^{-2}$
Platinum wire	$4.0 \cdot 10^{-4}$	Quartz	$7.4 \cdot 10^{-3}$
		Gold	$1.8 \cdot 10^{-3}$

The accommodation coefficient varied in the interval 10^{-6} to 10^{-1}. The lower limit is defined by the pressure instability in the system, and the upper by the diffusion of excited molecules to the sample surface. There was also an isothermal calorimeter in the system that was used in studies of temperature dependence of the accommodation coefficients, influence of adsorbed gases, and the course of catalytic oxidation reactions.

The general scheme of determination of the $O_2(^1\Delta_g)$ accommodation coefficient is similar to that of others in cylindrical or coaxial reactors. Catalytic insert had the form of a foil tube closely fitted to the reactor as well as a rod or strip positioned in the tube axis. Samples under investigation were cleaned mechanically and treated by the discharge flow. It was shown that longitudinal as well as transverse molecular diffusion could be neglected. Therefore, the accommodation coefficient has been determined from the expression which is valid in the kinetic regime

$$\epsilon = 4\,V\,k/\bar{v}\,S \tag{3.45}$$

where V is a volume per unit length of the reaction tube, k is the experimental rate constant of $O_2(^1\Delta_g)$ deactivation, $\bar{v}$ is the mean thermal velocity of molecules, and S is an insert area per unit length.

Results obtained are summarized in Table 3.4. The ϵ value on glass has the same order of magnitude of that measured previously.[171-174,179] Qualitative results of Grigor'ev[177] were also close to ours except for iron, where very weak deactivation was found.

The wide variety of objects studied does not allow us to accept that, for example, the largest value of the accommodation coefficient on silver is substantially greater than its least value on molybdenum, for the following reasons: (1) the measurement was performed only at room temperature and (2) the surfaces under study were, to a marked extent, oxidized by residual oxygen atoms and O_2 molecules in different states in the flow.

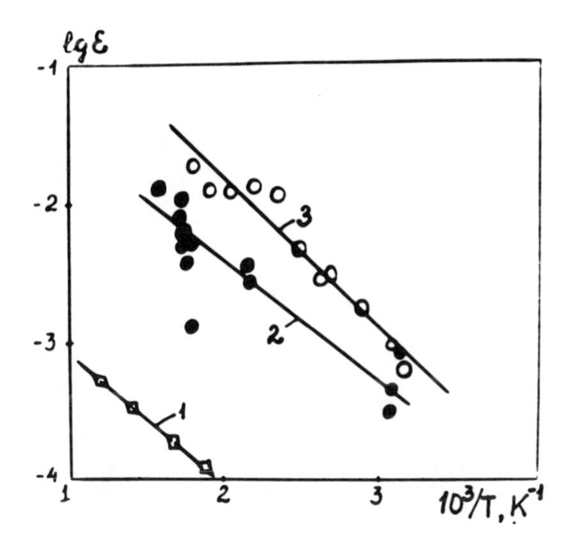

FIGURE 3.17. Temperature dependence of the accommodation coefficient ϵ of singlet ($^1\Delta_g$) oxygen on different samples: (1) quartz; (2) platinum wire; (3) deposited platinum film.

$O_2(^1\Delta_g)$ Deactivation on Quartz and Platinum

A knowledge of the temperature dependence of the accommodation coefficient is required for elucidation of the heterogeneous deactivation mechanism. The choice of platinum as an object of studies is explained, first, by the high stability of this material with respect to oxygen, which gives a good reproducibility of results, and second, by the high catalytic activity of platinum in most of the redox reactions. Deactivation on the quartz surface is interesting by itself. Besides, it is needed in connection with the fact that our reactor is made of this material.

Temperature dependence of the accommodation coefficient ϵ on quartz at a pressure of 66.5 Pa is presented in Figure 3.17. In the temperature interval from 523 to 828 K, this dependence can be described by the expression

$$\epsilon_{SiO_2} = (7.4 \pm 0.5) \cdot 10^{-3} \exp\left[\frac{(18.5 \pm 1)\text{kJ/mol}}{RT}\right] \tag{3.46}$$

Below 523 K the accommodation coefficient is so small that we failed to measure it, but the tendency of ϵ decreasing with the temperature decrease remains unchanged.

Two types of platinum were used in studies of singlet oxygen deactivation: the bulk, and the platinum film. The first had the form of spiral wire 0.5 mm in diameter with a total surface area of 34 cm³.

Platinum was cleaned by heating in air at 1273 K, then at 873 in a vacuum of 10^{-3} Pa, and, finally, in an oxygen flow. Figure 3.18 illustrates the

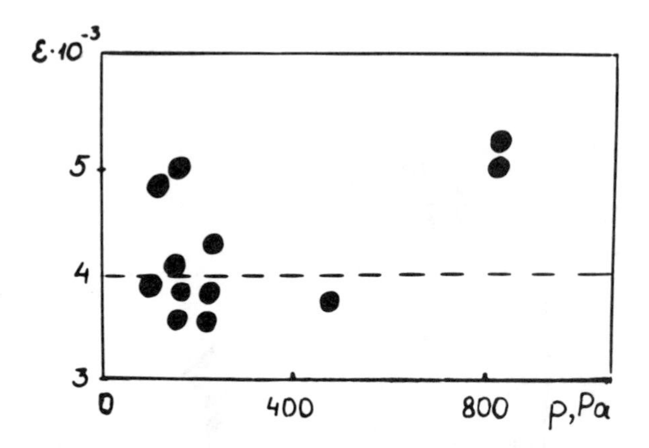

FIGURE 3.18. Pressure dependence of the accommodation coefficient ε of singlet oxygen for Pt wire at 523 K. Each experiment corresponds to an individual point.

reproducibility of results for the accommodation coefficient. At 523 K the ε value varies from time to time by approximately 50%. Because the experimental error is about 10%, the observed spread in ε values is connected with the irreproducible change of surface properties. In the interval from 0.6 10^2 Pa the pressure dependence of ε may be thought of as absent.

The results of measurements for the temperature dependence of the accommodation coefficient on Pt wire in the range 293 to 583 K (see Figure 3.17) can be described by the following expression

$$\epsilon_{Ptwire} = (0.25 \pm 0.2) \exp\left[-\frac{(17.6 \pm 2)\ \text{kJ/mol}}{RT}\right] \tag{3.47}$$

The ε value at the temperature of dry ice (196 K) turned out to be outside our measurement limits, but the general tendency of ε decreasing with the temperature decrease remains unchanged.

$O_2(^1\Delta_g)$ heterogeneous deactivation on Pt film deposited on the surface of the mobile quartz rod has also been measured. Because the rod was longer than the reactor, inserting the part with deposited platinum into the center of the reactor did not alter the gas-flow conditions. The temperature dependence of ε, in this case measured in the interval from 293 to 503 K at a pressure of 4 10^2 Pa, is well described by the expression

$$\epsilon_{Ptfilm} = (1.65 \pm 0.2) \exp\left[-\frac{(21.0 \pm 2)\ \text{kJ/mol}}{RT}\right] \tag{3.48}$$

Comparison of the results obtained for the singlet oxygen deactivation on Pt wire and film shows that the activation energies of the processes E_α are practically the same, but the preexponential factors differ sevenfold. Thus, the film appears to be several times more active with respect to the singlet oxygen deactivation. It seems to be connected with the fact that the geometric area of Pt film is substantially less than its real area. It is this fact that explains the overestimated value of the preexponential factor in excess of unity.

At the present time, it is quite unclear why E_α values in E_{exp} in Equations 3.46 and 3.48 for quartz and platinum are in such close agreement. In addition, it is difficult to explain the growth of the accommodation coefficient with increasing temperature. There was no theoretical and experimental work in the field of heterogeneous deactivation of electronically excited molecules before our studies. It seems that an excited molecule transits into the physisorbed state as a result of collision with the surface. Deactivation occurs in the physisorbed state during the lifetime τ. If the probability of the energy transfer unit time is denoted by w, the observed value of the accommodation coefficient is proportional to $w\tau$, i.e., it should decrease with increasing temperature. It seems likely that this mechanism fails to explain our experimental data.

The second mechanism assumes the existence of surface active sites that can strongly chemisorb excited molecules. The lifetime in the chemisorbed state is so long that deactivation occurs with the probability close to unity. At low temperatures these sites are occupied by chemisorbed impurities. As temperature increases, the number of unoccupied sites increases due to desorption; therefore, the deactivation rate grows. According to this mechanism, a number of active sites and, hence, the accommodation coefficient should be dependent on pressure, but this contradicts our data on singlet oxygen. We may suppose that almost all active sites are occupied, but a low activation energy of 21 kJ/mol excludes this possibility.

Both mechanisms under consideration explain the characteristic features of heterogeneous accommodation only formally. Therefore, their applicability to the processes of energy transfer is questionable.

The observed spread of values of the accommodation coefficient on platinum seems to be connected with the influence of the composition of adsorbed layer on the heterogeneous deactivation rate of singlet oxygen. Composition of the adsorbed layer may alter even at a constant temperature, depending on the impurity amount in discharge products, especially NO, CO, and water, which were detected using the mass spectrometer. This conclusion is confirmed also by the results connected with the influence of hydrogen, CO, and water on the heterogeneous deactivation rate of singlet oxygen on platinum and gold.

$O_2(^1\Delta_g)$ Deactivation on Gold

When the control over the surface state is absent, it is very difficult to distinguish between the temperature influence and that of adsorbed layer on the accommodation coefficient. For a separate study of these effects, the most convenient material is gold, having the weakest chemisorption activity. Until

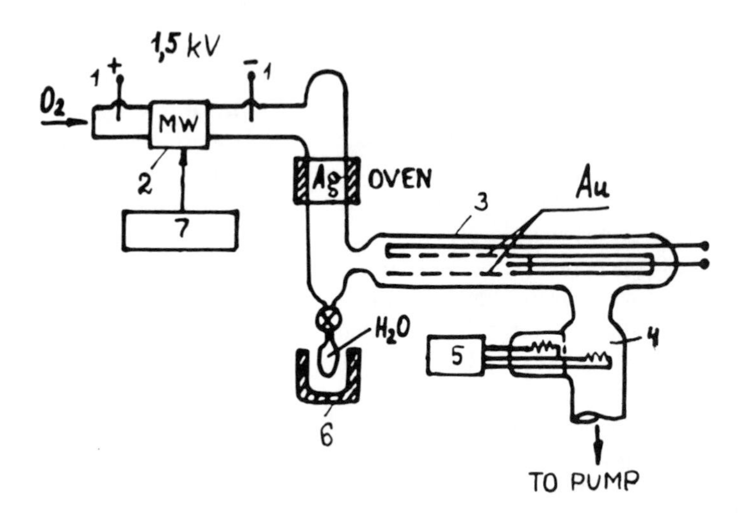

FIGURE 3.19. The scheme of the experimental setup for studies of the singlet oxygen deactivation on gold: (1) initiating direct current discharge; (2) UHF-discharge; (3) reactor; (4) detector; (5) registration unit; (6) source of water vapor; (7) modulator of the discharge generation.

our work there were no detailed studies in deactivation of singlet oxygen on gold. Thomas and Thrush[175] pointed out that the $O_2(^1\Delta_g)$ accommodation coefficient on addition of HCl to the oxygen flow appears to be related to HCl adsorption on gold. We will give the data concerning the deactivation of singlet oxygen on gold as well as the influence of water on this process.[138,178]

The sketch of the experimental setup is shown in Figure 3.19. Oxygen flow at a pressure of 200 Pa was purified from water by passing through a trap with liquid nitrogen. Singlet oxygen was generated using an ultrahigh frequency (UHF) discharge cell (2). A specially constructed thermometric detector (4) was used for the concentration measurements of excited oxygen molecules. It consisted of two tungsten wires, one of which was measuring and located in the flow, and the other was compensative and was protected from excited molecules by a glass-fabric screen, being out of the flow. Both wires were constituents of a bridge scheme and could be heated by a direct current up to 673 K. UHF discharge was modulated at a frequency of 1.6 Hz. The detection at this frequency was performed using a synchronized technique. Owing to modulation the signal-to-noise ratio increases, and the influence of heating elements of the setup on the detector is eliminated, which in turn, makes possible the studies of temperature dependence of the accommodation coefficient. To enhance the modulation depth the discharge power was modulated by 80% so that discharge was periodically drawn and quenched. For the sake of increasing the stability in this regime an additional initiating direct-current discharge (1) was employed. Oxygen atoms created in the discharge were removed using the glass tube covered by a thin silver layer on the inside.

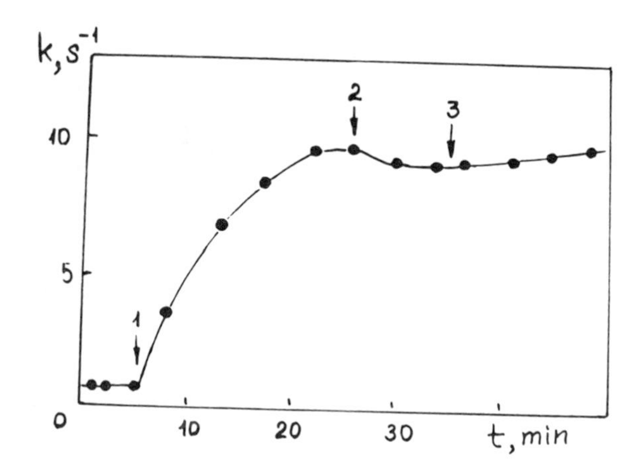

FIGURE 3.20. Heating and cooling effects for the rate constant of $O_2(^1\Delta_g)$ deactivation on the gold wire: (1) t = 5 min, heating to 353 K; (2) 26 min, cooling to 298 K; (3) 35 min, heating to 353 K.

Heating this tube up to 623 K resulted in the increase of detection signal approximately tenfold, which is connected, in our view, with the creation of singlet oxygen during the heterogeneous recombination of oxygen atoms. This conclusion has been confirmed by the direct spectroscopic detection of singlet oxygen.

The reactor had the form of a cylindrical glass tube 2.7 cm in diameter with a gold wire 0.03 cm in diameter and 11.3 cm in length, stretched along the tube axis. Oxygen pressure in all experiments was $2 \cdot 10^2$ Pa, the average flow velocity 0.6 m/s. In order for the rate constant of the singlet oxygen deactivation to be measured, the gold wire was screened by a thin tube moved with the help of an exterior magnet. The screening of the wire resulted in increasing the signal of singlet oxygen at the output of the reactor from I_{min} for the completely open wire to I_{max}, when the wire was absolutely screened. A knowledge of the contact time t allowed us to calculate the effective deactivation rate constant according to the formula $k = t^{-1} (I_{max}/I_{min})$. The effective rate constant depends on the accommodation coefficient, on the geometric characteristics of the reactor, and on the diffusion rate of excited molecules to the surface.

In order for ϵ to be defined using experimental values of the effective rate constant, the diffusion equation in the reactor must be solved. It has been solved using numerical integration. A set of functions $k = k(\epsilon)$ for different temperatures of the inner cylinder was plotted. It turned out that the deactivation rate constant depends on the temperature of preliminary treatment of the sample surface. The variation of the rate constant with rapid heating from room to higher temperature is shown in Figure 3.20. At the moment $t = 0$ the deactivation rate constant equals $0.75\ s^{-1}$, the corresponding value of the accommodation coefficient is $1.8 \cdot 10^{-3}$. In 5 min the temperature is increased

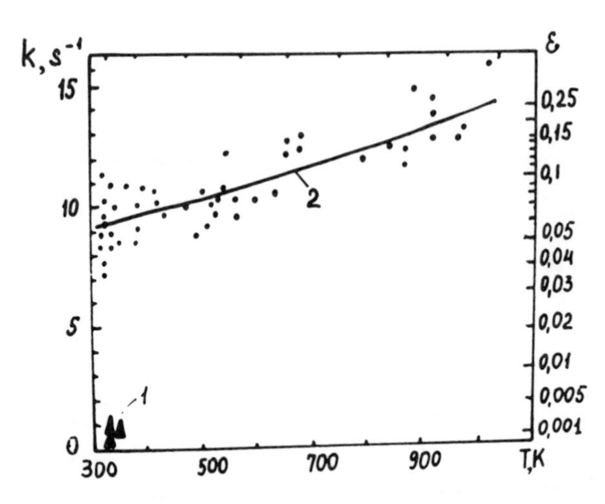

FIGURE 3.21. Temperature dependencies of the deactivation rate constant and accommodation coefficient for singlet oxygen on gold: (1) before heating up to 1073 K; (2) after heating.

to 353 K, and the rate constant starts to grow. This process continues ~10 to 15 min, and the rate constant reaches a value of 9.5 s^{-1}, the corresponding ϵ value being $5 \cdot 10^{-2}$. It should be noted that the time of heating and cooling the wire is of the order of 30 s, so a slow process of changing the rate constant does not contribute to the thermal heating of the wire. Upon turning off the heating at t = 26 min, the wire cools, but the rate constant does not return to its original value, i.e., the hysteresis is observed. A second heating up to 353 K (at t = 35 min) does not give rise to appreciable change of the rate constant.

Inasmuch as the hysteresis behavior of the deactivation rate constant could be connected with the impurity desorption which was initially present at the surface of the gold wire, it was heated at 1073 K for a period of 30 min in the oxygen flow. Such treatment resulted in a deactivation rate constant of the order of 10 s^{-1}. With the temperature increase, the rate constant grows slightly (see Figure 3.21). It is known[183-185] that heating gold in an oxygen atmosphere results in both carbon removal from its surface and the formation of a strongly bound oxygen layer. This effect is connected with the calcium segregation at the gold surface on heating. Eley and Moore[186] did not reveal oxygen chemisorption on the gold surface with the Auger-spectrometer control of the surface cleanliness. We dealt with the surface capable of oxygen chemisorption; therefore, a value of the deactivation rate constant k = 10 s^{-1} ($\epsilon \approx 0.1$) appears to refer to the oxidized surface.

Special examination has been carried out to check the influence of water additions in the oxygen flow on the heterogeneous deactivation of singlet oxygen. Initially at the moment t = 0 the deactivation rate constant equals 9.5 s^{-1}. On 5 to 10% addition of water in the oxygen flow (t = 5 min), the

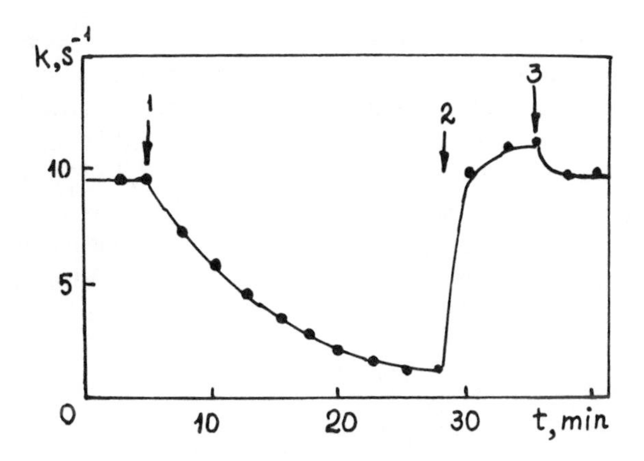

FIGURE 3.22. Effect of water adsorption and desorption on the rate constant of $O_2(^1\Delta_g)$ deactivation on gold: (1) t = 5 min, adding 10% water vapor in O_2 flow; (2) 27.5 min, heating of Au in dry oxygen at 773 K; (3) 35 min, cooling to 298 K.

rate constant starts to decrease and reaches a value of $\sim 1\ s^{-1}$ in 20 to 30 min. The accommodation coefficient decreases correspondingly from a value of $4.3 \cdot 10^{-2}$ to $2 \cdot 10^{-3}$, i.e., 20-fold. On cutting off the water supply, the deactivation rate constant does not change. Next, heating the wire at 773 K in dry oxygen at t = 27.5 min restores the rate constant to its original value of $10\ s^{-1}$ (Figure 3.22).

These data are evidence in favor of a strongly bound water layer created on the gold surface that does not desorb at room temperature. Assuming that the accommodation coefficient is additive, i.e.,

$$\epsilon = \epsilon_{H_2O}\Theta_{H_2O} + \epsilon_{AuO}(1 - \Theta_{H_2O}) \tag{3.49}$$

where ϵ_{H_2O} and ϵ_{AuO} are accommodation coefficients on water-covered and oxidized gold surface, respectively, and Θ_{H_2O} is the coverage by water, from 20-fold ϵ decrease on water adsorption, it follows that $\Theta_{H_2O} \approx 0.95$. The reason for the strong decrease of the accommodation coefficient on water adsorption seems to be connected with the change of the deactivation mechanism. At the clean gold surface $O_2(^1\Delta_g)$ energy is possibly transferred to conduction electrons when their orbitals overlap that of singlet oxygen. Adsorbed water prevents this overlapping, and $O_2(^1\Delta_g)$ energy is transferred to H_2O and adsorbent vibrations.

3.2.4. $O_2(^1\Delta_g)$ QUENCHING ON Pt SURFACE IN THE COURSE OF CATALYTIC CO OXIDATION: SDOSO METHOD

As shown previously, the rate of surface deactivation of singlet oxygen depends on temperature, the composition of adsorbed layer, and on physical and chemical surface properties. Carbon monoxide appears to be one of the inhibitors for deactivation of excited oxygen. To check this supposition we examined the influence of adsorbed CO layer on the deactivation rate at the Pt surface.[187] Indeed, platinum is an excellent catalyst of CO oxidation. Earlier, the mechanism of this reaction had been studied by us in detail on platinum surface cleaned in ultrahigh vacuum conditions,[188,189] so the possible deactivation channels for singlet oxygen are of interest as well.

CO oxidation on platinum is the subject of wide investigation.[190,191] As a result of joined activities, the following stage mechanism has been established

$$Z + CO \rightleftarrows ZCO$$

$$2Z + O \rightleftarrows 2ZO$$

$$ZO + CO \rightarrow CO_2 + Z$$

$$ZO + ZCO \rightarrow 2Z + CO_2$$

where Z is a surface site, and ZO and ZCO are chemisorbed oxygen and carbon monoxide, respectively.

Omitting details of the kinetic mechanism, we shall point out some characteristic features of this process. According to this scheme, the reaction proceeds predominantly through the Rideal mechanism. Gaseous CO reacts with chemisorbed oxygen practically without activation energy. Recently, this scheme was criticized by a number of researchers who proved using molecular beam methods that a pure impact mechanism in CO oxidation does not occur (see Section 4.4.2). These authors suppose that carbon monoxide reacts with oxygen from the preadsorbed state. Note that this question is of general importance in catalysis, and we shall discuss it later, but it is of no importance in our case because these mechanisms have the same kinetic behavior. It is of importance that the oxidation rate is governed by CO desorption from the surface when CO is in excess. Hence, by controlling the ratio between CO and O_2 in the gas phase, desired carbon monoxide and oxygen coverages can be obtained on the surface.

The revealed properties of singlet oxygen deactivation give promise that the study of $O_2(^1\Delta_g)$ heterogeneous deactivation in the course of catalytic reaction could enable us to keep permanent watch on the composition of the adsorbed layer *in situ*. Our method of observation is based on the study of surface deactivation of singlet oxygen (SDOSO).

Oxygen pressure in the flow was varied in the interval from 10 to 10^3 Pa. To remove the discharge products such as oxygen atoms, ozone, and $^1\Sigma_g^+$ oxygen, a silver wire placed in a heating glass tube downstream with respect to the discharge was used. The reactor had the form of a quartz tube 2 cm in diameter with a platinum wire 0.5 mm in diameter and 10 cm in length stretched along the tube axis. To eliminate the influence of temperature changes resulting from heat release in the course of the reaction, a special

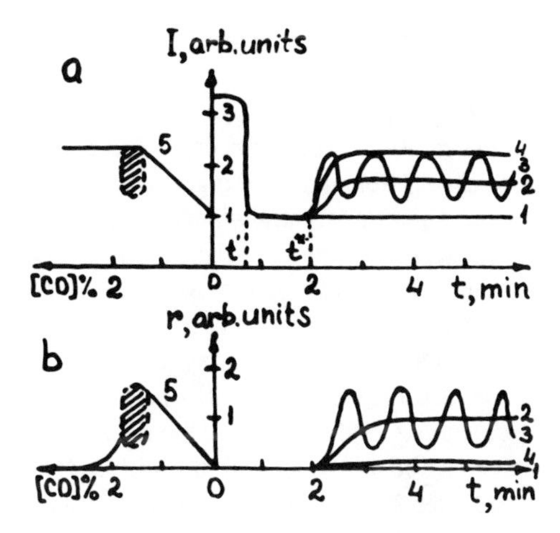

FIGURE 3.23. $O_2(^1\Delta_g)$ signal intensity (a) and reaction rate (b) as a function of time and CO concentration in the oxygen flow at $P(O_2)$ = 173 Pa: (1) [CO] = 0%; (2) 1%; (3) 1.5%; (4) 3%.

control apparatus for automatic temperature stabilization (isothermal calorimeter) was used. The singlet oxygen concentration was determined in a spectroscopic way. Carbon monoxide was mixed in the oxygen flow at the input of the reactor.

Figure 3.23 illustrates the influence of CO addition in the oxygen flow on the singlet oxygen deactivation on a Pt surface at 423 K. At the moment t = 0 the catalyst is at room temperature, and the $O_2(^1\Delta_g)$ signal has the maximum intensity. Increasing the temperature up to 423 K ($t = t'_1$) results in the drop of the signal, i.e., to the growth of the singlet oxygen deactivation rate (Curve 1). Further addition of a small amount of carbon monoxide in the oxygen flow ($t = t^*$) results in, first, appearance of CO_2 in the gas phase detected by a mass spectrometer, and second, the increase of both $O_2(^1\Delta_g)$ signal intensity and the reaction rate r (Curve 2).

A series of curves in the right-hand side of Figure 3.23 illustrates the influence of CO concentration in the gas phase on the function I(t) and r(t). The $O_2(^1\Delta_g)$ signal and the rate of the reaction as functions of CO concentration in the gas phase are represented by Curve 5. For increasing CO concentration up to 1%, the $O_2(^1\Delta_g)$ signal intensity and the reaction rate are unambiguous growing function (Curve 5). In the CO concentration interval from 1 to 2%, the $O_2(^1\Delta_g)$ signal and the reaction rate oscillate (Curve 3). The oscillation period is approximately 10 to 100 s. From these curves it is seen that the $O_2(^1\Delta_g)$ signal intensity and the oxidation rate oscillate out of phase: the reaction rate growth coincides with the decrease of the singlet oxygen concentration. The shaded regions in the left side of Figure 3.23 correspond to the unstable regimen. If the CO concentration is above 2%, the reaction rate becomes very slow and the $O_2(^1\Delta_g)$ signal intensity reaches its maximum value and remains unchanged on further increasing of CO concentration (Curves 4 and 5).

The contribution of gas-phase $O_2(^1\Delta_g)$ quenching by CO and CO_2 molecules is negligible under our conditions because CO addition to the oxygen flow does not lead to appreciable singlet oxygen signal change in the absence of the reaction. Hence, signal alternations in our experiments are caused by changes of the surface deactivation rate. Also, because CO oxidation experiments were carried out in isothermal conditions, the influence of the catalyst temperature on the SDOSO effect is excluded.

The deactivation probability of singlet oxygen on adsorbed oxygen layer is higher than that on adsorbed CO. It is this effect that is responsible for $O_2(^1\Delta_g)$ concentration change in the course of the reaction. Thus, the SDOSO probability depends on the coverage of platinum surface by reagents. The CO concentration increase over 2% does not alter the SDOSO probability because the platinum surface is completely covered by CO, and, therefore, the reaction rate is very slow. In the interval of CO concentrations from 0 to 1%, a partial substitution of adsorbed oxygen for CO occurs. Therefore, SDOSO probability decreases with the growth of CO concentration in the gas phase. Investigation of the singlet oxygen deactivation in the region of instability allows us to draw the conclusion that the described rate oscillations were connected with those of adsorbed particles, i.e., in fact, to exclude oscillation mechanisms which are related to the combined heterogeneous-homogeneous processes.

So, it has been shown that singlet oxygen deactivation is a surface-sensitive method which allows us to study the catalytic reaction *in situ*. The question about the microscopic mechanism of the singlet oxygen heterogeneous deactivation both on clean metals covered by adsorbates remains open. Kustarev[109] and Ryskin[178] suggested an exchange mechanism according to which the energy of excited oxygen molecules transfers to the metal-free electrons. Ertl et al.[190] have also noted the effect of adsorbed particles on the heterogeneous deactivation of metastable inert atoms.

3.3. KINETIC MECHANISMS OF HETEROGENEOUS DEACTIVATION OF EXCITED PARTICLES

The studies of pressure and temperature dependence of the accommodation coefficient and the effect of surface treatment enable us, in principle, to understand the stepwise scheme of the deactivation process, to check its applicability, and to determine the rate constants for different stages and their temperature dependence. This information is of great importance because it enables us to determine the lifetime of excitation. The suggested scheme of the process must be sufficiently proved; it should not have only a hypothetical character. When studying the heterogeneous deactivation of excited molecules, there are principally only two different mechanisms. In accordance with the terminology accepted in heterogeneous catalysis we call them "impact" (Rideal) and "adsorptive" (Langmuir-Hinshelwood). The first implies that an excited particle deactivates in the process of inelastic scattering at the surface, and the second proceeds via preliminary molecular adsorption on the surface. The deactivation probability per one collision can be given by

$$\epsilon = (1 - s)w_{imp} + snw_{ads} \quad (3.50)$$

where s is the sticking coefficient, w_{imp} and w_{ads} are deactivation probabilities through the impact and adsorptive mechanisms, respectively, and n is the number of molecular vibrations in the adsorption well during its lifetime on the surface.

Suppose that to a first approximation, $w_{imp} \approx w_{ads}$. Then taking into account that even at relatively high temperatures $n >> 1$, the second term in Equation 3.50 is assumed to be dominant. So, as a rule heterogeneous deactivation of excited molecules proceeds via the adsorptive mechanism. Kinetic treatment also faces problems, even though adsorptive mechanism is strictly proved, for example, in the elucidation of the question what microscopic mechanism is responsible for the deactivation. As we have already seen, for vibrationally and electronically excited molecules, ϵ is of the order of $10^{-1} \div 10^{-5}$. These values are defined by either a few adsorption sites or the low probability of losing the energy quantum during one collision with the surface. A choice between these possibilities is a complicated problem that requires the site identification to be solved. Such investigation has been carried out for heterogeneous deactivation of singlet oxygen on quartz.[178]

Severe difficulties are associated with the definition of the number of molecular vibrations in the adsorption well during its lifetime on the surface for two reasons. First, it is not clear whether the collision of a molecule with the surface is required for its deactivation, and second, at least in the case of electronically excited molecules, the definition of the molecular lifetime on the surface remains uncertain. For example, the so-called excimer deactivation mechanism has been proposed[155,178] for $Ar(^3P_2)$ and $O_2(^1\Delta_g)$ deactivation that takes into account the possibility of formation of a deep potential "well" for an excited molecule. In this case the lifetime of an excited molecule on the surface is defined by deactivation time rather than the depth of the well. In such a situation it is especially difficult to define the number of molecule collisions with the walls of the well during its lifetime.

Nevertheless, let us discuss kinetic mechanisms of the heterogeneous deactivation of excited molecules that take into account different surface initial states. To summarize the data on kinetic studies in heterogeneous deactivation of excited molecules, there are only three different kinetic mechanisms:

1. Physisorption or chemisorption takes place at sufficiently high temperatures, when the adsorbent surface is practically free from both excited and nonexcited adsorbed molecules.

2. Physisorption occurs on the surface that is almost completely covered by adsorbed molecules with a vibrational frequency only slightly different from that of a free molecule, i.e., exchange mechanism.
3. The surface is to a large extent covered by strongly bound chemisorption molecules or atoms having frequencies that are significantly different from that of a free molecule.

When considering these mechanisms, we accept that the Langmuir isotherm is valid and that the concentration of excited molecules is far less that that of nonexcited ones.

In the case of the first mechanism one can conceive the following scheme of the process

$$A^* + Z \underset{k_2}{\overset{k_1}{\rightleftarrows}} ZA^*; \qquad ZA^* \overset{k_3}{\rightarrow} ZA$$

Taking into account that the accommodation coefficient is equal to the ratio of the deactivation rate to a number of collisions of excited molecules with the surface, within a constant factor the ϵ value is given by

$$\epsilon = (k_1k_3/k_2)/(1 + k_3/k_2) \qquad (3.51)$$

For $k_3 >> k_2$, i.e., when the deactivation time from the adsorption state is far less than the lifetime, we obtain $\epsilon = k_1$, where k_1 is the sticking probability and $k_1 \sim 1$ in the case of physisorption. For $k_3 << k_2$ $\epsilon = k_1k_3/k_2 = ak_3$, where a is an adsorption factor of a physisorbed molecule and k_3^{-1} is the excitation lifetime. Two points should be emphasized. First, ϵ can diminish with growing temperature, and, second, it is independent of pressure.

The second mechanism can be reduced to the first. Indeed, at high surface coverage with physisorbed species, each new molecule collides with already adsorbed one. As a result of decreasing the binding energy for the next layer, its coverage is low, and we arrive at the first mechanism. Note, that in this case the nature of the surface is of no importance, and the deactivation probability is close to the gas-phase value.

There are two possibilities for the third mechanism: (1) deactivation occurs at a free surface, then the accommodation coefficient is inversely proportional to pressure and grows with increasing temperature; and (2) deactivation occurs through the physically adsorbed state with the participation of chemisorbed species. In the latter case, ϵ is independent of pressure and drops with increasing temperature. This was the situation that was observed for nitrogen and deuterium deactivation on the tungsten and copper surface.[79]

The work of Gershenzon et al.[92] is a good illustration of the above classification of microscopic mechanisms. These authors studied in detail the heterogeneous deactivation of vibrationally excited nitrogen on the molyb-

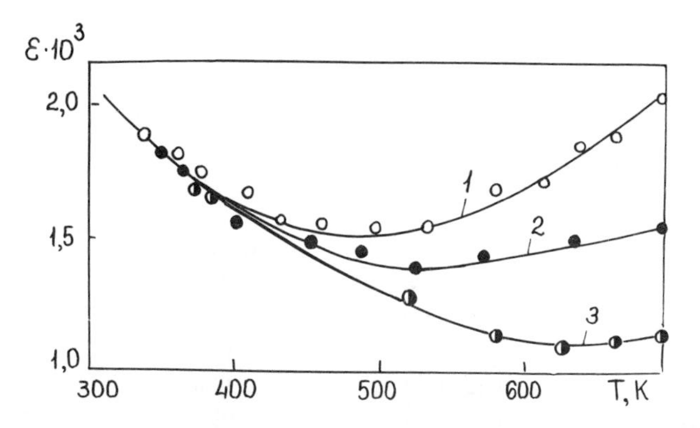

FIGURE 3.24. Temperature dependence of the accommodation coefficient of vibrationally excited nitrogen on molybdenum glass: (1) 2.5 10^2; (2) 5 10^2; (3) 8.5 10^2 Pa.[85,92]

denum-glass surface. Detailed description of this work allows us to understand what kind of information can be obtained from the kinetic studies of relaxation processes. Figure 3.24 demonstrates the temperature dependence of the accommodation coefficient.[92] With the increase from room temperature to 400 K, the accommodation coefficient drops and is practically independent of pressure. Then it grows, and the rising branch of the curve depends on pressure. The results are described by the following empirical equation

$$\epsilon\,(T,p) = 3.9\;10^{-4}\exp\left(\frac{1000}{RT}\right) + \frac{0.415}{P}\exp\left(-\frac{6000}{RT}\right) \qquad (3.52)$$

These authors suggest two possible mechanisms. A so-called "adsorption-phonon" mechanism is proposed to explain the inverse temperature dependence and independence of the accommodation coefficient on pressure at low temperatures. It reduces to the following kinetic scheme

$$N_2^* + Z \underset{k_{-1}}{\overset{k_1}{\rightleftarrows}} N_2^*Z \qquad N_2^*Z \xrightarrow{k_2} N_2Z$$

The term "adsorption-phonon" mechanism is, in our view, improper because one should distinguish between kinetic-deactivation and microscopic mechanisms. The latter elucidates the process of energy transfer from excited particles to the solid. The synthetic term "adsorption-phonon mechanism" may be misunderstood, but we still use it following Gershenzon et al.[92]

Assuming that the concentration of adsorbed vibrationally excited molecules is stationary and the corresponding coverage is small, the accommodation coefficient is given by

$$\epsilon = sk_2/(k_{-1} + k_2) \qquad (3.53)$$

where s is the sticking coefficient of N_2^*.

If the molecular lifetime on the surface $\tau_{-1} = 1/k_{-1}$ is far less than the relaxation time of the local vibration $\tau_2 = 1/k_2$, then we have

$$\epsilon = S\tau_{-1}/\tau_2 \tag{3.54}$$

Thus, a knowledge of $s\tau_{-1}$ allows us to determine τ_2. Gordon and Ponomarev[192] measured the value of $s\tau_{-1}$ for nonexcited nitrogen molecules. Assuming that this value does not appreciably change for excited molecules, Gershenzon et al.[92] estimated it to be $\tau_2 = 3.4\ 10^{-8}$ s at 300 K. Black et al.[98] arrived at a similar mechanism for nitrogen deactivation on Pyrex®. They estimated $\tau \approx 10^{-4}$ s. Such a discrepancy is explained by the fact that these authors[98] used the upper value of the interval for physisorption energies given in Reference 192, from 8 to 25 kJ/mol. The ascending branch of the curve in Figure 3.24 is attributed to the chemisorption mechanism.[92] According to this mechanism, deactivation occurs due to the chemisorption of nitrogen molecules, the process being limited by desorption of impurities from these sites. At low temperature, the chemisorption sites are occupied and do not contribute to the total process. In this case nitrogen deactivation is described by the following scheme

$$N_2^* + Z \xrightarrow{k_1} N_2^*Z \qquad N_2 + Z \underset{k_{-2}}{\overset{k_2}{\rightleftarrows}} N_2Z \qquad N_2^*Z \xrightarrow{k_3} N_2Z$$

On the assumption that the concentration of adsorbed vibrationally excited molecules is stationary, we have

$$\epsilon = \frac{4\sigma_o}{v} s_1 \frac{k_{-2}}{(\alpha s_1 + s_2)[N_2] + 4k_2\sigma_o/v} \tag{3.55}$$

where σ_o is the surface density of chemisorption sites in cm^{-2}, $\alpha = [N_2^*]/[N_2]$ is a part of excited molecules which does not depend on pressure, and s_1 and s_2 are sticking coefficients of excited and nonexcited molecules, respectively.

Gershenzon et al.[92] found that $\sigma_o \approx 10^9\ cm^{-2}$. Equation 3.55 well explains the observed temperature dependence on pressure. Unfortunately, the studied temperature interval is not sufficiently wide. It is obvious that further temperature increase must lead to the decrease of the accommodation coefficient as a consequence of lowering the lifetime of an excited molecule on the surface, providing the proposed model is valid.

Many researchers believe that deactivation of multiatomic molecules as well as of diatomic ones in some cases proceeds through the exchange mechanism, i.e., collision of a vibrationally excited molecule with an adsorbed resonance partner. Numerical estimations for such a process have not been performed. By analogy with deactivation processes in the gas phase, one would expect that a large number of inner degrees of freedom plays an important part in the deactivation process. The accommodation coefficient for diatomic molecules are 2 to 3 orders less than that for multiatomic ones, because the first have vibrational frequencies far higher than the Debye frequency of a solid.

Let us make a short historical review for relaxation mechanisms of vibrationally excited molecules.

Schäfer and Grau[193] calculated the molecular accommodation coefficient ϵ on the platinum surface. Assuming that the partial coefficient of translational and rotational energies are equal to unity, the accommodation coefficient of the ith vibrational mode of the CS_2 molecule has been found using the Euken formula similar to that of Feuer[195]

$$\epsilon_i = \tau/(\tau + \tau_i) \tag{3.56}$$

where τ is the molecular lifetime on the surface (τ is defined by the Frenkel formula $\tau = \tau_o \exp(Q/RT)$) and τ_i is the lifetime of the ith molecular mode on the surface. The calculation of the deactivation probability, defined as a value inversely proportional to the excitation lifetime on the surface, has been carried out to the first approximation of the perturbation theory. The change of the dipole interaction between the adsorbed molecule and a platinum atom on vibrational excitation served as a perturbation. It was also assumed that the platinum atom has the elementary charge e_o. The perturbation potential in this case is given by

$$\Delta E = \left(\frac{\delta\mu}{\delta x}\right)_{x=o} x\frac{e_o}{d^2}\cos\zeta \tag{3.57}$$

where μ is the molecular dipole momentum, x is the vibrational coordinate, d is the distance between the center of mass of the adsorbed molecule and the platinum atom, and ζ is the angle between the molecular axis and a straight line between the molecular center of mass and the platinum atom.

These authors modeled the platinum atom by a spatial oscillator. Numerical calculation of the deactivation probability has been carried out for all CS_2 vibrational modes, and the total number of accommodation coefficients was found. They coincided with the experimental ones at all temperatures studied.

Note that the accommodation coefficient is usually measured by indirect methods lacking sufficient accuracy. Nevertheless, more recent studies using the pulse and beam technique arrived at values very close to that of Schäfer and Grau. A weak dependence of the accommodation coefficient on the surface nature has been observed in this case; therefore, the model of Schäfer and Grau seems to be questionable.

Goldansky et al.[196] calculated the vibrational exchange probability in the collision of a diatomic molecule with a surface atom. The surface atom and incident molecule were assumed to be harmonic oscillators. The quantum one-dimensional task has been solved. The rotational motion has been neglected. The interaction potential was modeled by exponential repulsion. The frequency of molecular vibration was assumed to be $\nu_g > \nu$, where the latter is the vibrational frequency of the surface atom. The probability of one-quantum transitions has been found to the first order of the perturbation theory.

For molecules with a high-frequency vibrational quantum such as H_2, D_2, and N_2, the probability of the impact deactivation according to this theory is several orders less than experimentally measured values. Therefore, deactivation via the preadsorbed state seems to be more realistic. The temperature dependence of the accommodation coefficient has a better explanation in this case as well.[195]

When the frequency of molecular vibrations coincides with that of the solid ones, a very fast energy exchange is possible. It has been shown that the lifetime of such an excitation is $\sim 10^{-11}$ s, which is close to the lifetime values in solutions.

Hunter[197] proposed an interesting model for deactivation of vibrationally excited molecules. The calculation from the theory of Landau-Teller for the impact deactivation probability of CH_4, CO_2, and N_2O on polished NaCl surface and for N_2 on glass gives values which are substantially lower than experimental. Therefore, Hunter suggested that deactivation proceeds via surface migration of adsorbed molecules with the probability calculated again in accordance with the theory of Landau-Teller for each elementary molecular jump. To estimate a number of jumps, it was assumed that the activation energy of the surface migration is approximately half as much as the adsorption heat. Then a number of jumps can be estimated from the formula

$$m = \exp[(Q - E_m)/RT] \tag{3.58}$$

where Q is the adsorption heat and E_m is the activation energy for surface migration.

Because the parameters of interaction potential between the adsorbed molecule and the surface are unknown, their values were varied to fit the calculated and experimentally measured accommodation coefficients. Nevertheless, a disagreement between theoretical and measured values was found. For example, the calculated accommodation coefficient nitrogen on glass gives $\epsilon = 10^8$ compared with the experimental value of 10^{-3}. In the case of multiatomic N_2O and CO_2 molecules, the calculated deactivation probabilities per one jump are high enough; therefore, the accommodation coefficient practically coincides with the sticking coefficients.[197] This viewpoint differs from the widespread opinion that deactivation is explained by the exchange mechanism in these cases.

A number of microscopic relaxation mechanisms for vibrational energy is reviewed by Zhdanov and Zamaraev.[198] We list here the main microscopic relaxation mechanisms considered in various papers published before this one: energy transfer to (1) optical and acoustic phonons of the solid lattice,[199-203] (2) translational motion of the relaxing molecule as a whole,[203] (3) molecular rotation,[203] (4) vibrational modes of the crystal,[203] (5) adsorbed layer when the energy transfer to the lattice is less probable,[204] (6) conduction electrons in metals,[103,104] and (7) electron surface states for the semiconductor.[107]

Chapter 4

INTERACTION OF MOLECULAR BEAMS WITH SURFACE

In Chapter 3 we reviewed the results for vibrational and electronic molecular deactivation obtained by the flow technique. Strictly speaking, flow technique does not allow direct measurements of energy transfer velocities from vibrationally and electronically excited molecules to the solid. On measuring the vibrational accommodation coefficient ϵ and molecular lifetime on the surface τ_s, vibrational relaxation time is defined by the expression $\tau = \tau_s (\epsilon^{-1} - 1)$.[20]

As mentioned in Chapter 2, the molecular beam technique is a direct method for studies of energy transfer velocities from molecules to the surface. At the time we began our investigations in the field of molecular quenching on surfaces, only a few studies concerning molecular beam applications to this problem had been reported. In recent years the number of publications has grown rapidly.

4.1. DYNAMIC SCATTERING OF MOLECULAR BEAMS AND DESORBED MOLECULES

Energy exchange between beam molecules and solid surfaces is studied by spectral methods in which different energy states of reflected species are detected. The kinetic energy (translational temperature of scattered molecules T_t) is measured with the help of a time-of-flight (TOF) mass spectrometer. Molecular angular distribution is also an important source of information about energy exchange in collisions.

Elastically scattered molecules do not interact practically with the surface and are reflected specularly. Elastic scattering of molecular beams enables us to study directly the energy exchange mechanisms with the surface. Exchange of translational, rotational, and vibrational energy with the surface (V_s-T, V_s-R, and V_s-V exchange, respectively) can be registered in the process.

4.1.1. MONOENERGETIC ATOMIC BEAMS

Scattering of monoenergetic inert atom beams (He, Ar, Kr, and Xe) with a definite translational temperature and directed at a hot surface is the simplest example of V_s-T energy exchange.[205-215]

Measurements of the kinetic energy of atomic beams scattered from the hot surface reveal that the temperature of scattered atoms T_t is usually far less than the surface temperature T_s (translational cooling of atoms). Inert gas atoms scattered from graphite surface have $T_t << T_s$.[205] Of course, the temperature of scattered particles grows with the increase of T_s, but at high temperatures this growth, i.e., V_s-T exchange, is strongly retarded. At T_s = 2500 K the observed T_t is 550 K for He, 700 K for Ar, 800 K for Kr, and ~900 K for Xe, so energy exchange increases with the growth of atomic mass. Angular distribution of the scattered He beam is sharper and differs from the cosine profile more strongly than that of heavy atoms (see Figure 2.4.). TOF studies of Ar beams scattered from Ag(111) showed that in this case T_t is also less than T_s.[206] The V_s-T exchange probability differs also on the properties of a solid. Distribution of particles is sharper for heavy metals, but is more diffuse for light ones.[208] When an adsorbed layer is present, e.g., CO on Pt-Ni, the energy exchange of inert atoms with the surface is enhanced.

Many of these characteristic features can be explained by adsorption of species on solids giving rise to a natural increase of energy exchange coupling. General experimental features of atomic and molecular beam interactions with the surface are as follows. At high energies of beam particles (high translational temperature T_t), high surface temperatures T_s, and weak coupling (low adsorption heat Q), elastic scattering dominates (specular reflection). At low T_t, low T_s and large Q inelastic processes, in accordance with the "trapping-desorption" mechanism, take place. The term "trapping" is usually applied to physical adsorption, whereas the term "sticking", to chemisorption. According to Weinberg,[207] elastic (specular) scattering occurs at $D/kT_t \leq 0.25$ where D is the adsorption well depth, inelastic scattering takes place for $0.25 < D/kT_t < 2.5$, and the "trapping-desorption" scheme is valid for $D/kT_t > 2.5$. Hurst[210] observed two scattered Xe beams from Pt(111): the first was specular and independent of T_s, and the second was accommodated completely.

The probability of multiphoton excitation during atomic beam scattering increases rapidly with the growth of mass of an incident atom, along with the growth of energy and density of solid low-frequency vibrations — phonons.[212] One-phonon transition in He scattering was proved[213] using the TOF mass-spectrometric technique with a high velocity resolution of scattered atoms ($v/v \approx 0.8\%$). Surface phonon spectra of metals were also determined from He beam scattering measurements.[214,215]

Another mechanism of inelastic coupling of atomic beams with metals and semiconductors is connected with electron-hole pair generation. For instance, Xe beams were shown to excite electron-hole pairs in a semiconductor. The effect of electron-hole pair excitations is weak (<10%) for light-scattering atoms, e.g., He. The probability of electron-hole excitation in metal increases at low T_s and low-beam energies.

4.1.2. MONOENERGETIC MOLECULAR BEAMS: TRANSLATIONAL ENERGY CHANGE (V_s-T EXCHANGE) DURING SCATTERINC AND ADSORPTION

Due to a large amount of molecular degrees of freedom, new possibilities appear with the use of molecular beams. Their interaction with a surface can lead to changes not only in translational, but in rotational, vibrational, and electronic energy as well. Even the simplest molecules H_2, HD, and D_2 are shown to enter into V_s-R and V_s-V energy exchange along with the V_s-T one. At first, consider V_s-T exchange which is studied mainly with the use of angular distribution of scattered molecules.

Energy exchange depends strongly on surface cleanliness. For contaminated surfaces, i.e., covered by adsorbed molecules, thermal accommodation increases as a rule. Consider the results of studies connected with the energetic states of molecules desorbing from the surface.

A number of investigations were concerned with the angular distribution of molecules desorbing from a monocrystalline surface. For desorbing molecules as well as for scattered ones, the validity of Knudsen law (Equation 2.4) is the basic criterion to judge whether they have come to equilibrium with the surface. For the first time, a deviation of desorbed molecules from equilibrium was observed by Dabiri et al.[216] TOF studies of H_2 molecules desorbed from polycrystalline Ni revealed that they have a Maxwell-Boltzmann distribution but with $T_t \approx 0.45\ T_s$. By a similar method it was shown that for desorption of O_2 molecules from Ag(111) at 600 K, $T_t = 0.8\ T_s$.[206]

The energy distribution of hydrogen desorbing from Ni(111) and polycrystalline Ni depends on surface contaminations.[217] Partial sulfur coverage strongly increases directionality up to $\cos^4\varphi$. Nevertheless, the nickel side (110) is almost insensitive to impurities, and complete equilibrium and the validity of the cosine law is observed for H_2 desorption.

The main result in measuring angular distributions of desorbed H_2, D_2, and HD molecules is more or less strong directionality in accordance with the dependence $\cos^4\varphi$, where $1 \leqslant n \leqslant 9$. According to Goodman,[218] desorbed molecules are not in equilibrium in general ($T_t \neq T_s$), but should have a Maxwell-Boltzmann distribution with different temperatures for normal $T_\perp$ and tangential $T_\parallel$ directions to the surface. This formal model was checked for H_2, HD, and D_2 molecules desorbed from Ni when the velocity distribution had been measured by TOF mass spectrometry for all desorption angles.[219] It turned out that the energy does depend on the desorption angle in accordance with this prediction. In the T_s interval 949 to 1143 K the mean energy is proportional to T_s, being 70% greater than T_s for normal direction ($\varphi = 0$) and approximately equal to T_s for $\varphi = 90$.

The causes of excess potential energy taken away by desorbing molecules are associated with the existence of the preadsorbed state or precursor (see Section 6.1). The best-known model is that of Van Willigen[220] according to which a molecule transforms into the chemisorbed state through a certain activated state with the activation energy E_α. According to the principle of microreversibility (detailed balancing principle), the same is valid for desorption as well: molecules desorb passing through the activated state. Hence, an equation for angular scattering has been derived

$$I = I_o \frac{E + \cos^2\varphi}{(E + 1)\cos\varphi} \exp(-E\ tg^2\varphi) \qquad (4.1)$$

(where $E = E_\alpha/RT_s$), from which adsorption activation energy can be obtained.

The model has different modifications, for example, assuming the distribution of centers with different activation energies[221] or "discrete inhomogeneity" — the existence of both activated and direct (without activation) adsorption sites.[222] The derived equations, along with Van Willigen's Equation 4.1, more nearly approximate the angular distribution of desorbed molecules compared to $\cos^n\varphi$.

In more recent works,[223,224] angular distribution of desorbed molecules and molecular velocity distribution are calculated taking into account specific models of the potential surface. For instance, Harris et al.[223] showed that H_2 and D_2 molecules desorbed from Cu at low T_s are focused close to the normal direction with $E_t \approx 1$ eV. Rettner et al.[225] observed very narrow distribution for H_2, D_2, and HD molecules desorbed from Cu(111) $\sim\cos^{12}\varphi$. The increase of E_t is connected with the potential well shape "at the point of entry" (see Section 4.1.4), i.e., it has an identical origin to Van Willigen's Equation 4.1.

The adsorption probability increases with the increase of a number of atoms in the molecule. The accommodation probability increases also, and the angular distribution of the scattered beam is close to the cosine. West and Somorjai[226] investigated a large number of different molecules scattered from the pure cubic face Pt(100) (5 × 1) at $T_s = 450$ to 1200 K and $T_t = 300$ K (thermal beams). Almost all examined molecules: CO, N_2, O_2, NO, H_2, D_2, CO_2, N_2O, C_2H_6, and NO_2 were scattered with a maximum close to specular reflection, though the angular distribution width was different indicating an incomplete accommodation. Only ammonia exhibited the distribution close to the cosine. Similar behavior was observed for Pt(100) covered with graphite. With the increase of T_s, the distribution broadens, i.e., the energy exchange with the surface enhances. The authors[226] believe that if the thermal molecular translational energy is close to the vibrational energy of surface atoms, T-V_s exchange is hampered, and the molecule reflects specularly without adsorption. Low-frequency (0.02 eV) surface phonons on Pt(111) and on Pt + C are not localized and are excited completely; then the probability of T-V_s exchange is minimal.

Quite different behavior was observed for CO beam scattering from Pt(100) covered with an adsorbed CO layer and for C_2H_2 beams scattering from Pt(100) covered with an adsorbed C_2H_2 layer. Here, the distribution was equal to the cosine. In this case adsorbed molecules can form low-frequency surface vibrations excited by incident molecules, and effective energy exchange is achieved.

Campbell et al.[227] made out almost specular CO scattering from Pt(111) with a low adsorption probability (the sticking coefficient $s_o = 0.06$ at 300 K and $s_o = 0.025$ at 600 K). In contrast to this, CO scattering from Pt(111) occurred with the complete thermal accommodation independent of T_s and T_t in accordance to the cosine law.

Tenner et al.,[228] using the methods of angular scattering and TOF, showed that translational excitation of NO molecules scattered from Pt(111) and Ag(111) depends on molecular orientation in the beam. The molecule oriented by N-tip adsorbs with a greater probability than that of O-tip orientation. Energy accommodation of N-tip oriented molecules is correspondingly greater. Similar experiments were carried out with CHF_3 and tert-C_4H_9Cl molecules.[229] The final translational energy of molecules scattered from graphite (0001) is lower for H-tip orientation of fluoroform and Cl-tip orientation of tert-butylchloride.

4.1.3. ROTATIONAL ENERGY CHANGE (V_s-R EXCHANGE)

The vibrational energy of surface atoms can be transferred to rotational excitation of an adsorbed or scattered molecule when the energy of vibrational levels is close to the phonon energy of the solid. The probability of the energy transfer depends on molecular orientation relative to the surface.

Both angular distribution and rotational level populations of molecules of H_2, D_2, and HD molecules scattered from different surfaces were measured in a number of studies.[31-33,230,231] According to Palmer et al.,[230] characteristic patterns for H_2, D_2, and HD beam scattering from Pt(111) and Ni(111) differ markedly: specular reflection is observed for H_2 and a more diffuse one occurs for D_2 and HD. Rotational levels of HD and D_2 molecules are substantially lower than that of H_2 (~4 kJ/mol). Therefore, V_s-R exchange is not observed for H_2 molecules, but it is possible with one-phonon excitation for HD and D_2, giving rise to the broadening of angular distribution.

Rotational excitations of H_2, D_2, and HD molecular beams scattered from LiF, Pt, and Ag were examined.[232-234] Selective excitation of individual rotational levels of HD molecules scattered from Pt(111),[233] Ag(111),[234] and Ni(001)[235] was observed in studies of the scattered beam intensity. The HD molecule is a nonsymmetric rotator, and in contrast to H_2 and D_2, sharp peaks corresponding to $0 \rightarrow 1$, $0 \rightarrow 2$, 1–2, and so forth of rotational transitions are observed for certain angles of scattering. The authors believe that these modulations result from trapping of some part of the molecules into physically adsorbed states on the surface with the resonant excitation of rotational quanta.

Excitation of rotational states of NO molecules scattered from the surface is one of the most extensively studied cases.[234-256] The energy of desorbed NO molecules has also been investigated. Rotational distribution of molecules is measured with the aid of the LIF or REMPI technique (see Section 2.2.1). Spin-orbital states were also measured in the molecular beam.[236-239]

The translational temperature T_t of molecules desorbed from Ru(001) was measured using Doppler line broadening in the LIF spectrum.[239] For $j = 6^{1/2}$ (where j is the rotational quantum number) T_t practically coincides with T_s = 460 K. For high vibrational levels the spectral lines are less broadened which points to the lower kinetic energy E_t of such molecules.

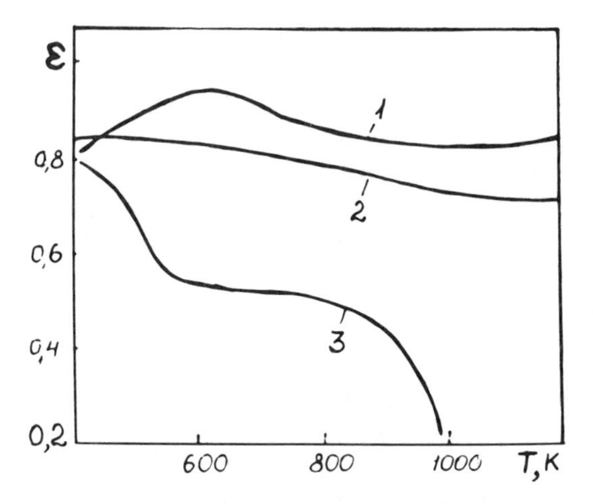

FIGURE 4.1. Temperature dependence of the accommodation coefficients of translational (1), vibrational (2), and rotational (3) energies for NO interacting with Pt. (From Assher, M. et al., *J. Chem. Phys.*, 78, 6992, 1993. With permission.)

The most characteristic feature revealed in studies of scattered and desorbed NO molecules is rotational cooling. NO molecules scattered from Pt(111)[238-242] and Pd[241] showed the Boltzmann distribution for rotational energy in most cases, but the rotational temperature T_r was almost always less than the surface temperature: $T_r < T_s$. Rotational energy accommodation is substantially hindered compared with vibrational energy accommodation and decreases with temperature increase. The dependencies of accommodation coefficients for translational, vibrational, and rotational energies on temperature are shown in Figure 4.1.[240] ϵ_r is approximately unity at 300 K and drops to a very small value at 1000 K. For Pt(111) T_s is only 70% of T_s. Spin-orbital temperature for the lowest rotational levels ($j = 0.5$ to 10.5) is also lower than T_s and becomes equal to it only for $j = 30.5$. Molecules excited in the lowest rotational levels are assumed to adsorb more easily than those excited in higher levels.[239]

The rotational cooling effect depends on the presence of other adsorbents. For instance, NO molecules scattered from Pt(111) at low temperature ($T_s <$ 300 K) exhibit complete rotational accommodation that is attributed to the presence of adsorbed NO molecules (or to C- or O-coverage).[238] At low temperatures (175 to 375 K) NO thermodesorption from Pt(111) was found to have equal temperatures for all states $T_r = T_{so} = T_s$ (where T_{so} is the temperature of spin-orbital levels of the NO molecule).[236] At a higher temperature ($T_s >$ 450 K) the surface of Pt(111) is free from adsorbed species, and $T_r < T_s$.[238] Rotational accommodation decreases with the increase of T_s, but translational accommodation is retained. For some cases the rotational temperature T_s tends to a limit which depends weakly on the surface nature. For NO scattered from different Pt faces, this limit is equal to 440 ± 20 K,[238] and 400 ± 20 K for NO scattered from Ge.[243]

NO scattering from Ag(111) is one of the most extensively studied cases.[61,228,244-246] It has been found that the rotational temperature depends linearly with the surface and translational temperatures[69]

$$T_r = aT_t + bT_s \tag{4.2}$$

where $a < 1$ and $b < 1$ are constants independent on T_r and T_s.

Some of the NO molecules scattered from Ni(100) have T_s according to Equation 4.2, and others have $T_r = T_s$.[247] Rotational temperature of NO molecules desorbed from Ru(001) is also lower than T_s.[248] For clean Ru(001) $T_r = 0.5\ T_s$; for oxidized Ru $T_r = 0.7\ T_s$.[249] It is likely connected with the fact that a major portion of NO molecules adsorb dissociatively on clean Ru and in molecular form on oxidized Ru. The rotational temperature of NO molecules scattered from graphite[250] is equal to the surface temperature for $T_s < 250$ K and is less than T_s at higher temperatures. Rotational cooling at $T_s > 400$ K is accompanied by the deviation from the Boltzmann distribution.

Rotational heating has been revealed for NO desorption from Ir(111).[236] Dissociative NO adsorption is likely to be the cause of this effect. Excess energy, released in recombinative desorption, N(ads.) + O(ads.) → NO(gas), can be transferred, in particular, to the rotational movement. The relationship $T_r < T_s$ is probably the criterion of the absence of the "chemical" desorption mechanism.

Theoretical treatment of rotational cooling in scattering and desorption processes reduces to different models of energy transfer between rotational and translational molecular degrees of freedom and to the choice of particular NO-surface potentials.[246,251-253] Bialkowski[251] proposed the model of the 2d-rotator in which the NO molecule rotates only in the plane parallel to the surface. A limiting value $T_r = T_s/2$ has been obtained in crude correspondence to the experiment[248] assuming weak coupling between molecular rotations and the surface and in accordance with the principle of equipartition of energy. Nevertheless, more complicated relationships are observed in other experiments. In a similar model, Tully[253] has found Boltzmann distribution of molecules over E_r and rotational cooling; to a first approximation, T_r is determined only by the rotational molecular constant but not the surface well depth. The theory fails to predict a limiting value of T_r.

Gadzuk et al.[254] proposed a model of a hindered rotator. When the NO molecule desorbs perpendicular to the surface, thermally populated bending vibration converts into rotation with angular momentum directed perpendicular to the surface ("overturning movement"). The desorption time is less than the molecular period of revolution; hence, energy exchange between the molecule and the surface does not occur.

It has been noted[253] that the mechanism of energy transfer between molecular rotational and translational motion is more appropriate in the case of NO scattering from Ag(111), whereas the model of a hindered rotator is suited to NO scattering from Pt(111).[254]

Tully et al.[246] studied two components of rotational energy: parallel $E_{r\parallel}$ and perpendicular to the surface $E_{r\perp}$ ("helicopter" and "cartwheel" models). Both components undergo rotational cooling: $E_{r\parallel} < 1/3kT_s$, $E_{r\perp} < 2/3kT_s$.

Experimental analysis of NO molecules by the LIF method with the linear polarization of the laser beam passing 2 mm above the Pt(111) surface has shown that for rotational quantum numbers $j > 12.5$ approximately, NO molecules desorbed from Pt(111) rotate parallel to the surface. These observations provide support for the hindered rotator model.[255] N-tip orientation of an NO beam colliding with Pt(111) and Ag(111) surfaces gave rise to higher T_r than O-tip orientation.[229]

When the potential is strongly anisotropic, the NO molecule interacts with the surface in a certain orientation. For very large rotational quantum numbers j (large E_r), fast rotation of an incident molecule averages the anisotropic component of the interaction between the molecule and the surface, which leads to interaction decrease and diminishes the sticking coefficient s. NO orientation resulting in rotational excitation gives weak energy exchange with a solid. Corresponding to the detailed balancing principle, the rotational temperature, $T_r = T_s \cdot s$, decreases with the increase of j, since s decreases with the increase of j.[256]

Rotational cooling ($T_r < T_s$) was observed as well for H_2 and D_2 molecular beam scattering from Cu(110) and Cu(111),[257] H_2 from polycrystaline Pd,[258] N_2 from Fe(111),[259] CO and CO_2 from Pd foil,[260] NH_3 and O_2 from Pt(111),[261] HF from LiF,[257] CO from LiF,[240] and I_2 from LiF.[262] In the last case translational and vibrational temperatures of scattered I_2 beam corresponded to T_s at $T_s > 300$ K, and $T_r < T_s$. During its lifetime (5 μs) determined using the beam modulation technique, the adsorbed I_2 molecule has no time to equalize T_r and T_s before leaving the potential well.

4.1.4. VIBRATIONAL EXCITATION (V_S-V EXCHANGE)

Few studies have been devoted to the vibrational excitation of scattered molecular beams. Assumptions of such an excitation (V_s-V exchange) are based on indirect evidence. Assher and Somorjai[263] studied NO beam scattering from Pt(111). Angular distribution data were used to calculate vibrational excitation and deactivation rates. Excitation occurs through the "trapping-desorption" mechanism.

Rettner et al.[264] revealed direct evidence of vibrational excitation of NO molecules scattered from Ag(111). Excited molecules were detected by the REMPI technique. The probability to excite into the $v = 1$ state depends weakly on kinetic energy of the incident molecules, but strongly on T_s with the activation energy 0.23 eV. It is just the energy needed to excite the NO molecule into the first vibrational level. Velocity distribution of NO molecules was found to be far from Maxwell one, and scattering of NO molecules was almost specular. This is a strong evidence for direct inelastic interaction rather than the "trapping-desorption" mechanism. The results were explained by the excitation transfer mechanism through electron-hole pairs in metal.[265,266]

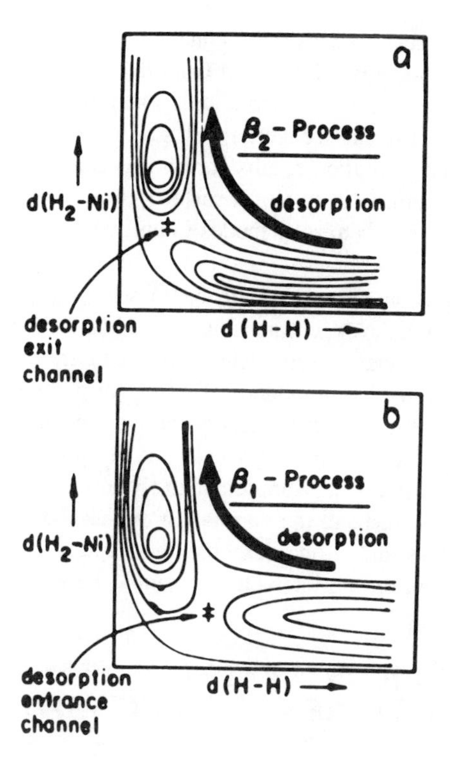

FIGURE 4.2. The surface of potential energy for H_2 desorption on Ni(111) from states β_2(a) and β_1(b). (From Russel, J. N. et al., *J. Chem. Phys.*, 85, 6186, 1986. With permission.)

Decrease of a number of vibrationally excited (v = 1) NO molecules scattering from graphite surface was observed using the REMPI method.[267]

Yates et al.[268] investigated H_2 thermodesorption from Ni(111) surface. Cosine-low angular dependence $\sim\cos\varphi$ was observed in the thermodesorption peak corresponding to the β_1-H_2 form at 290 K, and $\cos\varphi^{4.5}$ dependence was found for the β_2-H_2 form at 370 K. The authors[269,270] attributed this discrepancy to the peculiarities of the hydrogen activation mechanism during adsorption and desorption. Adsorption heat increases and activation energy decreases, with the increase of coverage Θ. Different types of potentials correspond to β_1 and β_2-H_2 states. At coverages $\Theta < 0.5$ (β_2-H_2 form) the desorption process is governed by passing through the "exit channel" (see Figure 4.2, a), and by passing through "entrance channel" at $\Theta > 0.5$ for the β_1 form (Figure 4.2, b). In accordance with this scheme, desorption of the β_2 form is accompanied by the translational excitation, and internal (vibrational or rotational) excitations are probably characteristic of the β_1 form. The H_2 molecule desorbs in a vibrationally excited extended state while passing through the "entrance channel" for β_1-H_2. The H_2 molecule behaves, as if it were not completely formed, and thermalizes in subsequent collisions with gas molecules. Quite different kinetic behavior is observed in this case for H_2 and D_2 (isotope effect).

Harris et al.[271] have drawn similar conclusions for H_2 and D_2 vibrational excitation during associative desorption from the Cu surface.

Paz and Naaman[272] revealed vibrational excitation of the NH_2-group in aniline scattered from surfaces covered with structured monolayers of organic species. Vibrational excitation results for two reasons: (1) the high ratio of molecular mass of aniline to that of a surface group contacting with it, and (2) the presence of low-frequency modes in aniline which are easily excited during the collision.

As a final note on monoenergetic beam scattering, we would like to emphasize that though the experiments discussed had little to do with ordinary catalysis, a set of important results, which are necessary for an understanding of elementary steps in adsorption and catalysis, were obtained. They include the absence or presence of an adsorption activation barrier; molecular orientation and the probability of energy exchange during the collision process; the possibility of translational, vibrational, and rotational excitation on account of interaction with a surface; the correlation of the above probabilities with the surface structure and properties; and the dependence of characteristic features of excitation on the interaction potential between a molecule and a surface.

4.2. DYNAMIC INTERACTION OF MOLECULAR BEAMS OF DIFFERENT ENERGIES WITH SURFACE

We looked at energy changes of the monoenergetic molecular beam after its collision with a surface. In recent years, much attention has also been paid to the ''inverse'' problem: what influence do the energy and internal excitations of molecules in the beam have on the molecular collision process? To solve this task, molecular beams with varied energy characteristics are required. We must be able to measure all molecular parameters in these beams, to determine the atomic constitution of a surface (Auger spectroscopy), its structure (LEED), and the positions of surface electron levels (XPS and UPS).

4.2.1. INFLUENCE OF TRANSLATIONAL ENERGY (T-V_s EXCHANGE)

Various nonmonotonic dependencies of the initial sticking coefficient s_o on the molecular kinetic energy were obtained. It turned out that the coefficient s_o depends on the translational temperature T_i (quadratic in the velocity of particles $\sim v^2$) and on the angle of incidence of the molecular beam φ_i as $\cos^2 \varphi_i$. Hence, only the normal velocity component of a particle is involved in the process. Adsorption probability drastically reduces with the decrease of φ_i. It has been found experimentally that

$$s_o = kE_i\cos^n\varphi \qquad (4.3)$$

where E_i is an average kinetic energy of a particle and $n > 2$. Kisljuk[273]

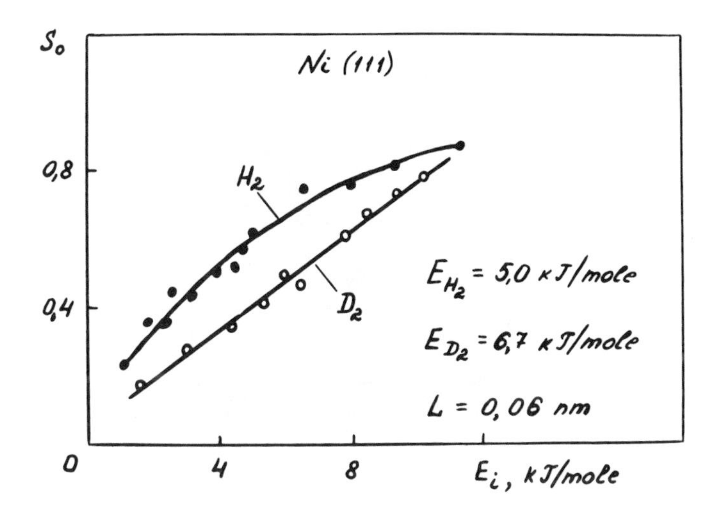

FIGURE 4.3. The sticking coefficient s_o for H_2 and D_2 on Ni(111) as a function of molecular beam energy. (From Hamza, A. V. and Madix, R. J., *J. Phys. Chem.*, 89, 5381, 1985. With permission.)

obtained theoretically a more complicated expression which covers a wide range of available data.

Rather weak dependence on T_s is observed in most cases. In the dependence like that of Equation 4.2, the first term evidently dominates. The validity of Equation 4.3 is repeatedly confirmed by experiment. For example, the probability of H_2 adsorption on the Ni(111) face is practically independent of T_s, but increases with the increase of E_i in accordance with Equation 4.3 from $s_o = 0.05$ at $E_i = 0.04$ eV to $s_o = 0.4$ at $E_i = 0.12$ eV.[231,279] According to Palmer et al.[230] and Robota et al.,[231] the hydrogen coefficient s_o on Ni(111) at 700 K depends on T_i with the apparent activation energy 8 kJ/mol. Similar independence of s_o on T_s is observed for the Ni(100) face, but s_o grows from 0.2 to 0.8 with the increase of $E_\perp$ from 0.7 to 7 kJ/mol.[275,276] The corresponding values of s_o for D_2 are lower and become comparable to that of H_2 only at high energies (see Figure 4.3). It is the authors' opinion that the quantum mechanical tunneling plays a part in dissociative H_2 adsorption in contrast to the D_2 case.

Palmer and Smith[33] observed the dependence of s_o for H_2/Pt(111) on the beam temperature with the corresponding activation energy close to 8 kJ/mol.

Much attention is given to H_2, D_2, and HD energy exchange on different Cu faces.[221,277-282] The probability of dissociative adsorption is negligible at $E_i < 0.2$ eV; then a rapid growth is observed up to $s_o = 0.05$ at $E_i = 0.4$ eV. The maximum value of s_o is observed on the close-packed Cu(111) face that is ascribed to the lower barrier for lateral diffusion of H atoms. It has been concluded that both translational and vibrational excitation of incident molecules is necessary to overcome the barrier for dissociative H_2 and D_2

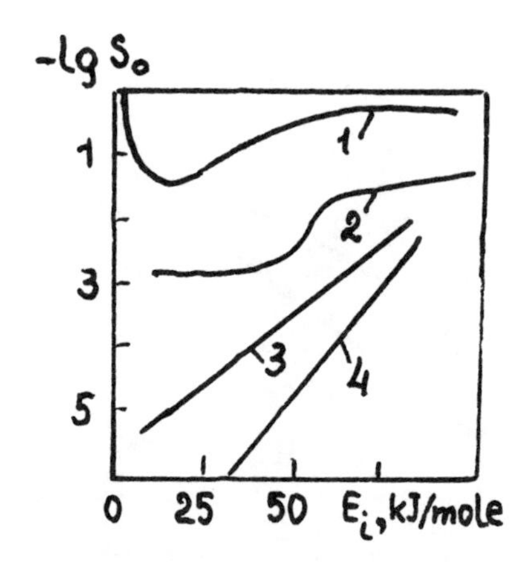

FIGURE 4.4. Energy dependence of the sticking coefficient s_o on W(110) at 800 K for O_2(1), N_2(2), CH_4(3), and CD_4(4).[285]

adsorption.[281-282] Higher s_o values for D_2 compared with H_2 failed to be explained.

Monotonic increase of s_o against E_i according to Equation 4.3 is accompanied by drastic changes in s_o at certain values of $E_\perp$. Abrupt changes of this dependence are usually explained by the activation barrier for adsorption in agreement with Van Willigen's Equation 4.1. Smooth dependencies are attributed to surface inhomogeneity, the presence of activation energy distribution for different adsorption sites. This distribution can be obtained by differentiating s_o with respect to $E_\perp$.

The N_2/W(110) system was studied in detail.[283,284] Over the range of E_i from 0 to 80 kJ/mol, the probability s_o of dissociative N_2 adsorption increased tenfold, and over the range of E_i from 80 to 200 kJ/mol, s_o increased 100 times, from $5 \cdot 10^{-3}$ to $5 \cdot 10^{-1}$. A sharp rise at $E_i = 50$ kJ/mol (Curve 2 in Figure 4.4) is in general agreement with the activation barrier models, but the permanent character of this rise indicates that the single barrier model is insufficient. The system N_2/W(110) exhibits one more characteristic properties: the sticking coefficient s_o depends on the total energy E_i, but not on $E_\perp$, according to Equation 4.3.

This new feature is explained by the possible influence of an exit potential barrier[285] (see Figure 4.2, a). Equation 4.3 describes the "direct" activated adsorption during one collision with the surface. The dependence of s_o on E_i corresponds to indirect adsorption, i.e., the "trapping-desorption" mechanism. The second dependence can be caused by the existence of a "hot" or "dynamic" precursor as well (see Chapter 6). Roughness of surface is an alternative explanation. It may exist from the very beginning or appear in the process because a particle strikes the surface many times until it passes through the potential barrier.

It should be noted that "direct" interaction corresponds to the Rideal mechanism which has been introduced in Section 3.3, and interaction mechanism with the excitation of phonons.

Invariant or even decreasing behavior of s_o with the increase of E_i is usually attributed to the existence of a long-lived mobile precursor (see Section 6.1).

For oxygen on Pt(111), s_o dependence on $E_\perp$ at 523 to 623 K corresponds to the adsorption activation energy of about 4 kJ/mol. Nevertheless, hot and cold O_2 beams behave identically at lower temperatures.[286,287] For O_2 beams on W(100), $s_o \approx 0.1$, at $E_\perp = 0.1$ eV and $s_o \approx 1$ at $E_\perp = 0.4$ eV. The value of s drops to $s = 0$ with the growth of oxygen coverage to $\Theta = 0.5$ at $E_\perp = 0.25$ eV; $s = 0.5$ for $\Theta = 0.5$ at $E_\perp = 1.3$ eV. The s dependence on $E_\perp$ allows us to calculate the value of the activation barrier for adsorption: 0.12 eV at $\Theta = 0$ and 0.34 eV at $\Theta = 0.5$. However, at very small energies (less than 0.03 eV) s increases with the increase of coverage degree instead of decrease, which points to more complicated processes during energy exchange (see Curve 1 in Figure 4.4). The model of mobile precursor has been invoked to account for this anomalous dependence of s on $E_\perp$ at low energies.[285]

A potassium monolayer on Pt(111) enhances the energy exchange for O_2 giving rise to appreciably greater values of s_o at the same E_i.[288] The authors explain this enhancement by charge transfer between K atoms and O_2 molecules with the formation of an O_2^--ion.

For a CO/W system it was found that s_o increases with increasing T_s and T_i.[289] Another dependence of s_o on $E_\perp$ was obtained for CO interaction with Ni(111) by Tang et al.[252] At $E_\perp < 17$ kJ/mol the sticking coefficient $s = 0.85$ and is independent of coverage, which is ascribed to molecular adsorption with a mobile precursor. It is the authors' opinion that the latter is a CO molecule approaching the adsorption site, Ni atom, with "improper" orientation, i.e., by a O atom instead of C atom, or parallel to the surface. The hindered rotator which has been mentioned in Section 4.1.3 in connection with rotational cooling is also a possible candidate for a precursor. The precursor lifetime τ and its rate of diffusion was estimated at 200 K using the Auger-spectrum pattern which spread out under molecular beam action: $10^{-5} < \tau < 10^{-1}$s. At high E_i the value of s_o lowers to 0.43 and remains further unchanged. The authors believe that the reduced value of s_o is caused by the fact that only molecules with the proper orientation, i.e., by a C atom, adsorb after the collision directly, which means overcome the adsorption barrier 17 to 30 kJ/mol and chemisorb just dissociatively without the precursor state; this scheme is supported by the Auger spectrum. A low value $s_o = 0.02$ independent of beam energy in the range 6.5 to 21 kJ/mol was observed for CO/Ni(100) interaction at $T_s = 500$ K.[290] Probably, only molecular adsorption took place.

A decrease of s_o with the growth of energy of the molecular beam was observed for CO and Ar at Pt(111)[291] and Ir(110).[292] The CO value of s_o at a clean Ir(110) surface diminishes from 0.8 at beam energy 8 kJ/mol to 0.35 at 140 kJ/mol. Over this energy interval, the sticking probability is independent of T_s at 200 to 500 K. This fact is explained by transformation of trapping-desorption interaction at low $E_\perp$ into direct inelastic scattering at high $E_\perp$. On Ir(110) covered with a CO monolayer, the dependencies of s_o on T_s are identical for CO and Ar. The value of s_o for Ar at the Ir surface covered with CO is much greater than s_o on a clean Ir surface. The CO adsorbed monolayer, probably, enhances gas-surface energy transfer.

So, different energy dependencies of the adsorption probability appear to be as illustrated by the example of CO adsorption on metals. The growth of s_o with the growth of $E_\perp$ is a characteristic feature of "direct" activated adsorption. The decrease of s_o with the growth of $E_\perp$ is observed in the case of molecular adsorption without activation or precursor.

Summarizing some results obtained for s_o, as a function of T_i and T_s, with the aid of molecular beams, Rendulic et al.[293] note that direct chemisorption without activation is characterized by the conditions: $ds_o/dT_i = 0$ and $ds_o/dT_s = 0$, whereas activated chemisorption gives rise to $ds_o/dT_i > 0$ and $ds_o/dT_s = 0$. The value of ds_o/dT_i for chemisorption passing through precursor is defined by the dissipation of the kinetic energy of adsorbed molecules into three main channels: excitation of solid phonons, excitation of molecular rotations, and generation of electron-hole pairs. All these mechanisms result in $ds_o/dT_i < 0$ at low temperatures. A minimum of s_o as a function of T_i (see, for instance, Figure 4.4 for O_2/W) gives an indication of change of chemisorption mechanism from precursor at low T_i to direct activated chemisorption at high T_i.

As it was already noted in Section 2.1.2, the study of ion-surface interaction using ion beams of varied energy provides practically the same information as in the case of molecular beams. Bykov et al.[41,42] investigated interaction of N_2^+ ions with the surface of polycrystalline Pt.

Adsorption probability of molecular ions s_o was determined by comparing the rates of the pressure drop in an omegatron detector, with and without a high-frequency electromagnetic field at resonant frequency. The difference between the slopes of these lines enables us to calculate s_o setting a certain value of σ_i, the cross-section of nitrogen molecules ionized by electron impact. In our experiments $\sigma_i = 2.87 \cdot 10^{-16}$ cm^2 taking into account that the ionization cross-section drops rapidly with decreasing energy, and the contribution of secondary electrons is negligible. The obtained dependence of the probability of N_2^+-ion adsorption as a function of energy is shown in Figure 4.5. In the framework of the theory of dissociative adsorption of diatomic ions each maximum is responsible for ion neutralization into the next electronically excited state. If such complicated behavior is not generated by an "apparatus effect", it may testify that not only Auger neutralization resulting in nonexcited N_2 molecule occurs, but the resonance N_2^+ takes place as well.

Later, Akazawa and Murata[294] investigated adsorption of N_2^+ and N^+ ions at Ni(100) and Ni(111) surfaces for lower kinetic ion energies: E_i from

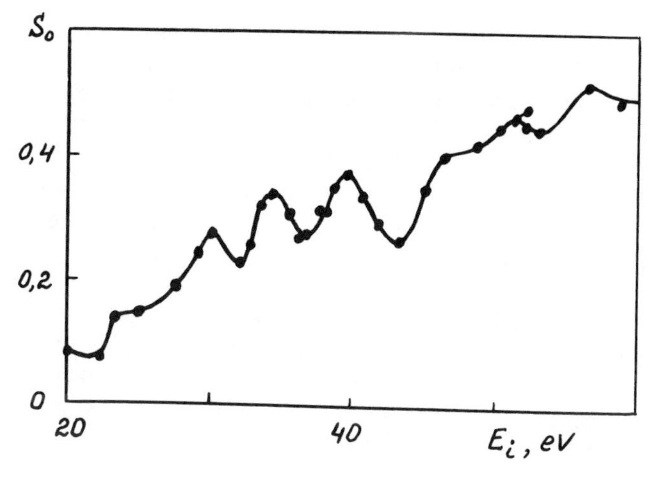

FIGURE 4.5. N_2^+ adsorption probability on polycrystalline Ni as a function of ion beam energy.[41]

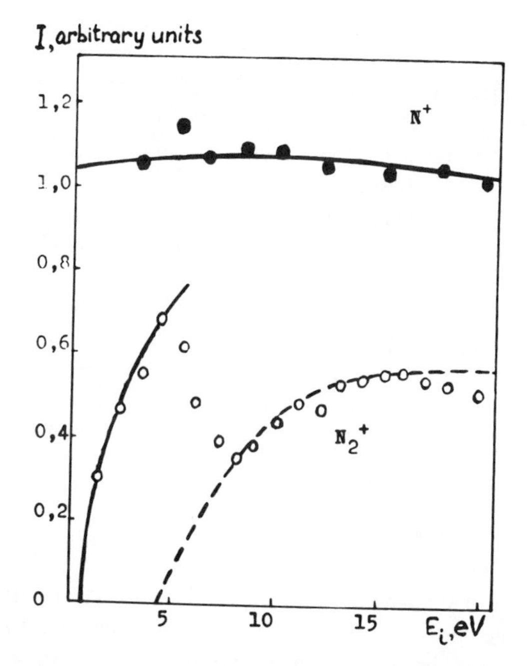

FIGURE 4.6. Nitrogen adsorption probability (using Auger spectra) on Ni(111) surface as a function of kinetic ion beam energy. (From Akazawa, H. and Murata, M., *J. Chem. Phys.*, 88, 3317, 1988. With permission.)

0 to 20 eV. The coefficient s_o does not depend on E_i for N^+, and a curve with a maximum and a minimum is observed for N_2^+ (Figure 4.6). The results are explained with the help of a model with different mechanisms of adsorption and ion neutralization. At low E_i, electronically excited molecules obtained

in the resonance ion neutralization adsorb dissociatively with low activation energy (~5 eV). At higher E_i, electronically excited molecules adsorb at the surface and dissociate further with high activation energy. Auger neutralization of N_2^+ ions occurs at closer distance with the increase of E_i, and the empty σ-orbital in N_2^+ overlaps with d orbitals in Ni.

Ion CO^+ beam interaction at the Ni(111) surface was studied by Rabalais and co-workers in the range 0 to 20 eV.[295] At low E_i Auger neutralization of CO^+ ions with 3d electrons of Ni can occur, giving rise to the formation of CO in one of the excited states with the inner energy up to E_{in} = I(CO) − φ[Ni(111)] = 8.7 eV, where I is the CO ionization potential and φ is the work function of Ni. At the lowest energies the exothermic CO adsorption without dissociation (ΔH = −1.55 eV) is the most probable. This beam energy can be transferred to a solid or produce CO desorption that results from a small s_o value at E_i = 3 eV. With the energy growth, an almost thermoneutral dissociation begins to proceed: $CO + Ni_s \rightarrow NiO_s + NiC_s$, which still requires activation energy. According to Auger measurements, O_s appears on the surface at E_i = 2.7 eV, and its surface concentration grows linearly with the increase of E_i. The concentration of nondissociated $NiCO_s$ on the surface decreases with the increase of E_i to ~7 eV. Thereafter, it remains approximately constant. The CO dissociation energy is calculated from the experimental data for ion beams and equals $E_{in} + E_\alpha$, where E_α = E_i = 2.7 eV. The maximal barrier height is estimated to be 11,4 eV which is rather close to CO dissociation energy −11.16 eV. Therefore, it was concluded that CO dissociation goes through the direct collisional mechanism of CO^+ ions with energy >2.7 eV rather, than catalytic CO dissociation at the Ni surface.

C^+ and O^+ ion beam interaction with Ni(111) leads to formation of the carbide NiC and oxide NiO, respectively, in the range 5 to 200 eV.[296,297] C^+ beam bombardment of NiO or O^+/NiC results in CO formation; the interaction efficiency drops with the growth of E_i in the first case and is enhanced in the second one. The reaction $O^+ + NiC \rightarrow CO$ leads to CO desorption and CO_2 formation.

4.2.2. INFLUENCE OF ROTATIONAL ENERGY (R-V_s EXCHANGE)

In a number of studies molecule-surface interaction was investigated with the aid of molecules excited in a certain rotational or vibrational status (R-V_s and V-V_s exchange). For this purpose molecules were excited in a hot supersonic beam or selectively excited using IR laser radiation. Scattered molecules were detected with the aid of LIF and REMPI techniques.

Taking into account the effect of rotational cooling after molecule-surface interaction (see Section 4.1.3), it would be natural to expect the influence of rotational excitation on the s_o value. The coefficient s_o is independent of the rotational energy E_R of excited molecules, if a potential well is sufficiently deep as, for instance, for NO adsorption on a Pt surface where Q = 0.58 eV.

For small Q values the coefficient s_o drops with E_R growth. This fact is attributed to rapid rotation of an excited molecule. Rettner et al.[298] revealed that NO scattered at Ag(111) is insensitive to the rotational energy of incident particles.

At low rotational energies the adsorption probability depends on molecular orientation at the instant of collision. Orientational dependence has been studied for NO scattered at Ag(111)[298] and Ni(100).[299] It turns out that s_o is 75% greater in the case of N-atom orientation compared with the O-atom case. Steric effects come into particular prominence for large molecules.

4.2.3. INFLUENCE OF VIBRATIONAL ENERGY (V-V_s EXCHANGE)

Dissociative adsorption of H_2 molecules in the v = 0 state and excited in the v = 1 vibrational level has been investigated on Cu(111)[300] and Cu(100),[301,302] faces. It turns out for all cases that vibrationally excited molecules adsorb substantially more easily than nonexcited ones. The calculation of the dissociation probability proved an appreciable gain with the growth of the vibrational quantum.[303] The vibrational deactivation of H_2 molecules was practically negligible at low-beam energies.

Crisa et al.[304] studied deactivation of vibrationally excited CO molecules using the molecular beam technique. Deactivation probability of the second vibrational level of the CO molecule interacting with polycrystalline silver decreases from 0.33 at 350 K to 0.20 at 440 K. The interaction mechanism involves electron-hole exchange during CO trapping followed by vibrational energy transfer to a rotational one or to the energy of surface phonons. An attempt to induce dissociative chemisorption of CO at Ni(111) surface using translational (up to 2 eV) or vibrational (up to 0.8 eV) excitation has not been successful.[305] The authors[305] suggest that the excitation in no less than the fourth vibrational level of CO is required (1.05 eV) to cause dissociative CO chemisorption at Ni(111). As we already mentioned in Section 4.2.1, low-energy ion beams are the most suitable to study dissociative CO chemisorption.

Up to 10% of NO molecules (v = 1) deactivate after collision with the Ag(111) and Ag(110) surfaces.[306] The interaction of vibrationally excited NO molecules (v = 1) with graphite surface was studied by Vach and co-workers.[307] At a surface temperature T_s = 600 K, almost all molecules remain in the excited state after collision with the surface. At T_s = 200 K, 20% of the excitation energy is lost, probably due to energy transfer to lattice phonons when a molecule is trapped near the surface. The authors observed even a small increase in NO rotational energy after interaction with the surface.

The probability of CO_2 dissociative adsorption ($\rightarrow$ CO + O) at Ni(100) enhances from $4 \cdot 10^{-4}$ to 0.15, i.e., 300 times, with the growth of the molecular beam energy from 8 to 40 kJ/mol.[308] A further growth of E_i to 100 kJ/mol has little effect on s_o. Molecular adsorption of CO_2 through a precursor proceeds possibly at $E_i \approx 8$ kJ/mol; at $E_i > 40$ kJ/mol the direct dissociative adsorption during collision of the CO_2 molecule with the surface takes place. Such a significant increase of the dissociative adsorption probability is likely indicative of the fact that the excitation of bending vibrational modes substantially increases the probability of CO_2 dissociation. This is experimentally confirmed using CO_2 beams with varied temperature. Temperature elevation from 300 to 1000 K at the source of the supersonic beam increases the value of s_o from 2 to 10 times. It is the authors' opinion[308] that bent CO_2 molecules participate in the dissociation reaction: $CO_2 \rightarrow CO + O$. The angle O–C–O is known to reach 133° in some metal-CO_2 complexes. The excitation of this mode promotes bending and the following rupture of the O–C bond.

Rather strong deactivation of vibrationally excited CO_2 molecules was observed at Pt[239] and Ag[309] in supersonic molecular beam experiments. It is unlikely that molecular vibrational energy transfers to lattice phonons because a multiphonon process with a greate number of phonons is required. Clary and De-Pristo[310] analyzed the results of Misevich et al.[309] and showed that CO_2 deactivation near metal surfaces occurs through energy transfer to electron-hole pairs or molecular rotation as a whole. The deactivation probability of vibrationally excited molecules at metal surfaces has typical values from 10^{-3} to 10^{-2}. Such values are associated with the trapping model and further energy transfer to an electron-hole pair. Energy transfer to surface plasmons is also possible.[311]

Rettner and Stern[312] showed that the dissociative adsorption probability s_o of an N_2 beam scattered at Fe(111) grows from 10^{-2} to 10^{-1} with increasing energy from 0.1 to 1.0 eV. To excite vibrationally, molecules were heated in a source up to 2000 K. It turned out that the influence of vibrational excitation on s_o is approximately half as effective as translational energy.

Most studies treating a possible involvement of vibrationally excited molecules in activated chemisorption are devoted to methane chemisorption. Stewart and Ehrlich[313] studied the dynamics of methane-activated chemisorption on Rh. A CH_4 molecular beam at 600 to 710 K bombarded a cold (245 K) rhodium point in the field emission microscope. The reaction rate was judged from the change of Rh work function because of the growth of a carbon layer due to CH_4 exposure. The activation energy estimated from the dependence of the reaction rate on T_i was close to 30 kJ/mol. The adsorption rate of CH_2D_2 molecules is three times, and that of CD_4 ten times, less than the CH_4 rate. It is the authors' opinion that such a significant distinction indicates a negligible contribution of translational and rotational energy of methane molecules. In contrast, vibrational excitation has a dominant role, probably enhancing the transition from precursor into the chemisorbed state. Using the Slater dynamic model,[314] the ratio of the rate constants of CH_4 and CD_4 was found to be

$$\frac{k(H)}{k(D)} = \frac{\nu(H)}{\nu(D)} \frac{\exp[-q^2/2\sigma^2(H)]}{\exp[-q^2/2\sigma^2(D)]} \tag{4.4}$$

where σ^2 is an average square of the breaking bond amplitude.

Equalizing a frequency ν to the bending frequency ν_4 of CH_4 gives a rate in agreement with the experiment.

Yates et al.[315] repeated this experiment using direct selective laser excitation of methane vibrational modes ν_9 and ν_4, but did not obtain a significant effect. According to their data, only $5 \cdot 10^{-5}$ of excited molecules chemisorb on the Rh(111) surface at 240 K. They suppose that more high-level excitation of CH_4 or some combination of translational and vibrational energy is needed to induce chemisorption of a methane molecule.

Rettner et al.[316] investigated interaction of the supersonic methane molecular beam with a W(110) surface. Vibrational energy was varied changing the source temperature, and translational excitation was quenched by adding foreign gases H_2, He, and Ar. CH_4 adsorption probability increased 10^5 times with increasing translational energy from 0.1 to 1.1 eV; the vibrational temperature elevation from 340 to 770 K resulted only in a tenfold increase of the sticking coefficient. The authors believe that translational energy of methane aids the tunneling of a hydrogen atom during chemisorption according to the semiempirical formulas for the sticking coefficient $s_o = \alpha \cdot \exp(\beta E_\perp)$, where $\alpha = 2 \cdot 10^{-6}$ and $\beta = 12.5\ eV^{-1}$, which corresponds to the potential barrier height 1.1 eV with a half width of 0.025 nm. It is due to tunneling that a strong dependence of the adsorption probability on mass of a tunneling particle and on its translational energy $E_\perp$ arises

$$s_o(E_\perp) \sim \exp\left[-\frac{\pi \Delta H}{h}\left(\frac{M}{V_o}\right)^{1/2}(V_o - E_\perp)\right] \tag{4.5}$$

where ΔH is a half width of the potential barrier and V_o is its height.

Rettner et al.[316] believe that the results of Stewart and Ehrlich[313] are explained by translational tunneling rather than vibrational excitation. The sharp distinction between $s_o(E_\perp)$ for CH_4 and CD_4 molecules and the sticking coefficient for O_2 and N_2 molecules is evident from Figure 4.4. The slope for CD_4 is even steeper than for CH_4.

More recently, Lo and Ehrlich[317] examined the chemisorption rate of methane on the W(211) molecular beam energy and found only a small difference between CH_4 and CD_4: the ratio of their rates is 2.5 at 800 K. The corresponding activation energies are 31 and 39 kJ/mol for CH_4 and CD_4, respectively. Estimations show that this difference is insufficient to explain the methane activation by H-atom tunneling from a hot molecular state. Therefore, the authors[317] returned to their previous explanation of CH_4 activation[313] through vibrational excitation.

The sticking coefficient of methane s_o at the Ni(111) surface grows linearly with the increase of $E_\perp$.[318,319] Nevertheless, for $E_\perp < 45$ kJ/mol CH_4, adsorption is not observed. This fact is attributed to the translational methane activation. It has been calculated that CH_4 deformation and its conversion into a triangular pyramid (with C atom at the center of the triangle formed by three H atoms, and the fourth H atom at the top) requires 68 kJ/mol.

Hence, the growth of s_o with the increase of $E_\perp$ is proportional to the degree of molecular deformation. Nevertheless, a strong isotope effect (CH_4 compared with CD_4) points to possible vibrational excitation or tunneling.[318]

A decrease of s_o with the increase of $E_\perp$ from 34 to 202 kJ/mol has been observed on Pt(111) at $T_s = 100 \div 140$ K.[320] At higher temperatures and beam energies (20 to 110 kJ/mol) s_o increases linearly with increasing $E_\perp$.[321,322] Madix and co-workers[321] assume that direct methane activation occurs when the H atom tunnels through the parabolic barrier with a half width of 0.013 nm. Translational and vibrational excitation enhance the tunneling probability. The ratio s_o (CH_4)/s_o (CD_4) equals 2.5, as in the case of W(211) surface.

One more work concerning methane activation should be mentioned.[323] Methane, adsorbed in molecular form at Ni(111), dissociates under argon atom bombardment at $E_\perp \geqslant 75$ kJ/mol. The probability of dissociative chemisorption grows exponentially with the increase of kinetic energy of Ar atoms in the range 75 to 160 kJ/mol. Such a "hammer mechanism" gives rise to vibrational excitation and molecular deformation, forcing the C atom of methane closer to the Ni atom. It is the authors' opinion[323] that such processes are even more possible at high pressures, and, probably, it is just these processes that are related to "pressure gap" which mean a large difference between the rates of catalytic reactions at low and high pressure.

Speculation about the possible role of vibrational excitation or tunnel effect in methane activation at metal surfaces has continued.[324] Luntz and Harris[325] suggested a quantum dynamics model of methane activation. All experimental results agree with the model of direct impact dissociation with the rupture of C–H bond during collision. The tunneling probability grows with increasing T_s; that is, thermal assisted tunneling owing to lattice thermal energy. Phonon absorption favors the passing of particles through the barrier because it results in a shorter tunnel length and a lower barrier height, thus, increasing s_o.

Consider the main results obtained for higher alkanes. The coefficient s_o decreases with increasing $E_\perp$ and T_s for ethane at the Ir(110) surface at $E_\perp < 40$ kJ/mol, which is trapping mediated chemisorption, but not direct dissociation.[326] The sticking coefficient is extremely low 0.03 at molecular beam energy 60 kJ/mol.[327] Changing T_s from 300 to 1400 K does not affect s_o (the activation energy is less than 400 J/mol), which seems to argue against the precursor model. In the energy range 60 to 160 kJ/mol, s_o increases linearly with increasing E_i up to 0.43 at 160 kJ/mol. Over this energy interval the activated dissociative chemisorption of C_2H_6 without a precursor occurs with $E_\alpha = 150$ to 160 kJ/mol.

Similar results were obtained for ethane at Pt(111).[328,329] At low-beam energies (less than 40 kJ/mol) and low T_s, s_o decreased with the increase of $E_\perp$. At the same time s_o increased with increasing coverage; i.e., ethane is more effectively trapped by the ethane adlayer but not by a clean metal surface. At high $E_\perp$, s_o grows with the growth of $E_\perp$; direct dissociative C_2H_6 adsorption occurs.

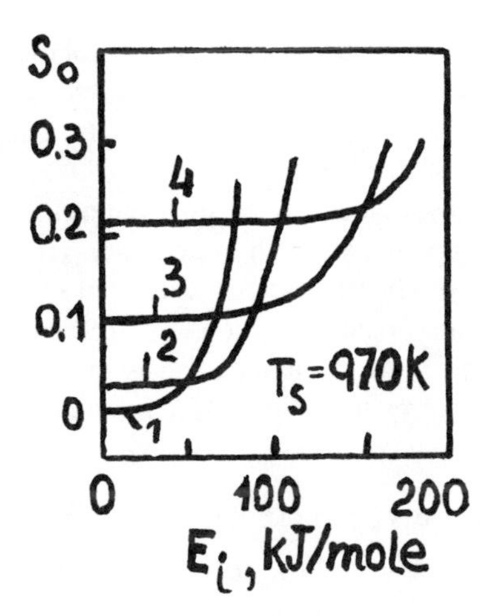

FIGURE 4.7. Energy dependence of the sticking coefficient s_o on Ir(110) for molecular beams of methane (1), ethane (2), propane (3), and butane (4). (From Hamza, A. V. et al., *J. Chem. Phys.*, 86, 6506, 1989. With permission.)

Hamza and Madix[330] studied s_o dependence of various paraffins C_1 to C_4 at the Ni(100) surface on molecular beam energy. The value of s_o did not practically depend on surface temperature T_s. At $E_\perp < 30$ kJ/mol, adsorption was not observed; at higher energies s_o grew linearly with increasing $E_\perp$. It should be noted that at the same temperature the sticking coefficient s_o of methane is greater than that of ethane, propane, and butane. These authors[330] calculated energy losses during molecular collisions with surface Ni atoms in accordance with classic mechanics, taking into account molecular and surface atom masses. On the assumption that the alkane molecule interacts with 5 Ni atoms, the dependence of s_o on $E_\perp - E_{loss}$ turns out to be identical for all alkanes. The interaction threshold 30 kJ/mol coincides with the activation barrier which was found for adsorbed alkanes at Ir(110) by Weinberg.[331]

In another study, Hamza et al.[332] studied the sticking coefficient $s_o(E_\perp)$ for alkanes on Ir(110). In this case (see Figure 4.7) $s_o = 0$ for methane at low $E_\perp$ (less than 40 kJ/mol); thereafter, s_o increased rapidly. A value of s_o is unchanged and equal to 0.03 for ethane up to $E_\perp = 60$ kJ/mol, 0.1 for propane up to 90 kJ/mol, and 0.21 for butane up to 110 kJ/mol. At higher energies a rapid increase of s_o was observed. It was concluded that at low $E_\perp$ molecular nonactivated adsorption through direct translational activation of the impinging molecule occurs (excluding CH_4). Further dependence $s_o(E_\perp)$ can be explained by the barrier distribution over the surface or by H-atom tunneling as well.

Halpern and Almitaz[333] studied dissociative adsorption of large hydrocarbon molecules C_nH_m ($n = 3$ to 8) at the Pt surface. At certain kinetic energy the rate of energy exchange (or deactivation) grows with increasing molecular size. Large molecules have longer lifetimes, desorb slowly, and chemisorb easily in the end.

This information is of considerable importance for one of the key problems of catalytic science, i.e., elucidation of activation mechanisms of methane at the surface. The results discussed show that methane dissociation probability is not less than that of other alkanes.

Summarizing the results considered in Section 4.2, it may be deduced that molecular beam experiments with varied energy have given conclusive evidence of the role played by translational energy in activated chemisorption. As to rotational and vibrational excitations, the currently available data are rather contradictory and make it difficult for us to draw a final conclusion. Vibrational activation seems to be less effective than translational in many cases.

4.3. EXCITATION OF CHEMICAL AND CATALYTIC REACTIONS ON SURFACE

We have presented a set of examples of dependence of the sticking coefficient on the molecular energy in the beam. In most cases an increase of the chemisorption probability vs. the kinetic energy ("translational temperature") of the beam has been observed. There are also direct investigations of the energy transfer between the molecules in the beam and the surface, which are indicative of the possible acceleration of the chemical and catalytic reactions.

4.3.1. EXCITATION OF REACTIONS ON SURFACE BY MOLECULAR BEAM ENERGY

A series of studies on the excitation of the noncatalytic monomolecular reactions induced by molecular interaction with a hot surface have been carried out by Rabinovich et al.[334-336] The reactions proceeding in the ordinary thermal way, but at somewhat elevated temperatures, were investigated. In the "one-collision" version the beam of molecules at thermal equilibrium falls on the surface at a higher temperature T_s. V_s-V energy transfer from the surface to the molecule occurs, and the latter undergoes transformations.

The isomerization of cyclobutane into butadiene[336] and 1-methylcyclopentane into isoprene[337] were studied in detail at a surface temperature of 500 to 800 K. The reaction probability was 10^{-4} to 10^{-6}. The excitation over the activation threshold of the monomolecular reaction (E_a = 120 to 290 kJ/mol) takes place, occurring practically independent of the surface character (Au, SiO_2). The vibrational accommodation coefficient was determined to be 0.95 to 0.96. The reaction probability of the nitromethane decomposition under the same conditions depends on the surface nature, i.e., the catalytic effect is observed.

Gerber and Elber[337] investigated molecular dissociation of IBr, ICl, and I_2 impinging at the inert MgO surface. I_2 dissociation on MgO(100) was shown to arise from the rotational excitation on the rigid surface that assisted the bond break. The probability of I_2 dissociation grows almost linear with the increase of the translational energy $E_{\perp}$ from 1 to 10 eV. Two maxima are present in the energy distribution of scattered atoms, the first atom of the I_2 molecule impinging on the surface acquires more energy than the second.

Botari and Greene[338] examined the decomposition of chlorine-substituted hydrocarbons, carbonyls of W and Mo, and dioxane after excitation in the supersonic beams.

McCarroll and Thomson[339] developed a molecular beam technique for the study of associative substitution of acetylene adsorbed on platinum by ethylene labeled with ^{14}C

$$C_2H_2 \cdot Pt + {}^{14}C_2H_4 = {}^{14}C_2H_2 \cdot Pt + C_2H_4$$

The beam temperature varied from 288 to 573 K, and the surface temperature was between 273 and 313 K. It turned out that only T_s affects the reaction rate, but T_i does not. Hence, the surface catalytic reaction take place.

Working on further use of ideas of this work,[339] Prada-Silva et al.[340] devised a molecular beam reactor with recycling. In this reactor, molecules from the supersonic beam struck the catalyst target; after that the reactor was evacuated, molecules were collected, and again directed toward a source. The integration over many cycles (molecule accumulation) brought the sensitivity with the help of a flame ionization detector to 10^{-9} to 10^{-10} per one collision. Adding inert gas into the space between the beam source and the target made it possible to retard translationally excited molecules and to study the influence of translational and vibrational energy on the interaction process independently of one another. In this manner the true nonequilibrium catalysis could be investigated (with different surface and gas temperatures).

With the use of the method developed by Haller et al.,[341-343] the isomerization of cyclopropane into propylene on mica was investigated. It turned out that the translational energy of C_3H_6 molecules in the beam had little effect on the reaction probability, but the vibrational energy and the surface temperature had a profound effect. The reaction rate vs. the vibrational energy has Arrhenius dependence, with the activation energy being 235 kJ/mol.

The Arrhenius dependence of the reaction rate on the vibrational energy and the surface temperature was observed, with the activation energy 235 and 88 kJ/mol, respectively. It was also found that $E_{\perp} + E_{\alpha} \approx 105$ kJ/mol, where E_{α} is the activation energy of the reaction itself. The meaning of these dependencies is not quite clear. The authors proposed that the reaction proceeds through the C–C bond break with the formation of a biradical followed by its conversion into propylene

$$\text{cyclo-}C_3H_6 + \text{Surface} \rightarrow \dot{C}H_2CH_2\dot{C}H_2 \text{ (ads.)}$$

$$\dot{C}H_2CH_2\dot{C}H_2 \text{ (ads.)} \rightarrow CH_3\text{–}CH{=}CH_2 \text{ (ads.)}$$

$$CH_3CH{=}CH_2 \text{ (ads.)} \rightarrow C_3H_6 \text{ (gas)}$$

For C–C bond breaking (the first step of the reaction) the vibrational excitation (235 kJ/mol) is needed. The biradical adsorption probability decreases with increasing translational energy $E_{\perp}$. The effect of $E_{\perp}$ influences, most likely, the following steps, i.e., C–H bond activation and adsorption, but this is not elucidated finally. The study of the same reaction on MoO_3[344] showed that the reaction rate was independent of T_s in the range 679 to 973 K, but depended on the beam energy with E_α close to 90 kJ/mol. The authors suggested that the vibrational excitation is necessary for the reaction to proceed. However, it should be noted that cyclopropane isomerization occurs as well in the absence of the catalyst, with the activation energy close to 260 kJ/mol.[349] Thus, it is possible that Haller et al.[341-344] dealt with a non-catalytic reaction. Besides, mica is not a good catalyst.

The conversion of butene-1 was studied by the same method on mica at T_s = 650 to 850 K for different beam energies.[341] Contrary to the above situation, the dependence only on surface temperature was revealed with E_a = 41.9, 48.1, and 48.2 kJ/mol for the isomerization to *cis*-butene, to *trans*-butene, and dehydration to butadiene, respectively. The authors drew the conclusion that in this case the reaction is truly catalytic, proceeds on the surface, and the energy barrier to adsorption is absent.

Molecular beams have been applied as well to study truly catalytic reactions. In studies of H_2-D_2 exchange (molecular beams H_2 + D_2, and HD desorption)[33,346] the variation of the angle of incidence led to an understanding of the active site arrangement in detail, for example, on a stepped surface of Pt monocrystal. By varying the translational temperature of H_2 and D_2 beams, it can be shown[33,347] that dissociative adsorption on Pt(111) needs the activation energy $E_\alpha \sim 4$ to 12 kJ/mol. On a stepped surface the reaction proceeds without activation.

As early as 1954, Shäfer and Gerstacker[89] studied experimentally the relation between the rates of vibrational relaxation and N_2O decomposition. It is clear from their results that the activation energy of N_2O decomposition on Pd (~100 kJ/mol) and on Ag (~90 kJ/mol) are rather close to E_α of N_2O vibrational deactivation on the same metals. Practically total accommodation was observed for platinum, E_a of N_2O decomposition being equal to 135 kJ/mol.

N_2O decomposition was investigated also by the molecular beam technique. It was shown that adsorption with a small activation energy is needed on platinum covered with carbon[348] and on W.[349]

The molecular beam studies of deuterium oxidation on Pt(111) were conducted by Smith and Palmer.[350,351] At the beginning, the Pt surface was covered with oxygen; then the modulated beam of D_2 molecules with varying energy was directed at it. With the use of hot D_2 beams, the activation energy was found to be $E_\alpha \approx 50$ kJ/mol. D_2O molecules desorbed without activation, which was confirmed by the cosine angular distribution.

To summarize the data it should be noted that there is strong evidence of the reaction rate dependence on translational excitation, whereas the dependence on vibrational excitation is less evident. However, the contributions of different types of excitations can be judged from the data on the reverse process, i.e., the energy distribution in the reaction products (see Section 4.4).

4.3.2. SELECTIVE EXCITATION

In catalysis, particularly in selective catalysis, the energy concentration in one of the bonds of the reacting molecule is required to break the bond. With the advent of lasers, the experimental opportunity for direct checking of this assumption has arisen. The selective excitation of one of molecular bonds becomes possible by the use of the laser with a proper frequency; thus, control over the processes of adsorption, desorption, and selective catalysis is realized. We have already referred in Section 2.2 to the difficulties associated with a distinction between laser-induced thermal and selective (quantum) effects. Nevertheless, a rather large number of papers have appeared in this field.[352-354] Relatively low-power lasers (1 to 30 W/cm^2) are usually used in order to prevent solid heating. Laser action on molecules near a solid is also applied: a laser beam is directed parallel to the surface to check a thermal effect. Then the dependence of product yield on laser intensity is examined in pulse experiments. The more rapid increase compared with a linear one often indicates the contribution of a thermal effect.

Chuang[354] pointed out that, taken alone, a distinction between thermal and laser heating in reaction yields is insufficient for proving the laser-induced selective effect since the laser thermal field differs from that of ordinary heating. For a clear demonstration of the selective effect, it is desirable

1. To prove the isotope dependence
2. To show that only one component is activated, other species reacting in the ordinary way
3. To find such an example of two parallel reactions, one of which can accelerate under the laser action
4. To produce the laser shift of the system from the equilibrium opposite to the thermal action
5. To demonstrate the laser effect for systems with a small amount of collisions, for example, for molecular beams
6. For reactions induced by one-photon absorption, to check for the strong specifity which should be different for different parts of the same absorption band

One of the first and, thus, often-cited studies on selective effect in heterogeneous processes was the paper by Basov et al.[355] in which the condensation rate of vibrationally excited CO_2 molecules on solid CO_2 surface was studied. For this purpose the mixture of isotope molecules $^{12}CO_2$ and $^{13}CO_2$ was exposed to electric discharge. As a consequence of the difference in vibrational temperature, the condensation rates of isotope-substituted molecules were also different. The ratio of condensation rates is defined by

$$\exp\left(\frac{E_\alpha}{T}\frac{\nu_1 - \nu_2}{\nu_1}\right)$$

where ν_i is a vibrational frequency of i molecule. $^{12}CO_2$ molecules condensed more rapidly. Therefore, the authors believed that they succeeded in separating the isotopes.

Attempts to separate different isotopes under conditions of the laser action on the processes of adsorption, desorption, and catalysis were also made by other researchers. Lin et al.[356] studied the interaction between BCl_3 and hydrogen on a titanium catalyst. In the absence of laser radiation, reaction outputs are B_2H_6, B_2H_5Cl, and $BHCl_2$; under the action of a CO_2-laser at ν = 10.55 nm, $^{10}B^{11}BH_2Cl_4$ and HCl were observed predominantly. The content of isotope-substituted $^{11}BCl_9$ molecules in the unreacted residue increased from 20 to 37%.

Petrov[357] showed that under laser action the diffusion flow of Br_2, SF_6, and $C_6H_5CH_3$ changed their passing through porous glass. This process is likely to be the main cause of separation of isotope-substituted molecules.

Experiments on laser excitation of SF_6 molecules are often carried out in homogeneous media, taking advantage of suitable absorption frequencies (942 cm^{-2}) for the CO_2 laser. Similar studies have also been made under heterogeneous conditions. For example, SF_6 vapor over silicon was exposed to a laser beam parallel to the Si surface.[358] The SF_6 molecule irradiated by CO_2 laser dissociates, and the etching of silicon surface attacked by fluorine atoms was observed.

Much research has been devoted to laser-induced selective desorption with the excitation of one of the vibrational frequencies of an adsorbed molecule or its bond to the surface. The general theory of laser-induced vibrations resulting in desorption of adsorbed particles was tested.[359-361] Laser action is likely to be the most effective during short (i.e., nanosecond) pulses. In that an amount of time the energy cannot dissipate and may be used for desorption of weakly bound, physically adsorbed molecules.[361] Dzhidzhoev et al.[362] studied SiO_2 in an ammonia atmosphere irradiated by a CO_2 laser with a frequency 950 cm^{-1} and power density of 10 W/cm^2. The rate of NH_2-group decomposition on SiO_2 was three orders higher than that without the laser. Nevertheless, a close look at the experimental conditions reveals that the pure thermal laser effect has not been ruled out.

CH_3F and SF_6 desorption under laser irradiation in the ν_3-vibrational molecular band was observed by Huang et al.[363] A CH_3F monolayer adsorbed physically on NaCl at 70 K was excited by a CO_2 laser. Desorption occurred at a radiation frequency 970 to 990 cm^{-1}. The observed width in photodesorption spectrum was ~10 to 15 cm^{-1}, i.e., substantially narrower than the corresponding band of IR spectrum. The yield of desorbed species was proportional to $I^{2.8}$, where I is the radiation intensity. Therefore, it was concluded that approximately three photons could participate into the process. Ammonia desorption from the surface of Cu(100)[364] and W(100)[365] was observed under the laser radiation corresponding to the stretching (3370 cm^{-1}) and asymmetric bending (1065 cm^{-1}) vibrations of the NH_3 molecule. Nevertheless, it was pointed out[364] that, most likely, this desorption is induced by the thermal laser effect.

When subjected to IR laser radiation, desorption of pyridine from KCl, Ni, Ag/SiO_2, and Ag(110),[366] methanol from metal surfaces,[367] and CCl_4 from a germanium surface[368] was observed. IR laser irradiation of Pt(111) with adsorbed water in the region of OH-vibrational band failed to cause water desorption.[369] CH_3OH photodesorption from a Cu(110) surface irradiated by laser in the absorption band of asymmetric CH_3 vibrations and valence OH vibrations was observed at 90 K.[370]

Chuang[366] recognizes the following characteristic features of ammonia and pyridine desorption under CO_2-laser action

1. Desorption takes place both from metals and dielectrics with quantum yield of $\sim 10^{-3}$ and desorption cross-section of $\sim 10^{-22}$ cm^2.
2. A clear dependence of selectivity on the vibrational frequency is absent. Isotope-substituted molecules desorb with the same efficiency, and the desorption yield is determined by IR absorption cross-section rather than by its vibrational frequency.
3. Desorption may be a two- or three-photon process.
4. The desorption yield increases with the increase of the thickness of an adsorbed layer, i.e., the energy is absorbed by adsorbed molecules.
5. For coverages close to monolayer, the maximum output is reached in the first few pulses; then it rapidly reduces; probably the most weakly-bound molecules desorb.
6. The translational temperature of desorbed molecules is usually low and close to T_s.

These effects are attributable to "nondirect" or "resonance" heating.[371,372] Initially, the thermalized molecule is successively excited by a laser to the first, second, and so forth, vibrational level up to a continuum which corresponds to the rupture of the adsorption bond. Phonon emission from intermediate levels occurs which results in heating of the sample. Theoretical calculations[373] are in good agreement with the experimental results, except that desorption temperatures are overestimated. V-V exchange between adsorbed molecules irradiated by laser was treated by Görtel et al.[374] It was shown that at coverages $\theta > 0.1$ to 0.3 V-V relaxation resulting from the dipole-dipole interaction can substantially increase laser-induced excitations of top levels in a harmonic system, thus, promoting desorption. At the same time the large-scale effect of V-V transfer is negligible, i.e., the laser radiation in one place of the surface does not induce desorption in the other.

Tro et al.[375] have shown that even nanosecond pulses are insufficient to prevent thermal effects, when subjected to a series of laser pulses (with the duration of 1 ns followed by the pause of 300 ns). C_4H_{10} desorption was observed in the absorption band of C–H bonds at 2875 and 2985 cm^{-1}. Nevertheless, the isotope effect was practically absent: C_4D_{10} desorption occurred with the same velocity in the absorption band of C-D vibrations. The vibrational energy rapidly thermalizes and heats the butene layer, resulting in its desorption.

Heidberg et al.[376,377] observed the desorption of ^{13}CO from NaCl at 20 K under selective laser radiation in the fundamental absorption band of CO (2107 cm^{-1}; the line width was 11 ± 3 cm^{-1}). The vibrational energy of adsorbed CO molecule far exceeded the energy of adsorption (16 kJ/mol). In this case, according to the authors,[376] the laser-induced resonance process is associated with photon absorption by the CO-bond rather than heating, and "vibrational predesorption" occurs

$$\text{NaCl–CO} \underset{}{\overset{h\nu}{\rightleftarrows}} \text{NaCl–CO } (v = 1) \longrightarrow \text{NaCl} + \text{CO(gas)}$$

Selective desorption of HD physically adsorbed on the LiF surface irradiated by IR light (9 to 15 nm) of weak intensity was detected at very low temperatures (1.5 to 4.2 K).[378] The desorption rate is proportional to the radiation intensity, independent of Ts (below 4.2 K) and increases rapidly with the increase of wavelength. The velocity distribution of desorbed molecules corresponds to 21 K, i.e., translational excitation is observed. Thermodesorption is absent under these conditions. It is the authors' opinion [378] that acoustic phonons are excited in LiF irradiated by an IR laser. Their lifetime (10^{-4} to 10^{-9} s) is sufficient to travel a distance from the point of their formation to the place of HD desorption.

Along with those of the IR region, visible and UV lasers have also been used to produce desorption through the excitation of solids. The theory of laser-induced charge transfer to the semiconductor surface was treated by Murphy and George.[379] To excite the silicon surface, the laser radiation in the Si absorption band of about 1.2 eV is required. One absorbed photon excites one Si surface state. The oxidation of silicon irradiated by the energy corresponding to the interband transition is really enhanced.[380] But the peak in the curve of the yield dependant on the incident quantum energy is not sharp, probably, due to Si spectrum broadening resulting from surface subbands in the region of interband transition.[381]

CO desorption from ZnO at 360 nm laser wavelength and from Ni and W metal surfaces has been studied.[382,383] Moiseenko et al.[384,385] investigated ZnO irradiated by a neodymium laser. Oxygen species O_2 and O were observed. The quantum yield in the region of the fundamental absorption band corresponding to the ZnO forbidden zone (λ = 354 nm) was an order of magnitude higher than it was outside this region (λ = 532 and 1064 nm), which is an indication of the selective laser effect. The kinetic energy of desorbed species measured by a TOF mass spectrometer was 0.8 eV (corresponding to the temperature ~7000 K).

CO_2 and H_2O desorption from oxide semiconductors irradiated by laser in the region of the forbidden zone[386] and CH_3Br desorption from LiF surface irradiated in the 222-nm F-center zone[387] were reported. Chlorine desorption from the metal surfaces of Cu, Ag, and Au irradiated by laser[388,389] was attributed to the plasmon excitation in a solid followed by ionization of surface species and ion reneutralization. Desorbed species Cl, CuCl, $CuCl_2$, Cu_3Cl_2, AgCl, and others had a high kinetic energy. The theory of laser-induced electron transfer from the metal bulk to the surface was considered by Murphy and George.[381] Frequencies from 0.7 to 4.4 eV can be used for excitation in metallic Na, and a photon of a given frequency can generate many surface states in contrast to the case of a semiconductor.

The interaction of the argon laser irradiation (514 nm, with the second harmonic of 257 nm) with mono- and multilayers of metal carbonyls adsorbed on Si(111) (7 × 7), results in carbonyl decomposition and CO desorption.[390,391] The second harmonic was required for $Mo(CO)_6$ selective decomposition; the base 514 nm frequency failed to decompose it, whereas both wavelengths resulted in decomposition of $W(CO)_6$ and $Fe(CO)_5$. Irradiation of 325 nm frequency at 90 K caused decomposition of $Mo(CO)_6$ adsorbed on Rh(100).[392]

Resonance NO desorption from Ni(100) surface irradiated by UV Ar-F laser in the band with the maximal absorption of about 6.4 eV took place at 170 K. Desorbed molecules, in accordance with LIF spectra and TOF data, were excited vibrationally in $v = 1$ and $v = 2$, and rotationally and translationally up to 2800 K, which argues for the true photodesorption. Irradiation at 5.0 eV gives broader molecular distribution due to thermal effects.[393]

NO desorption from irradiated Pt(111)[394] and Si(111)[395] revealed the dependence of desorption yield on laser wavelength (354 to 1907 nm), which points to a nonthermal desorption effect.[394] LIF spectra showed non-Boltzmann rotational and vibrational energy distribution in NO molecules.

The study of the system Pt(111) + O_2 (ads.) showed that at $\lambda < 295$ nm O–O bond excitation and atomic oxygen release occur, and at $\lambda < 420$ nm O_2 desorption takes place. It is the authors' opinion[396] that thermal heating does not describe the results as well as direct photoexcitation does. A TOF study of the velocity distribution of water molecules desorbed from Pd(111) irradiated by UV laser (6.4 eV) showed that in this case the desorbed molecules have also photochemical rather than thermal origin.[397]

Using very short 5-ns laser pulses at a very low power (0.1 to 1 mW/cm²) and energy 4.6 to 5.3 eV, Natzle et al.[398] observed selective NO desorption from Ag(111) at low temperatures (25 to 50 K). Desorbed molecules are excited rotationally and vibrationally; the population ratio of vibrational levels is $(v = 3)/(v = 2) = 0.85$. The authors are of the opinion that such a desorption is likely caused by $(NO)_2$-dimer photodissociation from the surface.

NO desorption from Pt foil irradiated by nanosecond laser pulses (λ = 532 nm) was observed by Burgess et al.[399] The authors believe that in metal the laser excites electron-hole pairs which have lifetime values greater than the pulse length (10^{-9} s). Therefore, desorption occurs.

When passing from nano- to pico- and femtosecond pulses, the probability of a selective laser effect increases further. Prybyla et al.[400] measured the yield and the final energy distribution of NO molecules desorbed from Pd(111) irradiated by 200-fs laser pulses with the energy 2.0 eV at 300 K. While NO temperature is close to Ts in nanosecond experiments, the concentration ratio $N(v = 1)/N(v = 0) = 0.3$ is observed for femtosecond pulses which correspond to vibrational temperature ~2200 K. Rotational temperature of NO molecules in $v = 1$ level was equal to 2600 K. The authors[400] suggested that for such a short time, electronic excitation involving 2π-valence orbitals of NO occurs due to electron or hole capture. This excitation, in turn, results in a vibrational one. The high desorption yield is attributable to the fact that several hot electrons can transfer their energy to each adsorbed NO molecule. It has been shown[401] that in the femtosecond region (pulse length 300 fs, or $3 \cdot 10^{-13}$ s) the selectivity of ion emission from protein layers containing tryptophan is greatly enhanced. Ions desorb with a high kinetic energy (fractions of an electronvolt) that is probably indicative of a biphoton ionization mechanism.

It would be most interesting to induce the catalytic reaction through selective laser action on one of the chemical bonds. A large number of experiments, especially with the NH_3 molecule, have been carried out in this field. NH_3 oxidation on platinum irradiated by laser was investigated by Sun and Quin.[402] Maximum yield of products $N_2 + H_2O$ was obtained at laser frequency 933 cm^{-1} coinciding with N–H-bond vibrations in gaseous NH_3. Belikov et al.[403] observed that the rate of decomposition of NH_3 irradiated by laser enhances 10% in the gas phase and 85% on Pt. The data obtained were attributed to the nonequilibrium photoexcitation of Pt oxide semiconductor film.

In collaboration with Dr. Ryskin we carefully examined and repeated this work. It turned out to be erroneous. The enhancement of NH_3 decomposition velocity likely resulted from laser heating of the catalyst surface. As to the theoretical treatment, we could argue that at studied temperatures 5% of NH_3 moleculas are in the first vibrational level, and the most effective using of laser pumping can lead only to a 50% population. So, if vibrational excitation of NH_3 promotes the reaction, its rate can increase only one order, but not six orders, as was observed.

Among other catalytic reactions with the laser-induced selective effect, HCOOH decomposition on Pt,[404] the interaction of C_2H_4 and NO_2 on Pt,[405] N_2O decomposition on Cu,[406] and dehydrochlorination of 1,2-dichlorethane on $BaSO_4$[407] should be mentioned. In the first case laser radiation increased the ratio of CO_2/CO in the HCOOH decomposition products. In the second case the argon ion laser increased the CO_2 generation rate four times. It is believed that this laser excites NO_2 molecules into the first electronic level; NO_2 electronic excitation results further in vibrational excitation, followed by dissociation into free radicals.

It was reported[408] that UV radiation accelerates the reaction $CO_2 + H_2O \rightarrow CH_4 + O_2$ catalyzed by Pt-$SrTiO_3$. The rate of increase in fragmentation of benzaldehyde, pyridine, and other aromatic amines on silver was observed[409] at laser frequency 350 to 410 nm corresponding to a surface-enhanced-Raman-spectroscopy (SERS) effect on Ag.

Taking into account all the data on laser-induced selective desorption and heterogeneous catalysis, it should be recognized that, up to now, the results are unconvincing, especially, for IR laser stimulation. Many publications have been disproved, and many results lack support from other researchers or provoke objections. More definite positive data have been obtained for electronic excitations of solids and adsorbed molecules irradiated by lasers in the visible region. Use of short ($\tau < 10^{-9}$ s) laser pulses at low powers and low temperatures appears to have the greatest promise in this field.

4.4. ENERGY DISTRIBUTIONS IN PRODUCTS OF CATALYTIC REACTION

Up until now, few firm data were available relative to the initial energy distribution effect of input reagents on the catalytic reaction process. In contrast, energy distribution in reaction products has received more attention. Exothermic reaction is the most convenient subject to investigate since released energy can be used for excitation of reaction products.

4.4.1. CATALYTIC RECOMBINATION OF ATOMS

Atomic recombination with the production of a diatomic molecule ($A + A \rightarrow A_2$) is the simplest example of an extremely exothermic catalytic reaction. Therefore, it was this reaction in which the first evidence of nonequilibrium energy distribution in reaction products was discovered.

In their early work[410] Wood and Wise pointed out that the recombination of hydrogen atoms occurs at metal wires with incomplete thermal accommodation. Melin and Madix[411] studied the transfer of energy to the surface during recombination of hydrogen and oxygen atoms at polycrystalline wires of Fe, Ni, Cu, Ag, Pt, and W. Atoms were generated in electric discharge and struck the catalyst surface. The accommodation coefficient was determined using the wire as an isothermal calorimeter and a catalyst simultaneously. In both cases (H + H and O + O) high accommodation coefficients $\epsilon \approx 0.5$ were obtained for metals of the copper group; the highest values were 0.95 for O/Ag and 0.87 for H/Ag. For other metals ϵ values were substantially lower (from 0.07 to 0.28) which is indicative of energy carryover by the reaction products. The values of recombination coefficients γ correlated with the values of ϵ as a rule. Dickens et al.[412] noticed the possibility of incomplete accommodation during recombination and estimated the lower limit $\epsilon = 0.5$. Catalytic reaction $2A \rightarrow A_2$ at active site Z was represented by the scheme

$$Z + A \rightarrow ZA \qquad \Delta H_1 = \chi \tag{4.6}$$

$$2Z + A_2 \rightarrow 2ZA \qquad \Delta H_2 = 2\chi - D = \Delta H_x \tag{4.7}$$

$$ZA + A \rightarrow Z + A_2 \qquad \Delta H_3 = D - \chi \tag{4.8}$$

where ΔH is the heat of an elementary step, χ is the atomic chemisorption heat, D is the bond energy in the diatomic molecule A_2, and ΔH_χ is the chemisorption heat of the A_2 molecule.

If the step in Equation 4.8 is exothermic, and $\chi = (D + \Delta H_\chi)/2$, then we have

$$\Delta H_3 = \frac{D}{2} - \frac{\Delta H_x}{2} > 0$$

Since $\Delta H_\chi \rightarrow 0$ when coverage $\Theta \rightarrow 1$, D/2 is the maximal energy which can be carried off by a diatomic molecule. The corresponding limit is $\epsilon = 0.5$. The values $\epsilon < 0.5$ are possible in the following cases: (1) the adsorption heat of A atoms is not completely transmitted to the wire; (2) Z sites are rapidly filled during dissociative chemisorption; (3) A atoms adsorb endothermally relative to the A_2 molecule, i.e., in the state of a mobile precursor.

A low value $\epsilon \approx 0.3$ during H-atom recombination at the Pt wire was also obtained by Kisljuk et al.[413] which points to significant energy carryover.

Recombination of nitrogen and deuterium atoms at different surfaces was studied in our work in collaboration with Kovalevsky.[101] It is suggested that vibrationally excited molecules $(ZA_2)^*$ can be generated at the surface

$$Z + A \longrightarrow ZA \tag{4.9}$$

$$ZA + A \longrightarrow (ZA_2)^* \tag{4.10}$$

$$(ZA_2)^* \longrightarrow ZA_2 \tag{4.11}$$

$$ZA_2- \longrightarrow Z + A_2 \tag{4.12}$$

The step in Equation 4.10 includes generation of vibrationally excited molecules followed by relaxation of vibrational energy. If this suggestion is valid, the whole process depends on the deactivation rate; hence, the corre-

TABLE 4.1
Accommodation and Relaxation Coefficients for N_2, H_2, and D_2 on Several Surfaces

Molecule	Surface	ϵ	γ	Molecule	Surface	ϵ	γ
N_2	Pyrex®	$1.5 \cdot 10^{-4}$	$3.0 \cdot 10^{-5}$	H_2	Pyrex®	$1.0 \cdot 10^{-4}$	$1.5 \cdot 10^{-5}$
	Quartz	$7.0 \cdot 10^{-4}$	$8.0 \cdot 10^{-1}$		Teflon®	10^{-5}	$3.0 \cdot 10^{-4}$
	Teflon®	$6.0 \cdot 10^{-4}$	$2.9 \cdot 10^{-5}$	D_2	Quartz	$9.5 \cdot 10^{-5}$	$7.0 \cdot 10^{-4}$
	Copper	$1.0 \cdot 10^{-3}$	$2.3 \cdot 10^{-3}$		NaCl	$6.3 \cdot 10^{-4}$	$4.1 \cdot 10^{-4}$
	Silver	$1.1 \cdot 10^{-2}$	$1.5 \cdot 10^{-3}$				

lation between accommodation coefficient ϵ and recombination coefficient γ must occur (see Table 4.1).

Molecular beam studies of atomic recombination confirmed the results obtained by Melin and Madix.[411] H_2 molecules[411] leaving the surface are really "more heated" than the surface.

Comsa et al.[219,414] revealed that H_2 molecules desorbed from the Pd surface after H-atom diffusion through the bulk and recombination were heated translationally. Zare and co-workers[61] studied H_2 molecular distribution over all states after recombination at Cu(100) and Cu(111) using LIF. *Ortho-* and *para*-states of the H_2 molecule were populated statistically. Rotational temperature was found to be 0.8 to 0.9 T_s. The ratio of populations of the first and zeroth vibrational levels of the H_2 molecule after desorption from the Cu(110) face was approximately 100 times greater than that of Boltzmann distribution. This value was 10 to 100 times less for Cu(111). In contrast, the presence of sulfur increases this ratio 10 to 100 times. The authors[61] proposed the mechanism of vibrational excitation of H_2 through the intermediate H_2^--ion state. The point of intersection of electron potential well with the Fermi level of Cu defines the degree of vibrational excitation of the H_2 molecule. Similar results were obtained for H_2 desorption from Pd.[62]

Lin and Somorjai[415] studied in detail the recombination process H + D → HD at the stepped Pt(557) face. H and D atomic beams struck the surface at T_s = 150 to 600 K. Translational temperature T_t of desorbing HD molecules was higher than T_s in all cases, and this distribution was broadly compared to Maxwell-Boltzmann one. The growth of T_t with decreasing T_s was surprising. For instance, at T_s = 300 K a mean value of T_t was about 500 K, and at T_s = 200 K it was T_t = 600 K (individual HD molecules had T_t = 1900 K). To explain their results, the authors[415] developed the following model. At low temperatures the H atom collides with a surface close to the chemisorbed atom and has enough time to react with it and to desorb without energy dissipation. At high T_s a coverage Θ is small. Incident H atoms find themselves far from chemisorbed ones and thermalize in the process of surface diffusion before the reaction with chemisorbed atoms. At 150 K, when $\Theta \approx$ 1, an HD signal was not detected. This is evidence that the direct impact H + Pt-D (Eley-Rideal mechanism) is ineffective and results only in beam scattering. It is obvious that the intrinsic, but not extrinsic, precursor participates in the reaction (see Section 6.1).

Similar characteristic features, i.e., lowered rotational and extremely increased vibrational temperatures, were observed in studies of hydrogen desorbed after recombination H + H from other surfaces: Fe,[416,417] Pd(100),[418] and Si(100),[419] The lowered T_R is attributed to retardation of motion of the H_2 molecule formed during desorption. Molecules excited in the fifth[417] and even the ninth[416] vibrational level were observed during recombination. This can be explained by an increased exothermic reaction mechanism (Eley-Rideal). It may be supposed as well that both H atoms are in a state with a very low adsorption heat.

Temperature dependence of the recombination coefficient γ of N atoms at W was accounted for by the formation of hot N_2 molecules.[420] It was assumed that excited molecules generate during recombination according to the Eley-Rideal mechanism, and thermally equilibrated molecules relate to Langmuir-Hinshelwood model. The data on energy carryover allows us to draw the conclusion about vibrational excitation.

The generation of translationally and vibrationally excited molecules N_2^* was observed using LIF during recombination of N atoms diffusing through an iron membrane.[66,421] On clean iron $T_v = T_s$, but on Fe covered with sulfur $T_v > T_e$. Rotational temperature was low. The results were attributed to vibrational excitation of N_2 molecules when passing through the potential barrier during desorption. Electronegative sulfur increases this barrier and the degree of vibrational excitation.

It is often assumed that molecules can be excited electronically after recombination of atoms. We already noted in Section 3.2 that the oxygen molecule has the lowest energy of electronic excitation. The generation of electronically excited $^1\Delta_g$ oxygen molecules at HgO,[422] Ni, Pt, Co, and other metals[429] and E_u^+ at Ni[424] was found during atomic recombination O + O.

The generation of electronically excited molecules during recombination of nitrogen atoms in the presence of oxygen at metals was observed by Harteck and co-workers.[128,129,425] Red luminescence was observed when passing a flow of $N_2 + O_2$ from discharge at a pressure close to 100 Pa over Ni, Co, and Ag surfaces. At first, electronically excited molecules in the metastable $A^3\Sigma_u^+$ state and excited in high vibrational levels are produced during recombination of N atoms at the surface (see the scheme in Figure 3.10). Leaving the surface, these molecules diffuse into the gas phase and, as a result of collisions with other particles, pass into the $B^3\Pi_g$ state. This transition must lead to $N_2(B^3\Pi_g)$ molecules vibrationally excited between the sixth and eighth levels because the terms $B^3\Pi_g$ and $A^3\Sigma_u^+$ cross each other in this region (see Figure 3.9). Molecules N_2 ($B^3\Pi_g$, $v^1 = 6$ to 8) illuminate immediately a photon in the first positive group band of N_2. The lifetime of active particles of $N_2(A^3\Sigma_u^+)$ responsible for the red illumination is 10^{-3} s; this time is necessary for vibrational relaxation, i.e., for transfer to the term $B^3\ \Pi_g$.

Blue luminescence, which is observed during passing of a mixture of N and O atoms over Co, Ni, and Cu surfaces, is explained by the formation of excited molecules of NO($B^2\Pi$) according to the scheme

$$N(^4S) + O(^3P) \xrightarrow{\text{Co, Ni, Cu}} NO^* (B^2\Pi)$$

$$NO(B^3\Pi) \longrightarrow NO(X^2\Pi) + h\nu$$

We have already mentioned in Section 3.2 that the generation of electronically excited molecules during recombination of N atoms proceeds on clean metal surfaces, namely, those where nitrogen adsorption is possible only in the molecular form. Oxide or nitride coverage deactivates a catalyst. Metals such as Ta or W, which adsorb nitrogen dissociatively, do not lead to the formation of excited N_2^* molecules desorbing into the gas phase.

4.4.2. CATALYTIC OXIDATION

Catalytic oxidation reactions are also highly exothermal. Therefore, the possibility of energy carryover in these reactions has also been studied extensively in recent years.

Heterogeneous CO oxidation is currently the only catalytic reaction for which the energy distribution of reaction products has been measured over all states.[426] Palmer and Smith[33,351] were the first to study it. A CO molecular beam struck a Pt(111) surface covered with oxygen atoms. The reaction product CO_2 desorbed with a strong directionality according to the law $\cos^6\varphi$. The authors suggested that hot CO_2 molecules desorb from the surface. The effect of enhanced reactivity of CO molecules around the periphery of a molecular beam points to a high mobility of CO molecules at the Pt surface covered with oxygen.

The same authors studied the oxidation of D_2/Pt(111)[427] and C_2H_4/Ag(111)[428] where the usual dependance $\sim\cos\varphi$ was found for the reaction products. Kinetic measurements showed that the lifetime of D_2O at the surface before desorption is 67 μs at 850 K and 1 ms at 600 K. It is obvious that this time is sufficient to result in complete thermalization. During oxidation of C_2H_2 and C_2H_4 at Pt(111), the distribution of reaction products was found to be wider than that of $\sim\cos\varphi$; the authors concluded that cooling of the reaction products may occur.

According to Comsa and David,[429] CO_2 angular distribution during CO oxidation at Pt corresponds to the expression

$$I_{CO_2} = \cos\varphi + (1 - a)\cos^7\varphi \tag{4.13}$$

where the first term describes the molecules desorbing in thermal equilibrium and the second, the molecules with excess kinetic energy. Parameter a depends on T_s and coverages Θ_o and Θ_{CO}.

A similar two-channel model was confirmed by other scientists.[430,431] Becker et al.[430] studied the CO molecular beam interacting with Pt(111) in an O_2 atmosphere ($\sim 10^{-4}$ Pa). Along with angular distribution velocity, the distribution of CO_2 molecules was measured using TOF mass spectrometry. The temperature of the Pt surface was 880 K. The temperature T_t of molecules desorbing normally to the surface was 3560 K, and at an angle of 45°, 2140 K. This magnitude of T_t allows us to estimate a part of the energy carried off by CO_2 molecules, which is equal to 10% of the reaction energy, or 30 kJ/mol. Two-channel distribution similar to that of Equation 4.13 was observed for CO oxidation at Rh(111).[432] The peak relating to "hot" molecules was sharper than that of Pt, $\sim \cos^9 \varphi$.

Ertl et al.[431] found that a part of the thermally accommodated molecules increases for stepped Pt surfaces, and in the case of the oxide underlayer formation after O_2 exposure at 700 K, it decreases with increasing Θ_o, Θ_∞, and T_s.

Matsushima et al. studied angular distribution of CO_2 molecules using temperature-programmed heating at Pd(111)[433,434] Pd(110),[435] Pd(100),[436] Pt(111),[437] Pt(110),[438,439] and Ir(110).[440] Several maxima are observed in the thermodesorption spectrum: β_1, β_2, etc. Sharp distribution was observed in all cases. The dependence $\cos^{10}\varphi$ was observed for CO_2 molecules desorbed after the reaction CO(ads.) + O_2 (ads.) at Pt(111), and $\cos^9\varphi$ for reaction CO(ads.) + O (ads.) on the same surface. In general, it may be deduced that sharper distributions are observed for more close-packed faces. For instance, CO_2 desorption from the β_1 state at Pd(110) showed the angular dependence $\cos^{20}\varphi$ at a desorption temperature of T_d = 250 K, $\cos^{12}\varphi$ from the β_2 state at T_d = 360 K, and $\cos^{10}\varphi$ from the β_4 state at T_d = 440 K.[435] The estimation according to the Van Willigen Equation 4.1 gives a value of excess translational energy of CO_2 molecule close to 17 kJ/mol after desorption from Pd(111).[433] Figure 4.8 illustrates the model of the activation complex for reaction CO + O at Pd(111) which is responsible for a narrow angular dependence of desorbing CO_2 molecules. The angular distribution is conditioned by the peculiarities of CO_2 molecular abstraction from the surface. In the activation complex, CO and O particles are located closer to the surface than CO_2 molecules, but farther than CO and O adsorbed separately. The activation complex is confined in motions parallel to the surface. At low CO_2 coverages this limitation is weaker than at high ones which results in a less sharp angular distribution.

Such limitations are absent for motions perpendicular to the surface. The CO_2 molecule passes the top of the barrier as though it were not formed completely and possesses an excess of translational and possibly vibrational energy. It does not have enough time after reaction CO + O to sink to the bottom of the potential well, i.e., to come to equilibrium with the adsorbed CO_2 molecule. Its further thermalization occurs through subsequent collisions with gas molecules. Matsushima[434] studied CO_2 desorbing from different sites

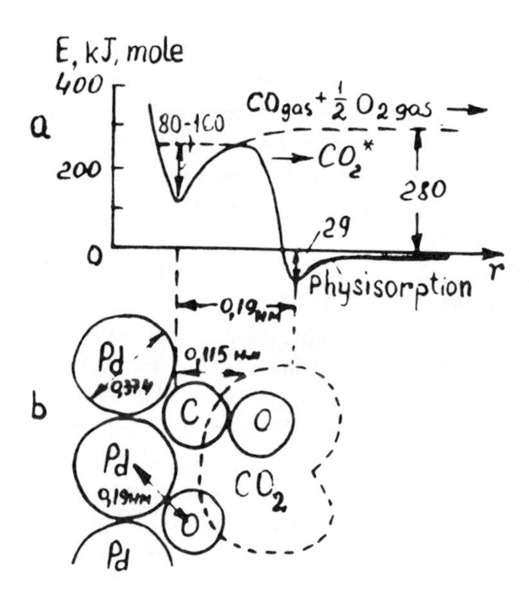

FIGURE 4.8. The model of activated complex for the reaction CO + O(ads.) → CO_2(gas) on Pd(111).[434] Energy profile (a); the structure of complex (b) (dashed line denotes the position of physisorbed CO_2). (From Matsushima, T., *J. Chem. Phys.*, 91, 5722, 1990. With permission.)

of formation at Pd covered with potassium. Most of CO_2 molecules at Pt(111) + K are bound in carbonate chemisorbed forms. These forms are completely thermalized before desorption and obey the $\cos\varphi$ law. At the clean Pd surface and at low temperature, CO_2 has no time to bind in carbonate and desorbs with a large excess translational energy according to the $\cos^n\varphi$ law where n varies from 6 to 30.

Haller et al.[426,441-444] studied the distribution of inner molecular energy of CO_2 molecules after the reaction CO + O_2 at Pt, Pd, and Rh. Excitation of vibrational and rotational levels of CO_2 was found using IR spectroscopy with Fourier analysis (the resolution was 0.012 cm^{-1}). About half the reaction heat is transferred into excitation of these levels, and the rest dissipates into phonons. Boltzmann distribution was observed for vibrational and rotational energies, but with different temperatures. At a high surface temperature T_s and correspondingly low values Θ_{CO} the energy of symmetric stretching vibration of CO_2 (the observed value T_v = 3150 K) increases compared with the energy of other vibrational levels (T_v = 2230 K for antisymmetric stretch levels, and 1820 K for bending levels), and translational temperature T_t decreases. At low T_s and high Θ_{CO} the energy of stretching CO_2 vibration is low, and T_t is high (higher than T_s). The authors[426,442] succeeded in measuring the independent influence of Θ_{CO} and Θ_o at constant T_s. For this purpose Pt and Pd surfaces were exposed to a stationary O_2 molecular beam and a pulsed beam of CO. For a time between pulses Θ_{CO} changes because of the CO reaction with the O_2 beam. It turned out that the Θ_{CO} change weakly affects the inner energy distribution, but vibrational and rotational temperatures increase with the growth of Θ_o. In accordance with the results obtained, the activation complex can be represented by a CO_2 complex perpendicular to the surface for Pt and Rh, and slightly bent for Pd (at an angle intermediate between 90 and 180°) corresponding to the stronger backdonation capability of Pd.[426]

Brown and Bernasek[445] reported that at T_s = 650 to 1100 K, CO oxidation on Pt results in strong vibrational excitation of CO_2. The vibrational excitation, especially the asymmetric stretching modes, enhances with decreasing Θ_o. The vibrational excitation of this mode was detected by measuring IR luminescence of CO_2 (ν = 2349 cm^{-1}). These data seem to be inconsistent with those of Haller and Coulston.[426] Kori and Halpern[446] showed that the great majority of CO_2 molecules excite vibrationally during CO oxidation on a Pt surface with the maximum at v_{av} = 9 and the uppermost value v_{up} = 16. To calculate the vibrational energy distribution in the reaction R-B + A → AB + R (Pt-CO + O → CO_2 + Pt), they used the formulas[447]

$$v_{\alpha v} = v_{up}\,(1 - \gamma/\lambda) \qquad (4.14)$$

where $\gamma = s + 9/2$, s is the number of oscillators, and λ is a parameter characterizing the deviation from the initial distribution (λ = 15.2). A very small value of s was obtained for v_{av} = 9 and v_{up} = 16; this means that the CO_2 molecule is bound to only one Pt atom.

The same authors[448] studied the oxidation of carbon created as a result of the methane decomposition on Pt foil. According to IR Fourier spectra, CO molecules formed on oxidation are excited up to the seventh vibrational level which far exceeds the equilibrium distribution (see Figure 4.9). Using Equation 4.14, it can be shown that the C atom is bound to two or three Pt atoms in the activation complex. The first level is populated somewhat greater than it should be according to statistics, and the higher levels lower, because they are more effective in the energy exchange with the conductivity electrons. A characteristic length of the energy-exchange region is several nanometers from the adsorption site.

The translational excitation of CO molecules created on carbon oxidation on Pt foil was found by Kisljuk et al.[449] As a result of the reaction between carbon and oxygen adsorbed on Pt, the angular distribution of desorbed CO molecules was found to be $\sim\cos^{1.6}\varphi$. If oxygen is not removed from residual gases in the chamber, a narrower distribution $\sim\cos^{2.3}\varphi$ is observed. At last, the $\cos\varphi$-law dependence is observed for CO molecules which were preliminary specially adsorbed on Pt surface. Thus, in this case too, a molecule of CO desorbed just on reaction C + O differs from that of adsorbed earlier on the surface. On recombination C + O, the CO molecule leaves the surface immediately bypassing the chemisorption state.

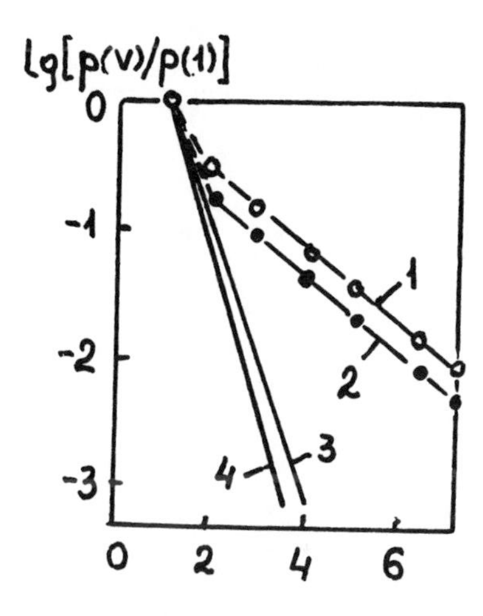

FIGURE 4.9. The distribution of CO molecules over vibrational levels observed for the carbon oxidation on Pt foil at 1400 K(1), at 1000 K(2), and that calculated (for equilibrium occupation) at 1400 K(3) and at 1000 K(4)[448] (the vibrational level numbers are along the x axis, the logarithm of the ratio populations between the nth and ground levels are along the y axis). (From Kori, M. and Halpern, B. L., *Chem. Phys. Lett.*, 98, 32, 1983. With permission.)

A striking result has been obtained by Somorjai and co-workers.[450] Using the TOF technique, they observed translationally cold D_2O molecules (T_t = 283 to 470 K) on reaction of $D_2 + O_2$ on the Pt surface at high temperature (T_s = 664 to 913 K), see Figure 4.10. The angular D_2O distribution was identical for reactions $D_2 + O_2$ and $D + O_2$, which points to the dissociative D_2 adsorption. The reaction proceeds through the formation of intermediate hydroxyls on the surface

$$O(a) + D \rightarrow OD(a);\ 2OD(a) \rightarrow D_2O + O(a)$$

The last step is exothermic with a reaction heat of 160 kJ/mol.

Taking into account the activation energy of this reaction ~70 kJ/mol, approximately 90 kJ/mol can be distributed over the reaction products. Nevertheless, the D_2O molecule seems to be thermalized on creation and remains on the surface for a long time. One may speculate that the desorption rate is greater than the rate of the thermal excitation of D_2O on the surface; then the desorption process will occur owing to molecules populating higher vibrational levels and will disturb the equilibrium. As a result, the velocity distribution of molecules will be shifted to lower values than would be for the thermal equilibrium with the surface.

These authors[450] underline that in gas-phase reactions, nonequilibrium effects play an important role for $E_\alpha/RT < 5$ to 10. In the case given $E_\alpha/RT = 9$ to 17, i.e., their possible influence is questionable. Another explanation of cold nonequilibrium D_2O molecules concerns the concert mechanism of

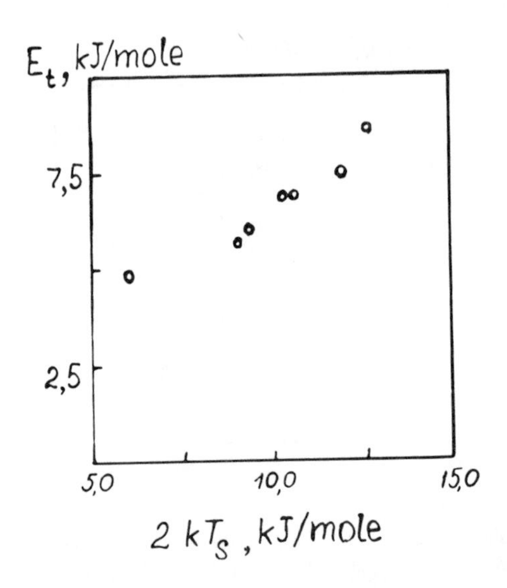

FIGURE 4.10. Translational energy of molecules formed in the reaction of D_2 oxidation on Pt(111) as a function of temperature of Pt surface. (From Ceyer, S. T. et al., *J. Chem. Phys.*, 78, 6982, 1983. With permission.)

the formation of the bond Pt-OD . . . D and D_2O desorption. In this case the distribution of D_2O molecules is substantially determined by the distribution of binding energies of Pt-OD bonds. At long reaction times T_t will correspond to T_s. At short times only a small part of vibrational levels of Pt-OD is effective, and the observed translational energy of D_2O is less than the equilibrium one.

At T_s = 1227 to 1479 K H_2 oxidation on Pt(111) and on the polycrystaline Pt surface results in the desorption of radicals · OH.[451,452] Their vibrational temperature T_v = 1.48 T_s, and rotational temperature T_r = 0.85 T_s and does not practically depend on Θ and on the ratio Θ_O/Θ_H on the surface, i.e., insensitive to Pt-OH bond strength. These authors[451] draw the conclusion that · OH desorption occurs through the freely rotating long-lived precursor. A weak rotational cooling is due to dynamic effects during desorption, probably due to the interaction of oxygen π-orbitals with the surface.

In another work[453] it was found that the angular distribution of OH radicals desorbing during oxidation of H_2 on the Pt surface corresponds to the $\cos\varphi$ law. Then they cool in collisions with gas-phase molecules. The authors note that a knowledge of rotational redistribution of this kind may be of great help in the understanding of the catalytic promotion of gas-phase reactions.

The relationship T_r = 0.45 T_s was found for NO desorption from Pt(111) on oxidation of NH_3.[238,261,454]

In several cases the generation of electronically excited molecules is observed. For example, on exposure of O + NO, the chemiluminescence on

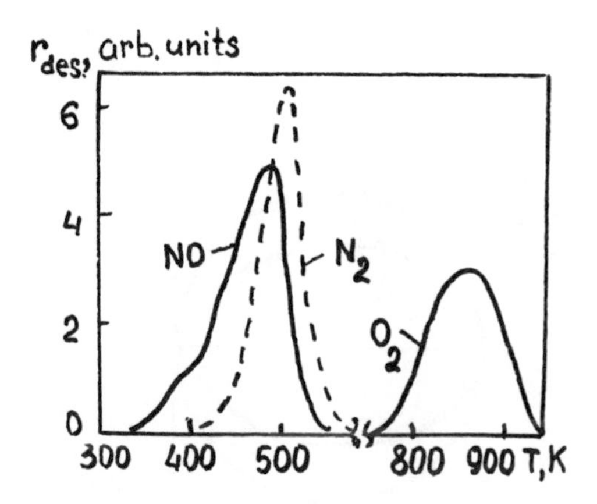

FIGURE 4.11. Thermodesorption spectrum for NO decomposition on Pt(100).[458]

Ni, Co, Pt, and some other metal surfaces was ascribed to the formation of electronically excited NO_2^* molecules.[455,456]

The generation of singlet oxygen $^4\Delta_g$ was obtained in the process of butene oxidative dehydrogenation over an Li-Sn-P-oxide catalyst.[457] The presence of $O_2(^1\Delta_g)$ was detected by watching decolorization on rubrene because of its selective oxidation. These authors suggest that singlet oxygen is produced through an intermediate peroxide complex which is bound to the Li atom. The removal of Li from the catalyst results in its deactivation with respect to the $O_2(^1\Delta g')$ generation. A large number of other catalysts including the catalysts of total (Cr_2O_3, Fe_2O_3, and others) and partial (MoO_3, V_2O_5) oxidation were examined, but all attempts to generate $O_2(^1\Delta g)$ have failed.

4.4.3. OTHER CATALYTIC REACTIONS

Savkin et al.[458] studied NO decomposition on polycrystaline platinum with the dominating (100) face using a TPD technique. The thermodesorption spectrum is shown in Figure 4.11. It turned out that nitrogen is liberated at a somewhat higher temperature than NO, and oxygen at an appreciably higher temperature. Angular distribution studies showed $\cos\varphi$-dependence for NO, $\cos^{1.5}\varphi$ for O_2, and $\cos^4\varphi$ for N_2 (see Figure 4.12). Recalculating the last value taking into account the surface roughness gives $\cos^{16}\varphi$. The estimation of the adsorption energy of NO using Equation 4.1 yields 13 ± 20 kJ/mol. The authors suggest the following mechanism

$$NO + Z \rightleftarrows ZNO \tag{4.15}$$

$$ZNO + Z \rightleftarrows ZN + ZO \tag{4.16}$$

$$2ZN \rightleftarrows N_2 + 2Z \tag{4.17}$$

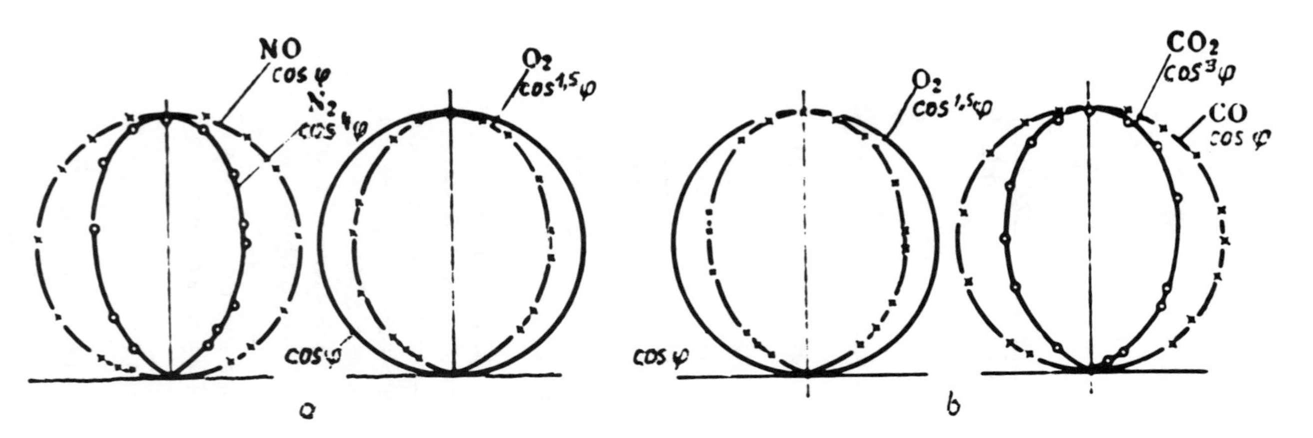

FIGURE 4.12. Angular distribution of particles desorbing from Pt surface during NO decomposition (a)[418] and formed during interaction between N_2O and CO (b).[36]

where Z is the active site. The potential curve in accordance with this scheme and experimental data obtained are presented.[445] Nevertheless the Equation 4.16 is questionable because nitrogen does not adsorb on platinum at a detectable rate. Therefore, the authors give a more probable scheme

$$2ZNO \longrightarrow N_2 + 2ZO \tag{4.18}$$

$$ZNO + ZN \longrightarrow N_2 + ZO + Z \tag{4.19}$$

The N_2 molecule is released in the translationally excited state, and the O_2 molecule is created only at a higher temperature and is desorbed at $T_t \approx T_s$.

In another work[36] the same authors used a TPD angular-resolved technique to study the reaction

$$N_2O + CO \longrightarrow N_2 + CO_2$$

on polycrystaline platinum. The reaction product, CO_2, behaves identically both in the case of the interaction of N_2O with CO, and O(a) with CO. This is indicative of the following reaction mechanism

$$N_2O \rightarrow O(a) + N_2 \qquad O(a) + O(a) \rightarrow O_2$$

$$CO \rightleftarrows CO(a) \qquad CO(a) + O(a) \rightarrow CO_2$$

The angular distribution of desorbing CO molecules obeys $\cos\varphi$ law, that of O_2 — $\cos^{1,5}\varphi$, and $\cos^3\varphi$ for CO_2 (see Figure 4.12 b). On account of the surface roughness, $\cos^6\varphi$ dependence was obtained for the latter case.[36] The cosine powers are lower than that of the above works of Matsushima et al.[437-439] on Pt surfaces, but still indicate appreciable deviation from the equilibrium. Using the Van Willigen formula (Equation 4.1), these authors obtained for the inverse process, i.e., CO_2 adsorption, the activation energy value of 10.5 ± 1.7 kJ/mol.

Using LIF, NO molecules formed on the reduction of NO_2 on Ge surface were found to be significantly deviated from the equilibrium rotational level populations. Nevertheless, an excess of the translational and vibrational energy in the desorbed molecules was not found in this case.[459] The NO molecule seems to pass from the strongly bound state into the weakly bound precursor. On desorption, the rotational cooling owing to the rotational energy transfer to the energy of translational motion is observed. The rotational cooling of NO has already been mentioned in Section 4.1.3.

Molecular nitrogen and hydroxyls are desorbed in the reaction of NO_2, reduction by H_2 on Pt(111) at 1078 to 1433 K.[460] OH rotational cooling has been observed. The ratio T_R/T_s, i.e., the rotational accommodation, increases with the increase of oxygen coverage on Pt surface.

Also studied was the exothermal reaction of the catalytic hydrazine decomposition. Savin and Merrill[461] found an extremely sharp angular distribution of nitrogen $\sim\cos^n\varphi$ with $n > 100$ on N_2H_4 decomposition on Ir(110) at 250 K. All nitrogen molecules have been detected within 7° of the normal to the surface and are heated translationally. At 600 K the desorbed molecules ($N_2 + H_2$) are likely to be thermally equalized. To explain their low-temperature results, these authors take advantage of the mechanism of Wood and Wise[462] according to which N_2H_4 molecule from the gas phase interacts with the doubly adsorbed complex

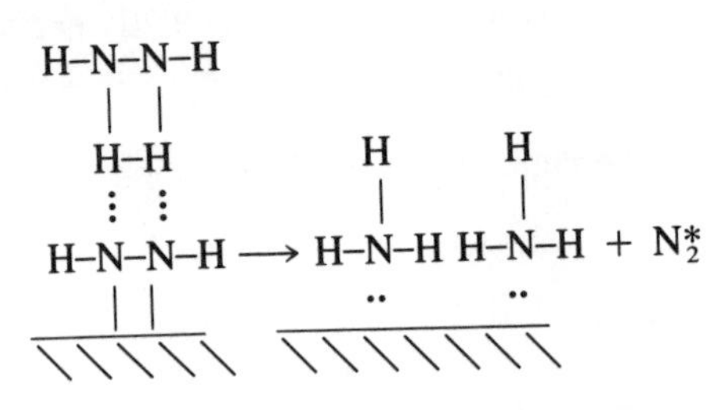

which transforms to the excited nitrogen molecule desorbing from the surface and two adsorbed ammonia molecules.

N_2H_4 decomposition on Pt, Ir, and W at $T > 900$ K results in the nonequilibrium energy distribution of created molecules.[463] Up to 30 to 40% of the energy of the exothermal reaction is transmitted to the molecular translational energy. At $T_s = 1520$ K the thermal equilibrium is reached. The vibration excitation on account of the reactions in Equations 4.26 and 4.27 (see below) is also possible.[464]

In connection with the results of the work[461,463] where excited molecules were observed on hydrazine decomposition, it is worth mentioning our work[465] concerning the catalytic N_2H_4 decomposition using a separate calorimeter technique. The thermocouple in the gas phase had a temperature value higher than that of the catalytic reaction on the surface which proceeds further by the chain mechanism in the gas phase. Excited particles are likely to be generated in the process giving rise to the thermocouple heating in the gas phase.

The separate calorimeter technique is considered the basic method for detection of the chain homogeneous extension of the heterogeneous catalytic reaction. Above, it has been shown that in catalysis, especially in the exothermal catalytic reactions, it is not unusual to find the desorption of excited molecules instead of radicals. These molecules can further deactivate heating the gas bulk. Therefore, all the data obtained with the aid of the separate calorimeter technique should be revised in the context of whether they prove the chain-reaction mechanism involving radicals in the gas bulk. It is really so only in the case of the simultaneous measurement of the concentration of free radicals.

Translational and vibrational temperature of particles desorbed from the catalyst surface is extremely high, and succeeding reactions could very likely occur in the near-surface gas layer. We are not acquainted with papers where these processes in the heterogeneous catalysis were taken into account.

The endothermic reaction of the catalytic decomposition of ammonia over Pt is of special interest relative to the creation of excited particles. Foner and Hudson[464] studied the distribution of N_2^* molecules generated in this reaction at 1233 K. Most of the N_2 molecules have a excitation energy less than 1.3 eV, but individual ones up to 2.4 eV which corresponds to the ninth vibrational level. Ammonia is known to decompose according to the scheme

$$NH_3 \rightleftarrows NH_3(a) \tag{4.20}$$

$$NH_3(a) \rightleftarrows NH(a) + 2H(a) \tag{4.21}$$

$$NH(a) \rightleftarrows N(a) + H(a) \tag{4.22}$$

$$N(a) + NH(a) \rightleftarrows N_2 + H(a) \tag{4.23}$$

$$N(a) + N(a) \rightarrow N_2 \tag{4.24}$$

$$2H(a) \rightarrow H_2 \tag{4.25}$$

To explain their results, Foner and Hudson suggested two additional steps

$$NH(a) + NH(a) \rightarrow N_2^* + H_2 + 1.75 \text{ eV} \tag{4.26}$$

$$NH(a) + NH(a) \rightarrow N_2^* + 2H(a) + 2.4 \text{ eV} \tag{4.27}$$

It is seen that only the step in Equation 4.27 could explain the vibrational excitation of N_2 molecule to the ninth level. The energy pathway of NH_3 decomposition over Pt catalyst is shown in Figure 4.13.[464]

Weinberg and co-workers[466] proposed another explanation of these results. They assumed that the observed vibrational excitation of the N_2 molecule might be ascribed to the nitrogen atomic recombination if N_2 dissociative adsorption had the activation energy of more than 85 kJ/mol.

Significant deviations from the equilibrium energy distribution are often observed in the molecular products of the catalytic reaction. This may be indicative of the nonequilibrium conditions both in the last stage, i.e., the desorption from a weakly bound state, and in preceding stages as well. In accordance with the principle of microreversibility, the translational and

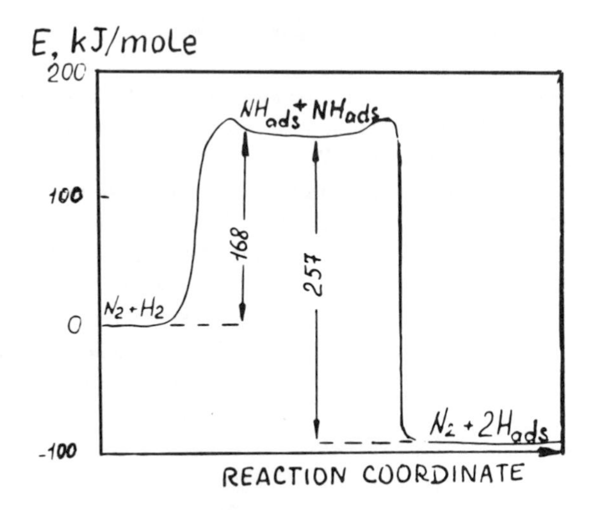

FIGURE 4.13. Energy profile for NH_3 decomposition on Pt. (From Foner, S. N. and Hudson, R. L., *J. Chem. Phys.*, 80, 518, 1984. With permission.)

vibrational excitation of initial reagents can also have an appreciable effect on the catalytic process.

All studies discussed including molecular beam studies, were carried out at low pressures. At high pressure excited molecules deactivate rapidly in the gas phase. Therefore, they are likely to have a little effect on the stationary catalytic reaction rate in most cases. Systematic investigations in this field are lacking. Such long-lived molecules like singlet oxygen $O_2(^1\Delta_g)$ might significantly affect the catalytic reactions.

To summarize the results, it should be noted that the generation of excited molecules appears clearly established for many exothermal and several endothermal catalytic reactions. The excitation of different degrees of freedom depends on T_s, the coverage, the structure of the activation complex, and the energy profile of the reaction. An excitation is most often observed in the case of desorption from the weakly bound preadsorbed state, i.e., precursor, when a small potential well exists on the top of the potential barrier.

Chapter 5

LIFETIME OF EXCITED MOLECULES ON SURFACE

The value of the lifetime of an excited molecule on the surface is of great importance for the catalytic reaction process. When this time exceeds the period between molecular collisions with a surface atom (about 10^{-5} s at 1 kPa), it would appear reasonable to expect a significant influence of excited molecules. It may be shown that this influence would be observable even for shorter lifetimes.

Until recent years there were few experimental works concerning the lifetime measurements of excited molecules on the surface. Now their number is rapidly increasing. In collaboration with Kozhushner[468] we estimated theoretically the molecular lifetime in a vibrationally excited state.

5.1. GENERATION PROBABILITY AND LIFETIME OF MOLECULES IN VIBRATIONALLY EXCITED STATE

We have already listed a number of works, in which vibrationally excited molecules were observed in the process of catalytic reactions. Chemisorption is a necessary stage of any catalytic transformation, and it is exothermic. The molecular excitation probability in the exothermic steps of catalytic processes depends substantially on the deactivation mechanism of the vibrational and electronic energy. Clearly, when the deactivation process is strongly hindered, the yield of excited molecules can be appreciably greater. Since the vibrational quantum of an excited molecule far exceeds the energy of solid phonons, deactivation becomes multiphonon and, hence, unlikely.[21,120,468]

Consider the excitation probability of molecular vibrations at the cost of the energy released in the adsorption process. The translational energy of a molecule can transfer to the energy of its inner vibrations only if the adsorption energy D_α is greater than $\hbar\omega_o$, where ω_o is the frequency of inner molecular vibrations. In order for the inner vibrational excitation to become possible in the case of $D_\alpha < \hbar\omega_o$, the closest surface atom (or molecule) must accumulate the vibrational energy $\hbar\omega_o >> k_BT$. We are considering a case of moderate temperatures, whereas $\hbar\omega_o$ for most diatomics is about several thousands of degrees; thus, such a situation is unlikely. In addition, the energy fluctuation may cause the desorption of an unexcited molecule; therefore, the ratio of the desorption probability w_d to the excitation probability w_e is

$$w_d/w_e \approx \exp\left[\frac{\hbar\omega_o - D_\alpha}{kT}\right] >> 1 \qquad (5.1)$$

i.e., the desorption proceeds more rapidly than the excitation. Thus, it is not unreasonable to treat the generation of vibrationally excited molecules only for sufficiently large adsorption energies.

Consider possible mechanisms of the vibrational excitation of the adsorbing molecule. Two ways may be suggested. First, a molecule excites vibrationally during the collision with the surface and immediately transits into a sufficiently deep vibrational level in the adsorption well. Then adsorbed molecules populate lower levels corresponding to the surface temperature. Second, a molecule transits into a higher level on the collision, and then can be excited during subsequent transitions between vibrational levels in the adsorption well.

The probability of adsorption in the higher level of the well is substantially greater than the probability to be adsorbed in a deep level because in the first case the energy loss is on the order of k_BT. Therefore, the experimentally measured adsorption probability w_a is approximately equal to the higher-level excitation probability w_e. We proceed for calculation of w_e and relaxation probability w_R from the assumption that the major contribution is made by the region with the most rapid change of the interaction potential, i.e., the region where repulsion forces dominate.

In the first case discussed above the probability may be calculated as the vibrational excitation probability of a molecule colliding with a wall. The total probability of molecular transition from any adsorption level is proportional to the molecular lifetime on this level. This lifetime is defined by the thermal deactivation rate, i.e., the rate of transition between close adsorption levels with the energy transfer to lattice phonons.

The total relaxation probability w, which we are interested in is defined by

$$w = \sum_{D_a + \hbar\omega_0 < E < 0} \int_0^\infty w_e(E)N(E,t)dt \tag{5.2}$$

The summation is performed over all energetic levels from which the vibrational excitation transition is possible. Here $w_e(E)$ is the transition probability from the adsorption level with energy E to the level with energy $E - \hbar\omega_o$, the inner vibrational excitation $\hbar\omega_o$ being excited simultaneously. Exact equality $n\hbar\omega = \hbar\omega_o$ cannot always be fulfilled (here ω is the vibrational frequency of an adsorbed molecule and n is an integer). However, since adsorbed levels are broadened due to the interaction with phonons, and there are also other molecular degrees of freedom (the motion parallel to the surface and rotations), the resonance may be thought of as always being reached. The transition probability weakly depends on the energy transferred to these additional degrees of freedom. The value of N(E,t) denotes the population of a level with the energy E at a time t. Time is counted off from the instant of molecular trapping in one of the higher levels. N (E,t) $\leqslant$ 1, because it refers to one particle.

The population N (E,t) obeys the kinetic equation in the energy space

$$\frac{\partial N(E,t)}{\partial t} = \bar{n}(T)(\hbar\bar{\omega})^2 \frac{\partial}{\partial E}\left[w_R(E)\frac{\partial N(E,t)}{\partial E}\right] - (\hbar\bar{\omega})\frac{\partial}{\partial E}[w_R(E)N(E,t)] \tag{5.3}$$

where $w_R(E)$ is the transition probability between two close adsorption levels with the energy E and E $-$ $\hbar\bar{\omega}$; $\bar{\omega}$ is a characteristic phonon frequency for such a process (for a simple crystal lattice $\bar{\omega} \approx \omega_D$ and ω_D is the Debye frequency; when the high-frequency optical phonons with a frequency ω_{opt} play an important part in such processes, $\bar{\omega} \approx \omega_{opt}$); $\bar{n}(T)$ is the phonon distribution function depending on the lattice temperature T

$$\bar{n}(T) = [\exp(\hbar\omega/kT) - 1]^{-1} \tag{5.4}$$

Equation 5.3 has been derived under the following assumptions: (1) the energy difference between two adjacent levels is far less than the adsorption energy D_α

$$\hbar\omega << D_\alpha \tag{5.5}$$

(since $D_\alpha > \hbar\omega_o >> \hbar\bar{\omega}$, the Inequality 5.5 is always fulfilled); (2) deactivation occurs at transitions between close levels with the creation or absorption of only one phonon. Note that we need not know N(E,t) for the calculation of w according to Equation 5.2, only $\int_0^\infty N(E,t)dt$ is necessary. Then on performing Laplace transform over time,

$$\bar{N}(E,w) \equiv \int_0^\infty N(E,t)\exp(-wt)dt \tag{5.6}$$

we are needed in $\tilde{N}(E,o)$. Therefore, the Laplace transform should be carried out in Equation 5.3. Equation 5.3 fails to be valid for the uppermost levels ($|E| < \hbar\bar{\omega}$) because of the absence of levels with E > O, and molecular desorption should be taken into account. Assuming all the coefficients to be independent of energy in this small energy region and equal to their values at E = O, we have

$$\frac{\partial N(O,t)}{\partial t} = -\tilde{w}(0)\bar{n}N(0,t) - w_R(0)N(0,t) + w_R(\bar{0})\bar{n}(\hbar\omega) \times \left[\frac{\partial N(E,t)}{\partial E}\right]_{E=O} \tag{5.7}$$

$$N(0,0) = 1 \tag{5.8}$$

where $\widetilde{w}(0)$ is a molecular desorption probability which is possible only for $E = 0$ and at $T > O$; Equation 5.8 is the initial condition.

On performing Laplace transform on Equations 5.3, 5.7, and 5.8, we have

$$\begin{aligned} w\overline{N}(E,w) &= \overline{n}(T)(\hbar\overline{\omega})^2 \frac{\partial}{\partial E}\left[w_R(E)\frac{\partial\overline{N}(E,w)}{\partial E}\right] - (\hbar\omega)\frac{\partial}{\partial E} \\ &\times [w_R(E)\overline{N}(E,w)] \\ w\overline{N}(0,w) &= 1 - \widetilde{w}(0)\overline{n}(T)\widetilde{N}(0,w) - w_R(0)\widetilde{N} + w_R(0)\overline{n}(\hbar\omega) \\ &\times \left[\frac{\partial\widetilde{N}(E,w)}{\partial E}\right]_{E=O} \end{aligned} \tag{5.9}$$

Equation 5.9 has an easy and exact solution at $T = 0$ when $\bar{n}(T) = 0$. Then

$$\widetilde{N}(E,0) = 1/w_e(E) \tag{5.10}$$

and substituting in Equation 5.2 the summation over E by the integration, we find

$$w = \frac{1}{\hbar\omega}\int_{-D_\alpha+\hbar\omega_o}^{0} \frac{w_e(E)}{w_R(E)}\,dE \tag{5.11}$$

Equation 5.9 cannot be solved analytically at $T \neq 0$. Because we seek only approximate solutions of Equation 5.9, assume $w_R(E)$ to be constant, i.e., substitute $w_R(E) \rightarrow \overline{w}_R$. Inasmuch as the energy change in the deactivation process is small $\hbar\overline{\omega} << D_\alpha$, it is evident from the formulas for transition probability with the exponential potential,[120] the dependence $w_R(E)$ is not sharp $w_R(E) \sim |D_\alpha - E|$. Besides, according to Equation 5.2, a knowledge of populations only in the interval $-D_\alpha + \hbar\omega_o < E < O$ is required. Taking into account the aforesaid, we can easily obtain

$$\widetilde{N}(E,0) \approx 1/[\widetilde{w}(0)\overline{n}(T) + \overline{w}_R] \tag{5.12}$$

In order for the passage to the limit T = O to retain validity with the same accuracy, we can write

$$\tilde{N}(E,0) \approx 1/[\tilde{w}(0)n(T) + w_R(E)] \tag{5.13}$$

Finally, the probability w at arbitrary temperatures can be expressed as

$$w = \frac{1}{\hbar\omega} \int_{-D_\alpha + \hbar\omega_o}^{0} \frac{w_e(E)}{\tilde{w}(0)n(T) + w_R(E)} \tag{5.14}$$

Now let us calculate the probability of molecular transition into the vibrationally excited state. Suppose this task to be one-dimensional, i.e., molecular motion only perpendicular to the surface is taken into account. The system Hamiltonian can be written as

$$\hat{H} = \hat{H}_{ph} + \hat{T} + V(R,r,x) + \hat{H}_m \tag{5.15}$$

where $\hat{H}_{ph}$ is the Hamiltonian of the crystal (phonon) vibrations; $\hat{H}_m$ is the Hamiltonian of the inner molecular vibrations; $\hat{T}$ the operator of kinetic energy of a molecular motion as a whole; V(R,r,x) the interaction potential between the molecule and a surface atom depending on R, the distance between the centers of the molecule, and the atom equilibrium position; r the atomic declination from the equilibrium; and x the molecular vibrational coordinate, i.e., the declination of atoms in the molecule from their equilibrium positions.

Assume the characteristic length L for the potential V (the distance at which V varies significantly) to be far greater than the ground-state vibrations of either the surface atom and molecular atoms, i.e.,

$$x/L,\ r/L << 1 \tag{5.16}$$

These relations allow the application of the perturbation theory in this task. Then Equation 5.15 with the accuracy to the first order of small parameters x/L and r/L can be written as

$$\hat{H} = \hat{H}_o + V_i \tag{5.17}$$

$$\hat{H}_o = \hat{H}_a + \hat{H}_m + T + V(R,0,0) \tag{5.18}$$

$$V_i = r \frac{\partial V(R,r,x)}{\partial r}\bigg|_{\substack{r = o \\ x = o}} + x \frac{\partial V(R,r,x)}{\partial x}\bigg|_{\substack{r = o \\ x = o}} \tag{5.19}$$

The eigenfunctions of the Hamiltonian $\hat{H}_o$ which is separable for the variables R, r, and x are known

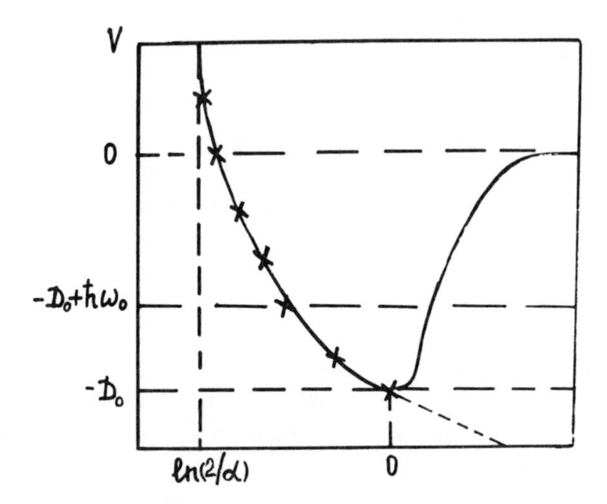

FIGURE 5.1. The scheme for exponential approximation of the repulsion branch of the Morse potential.

$$\psi = \psi_R(r)\,\psi_m(x)\,\psi(R)$$

A motion of the molecule as a whole in the R coordinate is finite if the molecular energy E_{transl} is less than zero. The molecule is moving in the potential well V(R) which, for the sake of definiteness, we shall choose as Morse potential

$$V(R) = D_\alpha[\exp(-2\alpha R) - 2\exp(-\alpha R)] \tag{5.20}$$

where the equilibrium position is at $R = 0$, and α is a constant for a given molecule.

For higher energy levels in this well the molecular motion is quasiclassic; transitions occur mainly in the region of strong repulsion. Here the true potential Equation 5.20 can be substituted by the pure exponent with good accuracy (Figure 5.1). The transition probabilities for this potential are well known,[120] and the relation $w_e(E)/w_R(E)$ can be easily obtained.

So, we write

$$V_1(R) = U_1 \exp(-\beta R) - U_2 \tag{5.21}$$

where parameters U_1, U_2, and β are defined by the following conditions

$$V_1(R) = V(R);\ dV_1/dR = dV/dR;\ V_1(R') = V(R')$$

Here $R = -\ln 2/\alpha$, and R' is a classic turning point at $E_{const} = -D_\alpha + h\omega_o$. A good approximation for $V_1(R)$ in a rather wide range of $\hbar\omega_o/D_\alpha$ (from

0.2 to 0.7) is obtained at $\beta = 3.5\alpha$, $U_2 = D_\alpha$, $U_1 \approx 2^{-3,5} D_\alpha$. The potential (Equation 5.19) which determines the transition probabilities on substitution V(R) by $V_1(R)$ has the form

$$V_1 = AV_1(R)\beta r + BV_1(R)\beta x \tag{5.22}$$

where A and B are coefficients on the order of unity which depend on molecule-to-surface orientation and on the mass ratio.

Then we have

$$w_R(E) = \frac{2\pi}{h}\left| <\Psi_{\Phi_1(r)} A\beta r \middle| \Psi_{\Phi_o(r)}> \right|^2 \left| <\varphi_E - \hbar\bar{\omega}(R) \right|$$

$$\times\ V_1(R)|\varphi_E(R)>|\rho^2(E) \tag{5.23}$$

where $\Psi_{\Phi_o(r)}$ $\Psi_{\Phi(r)}$ are vibrational wavefunctions of a surface atom corresponding to the zeroth and first levels of the crystalline oscillator; $\varphi_E(R)$ is a finite translational wavefunction of the molecular translational motion which differs from the infinite one only by the normalization factor depending on ω, the frequency of molecular vibrations as a whole; and $\rho(E)$ is the density of states involving both the phonon density of states and the molecular one.

The probability $w_e(E)$ can be represented in a similar way, but $\Psi_o(r)$ should be substituted by $\varphi_m(x)$. Strictly speaking, $\rho(E)$ is not constant, but we neglect this change. Now we take advantage of the formula for the relaxation transition probability

$$w_R \sim <r>^2 \left| \frac{1/4\pi^2(q_i^2 - q_f^2)}{ch\pi q_i - ch\pi q_f} \right|^2 sh\pi q_i sh\pi q_f \tag{5.24}$$

where $q = \sqrt{8m/(D_\alpha - E)}/(\hbar\beta)$; m is molecular mass; i and f indexes refer to the initial and final molecular energies; and $\langle r \rangle$ is the amplitude of zeroth atomic vibrations or the inner molecular one.

Taking into account that $\hbar\bar{\omega} << (D_\alpha - E)$ and $\pi\sqrt{8m(D_\alpha - \hbar\omega_o)}/(\hbar\beta) \geqslant 1$, and that $\hbar\bar{\omega}_o/D_\alpha \leqslant 2$ as well, for the ratio w_e/w_R we derive

$$\frac{w_e}{w_R} = \frac{<x_o>}{<r_o>^2} x^2 e^{-x} \tag{5.25}$$

$$x = \frac{\pi\sqrt{2m\hbar\omega_o}}{\hbar\beta}\left[\frac{\hbar\omega_o}{D_\alpha}\right]^{1/2} \tag{5.26}$$

Substituting Equation 5.25 in Equation 5.13 and taking into account that $\bar{w} \approx w_R$, we obtain the following expression for molecular excitation probability during adsorption

$$w \approx \frac{<x_o>^2}{<r_o>^2} \frac{D_\alpha}{\hbar\bar{\omega}} x^2 e^{-x} \frac{1}{1 + \bar{n}(T)} = \frac{m_A}{\mu} \frac{D_\alpha}{\hbar\omega_o} x^2 e^{-x} \frac{1}{1 + \bar{n}(T)} \tag{5.27}$$

where m_A is the crystal atom mass and μ is the reduced molecular mass.

Special attention must be given to a very interesting isotopic effect which is evident from the analysis of Equation 5.27. According to Equation 5.26, x mainly depends on the molecular mass. Nevertheless, the value $m\hbar\omega_o$ involved in Equation 5.26 is invariant relative to the isotope substitution of both atoms in the molecule: $x(H_2) = x(D_2)$, but $x(HD) = 1.06x(H_2)$. Therefore, $w(HD) < w(H_2) = w(D_2)$. The inverse relationship takes place for the rotational excitation.

The probability of formation of the vibrationally excited CO molecule on a platinum surface estimated using our data[189,470] has a minimal value 10^{-5}. The adsorption lifetime of such a molecule estimated with the same data cannot be less than 10^{-5} s. It should be noted that these minimal evaluations of w and τ are strongly dependent on the value of parameter α. To make more accurate calculations, experimental results on molecular beam scattering at surfaces are required. This enables us to restore the true interaction potential.

The lifetimes of vibrationally excited impurities in solids have been the objective of much research. Jortner et al.[471-473] used the Morse potential to the model molecular potential well. Omitting details, the relaxation transition probability to the first order of the perturbation theory over the nth anharmonic term has been considered in these works (if we suppose $\omega_o = n\bar{\omega}$, where $\bar{\omega}$ is a characteristic phonon frequency).

But this multiphonon approach may include as well anharmonics of lower orders (cubic or fourth power) taking into account higher orders of the perturbation theory. The question of which contribution to n-multiphonon transition is more important — that of the corresponding n-power anharmonic term to the first order of the perturbation theory or the lower anharmonic terms to higher orders — is rather complicated and is substantially determined by the bond character of the impurity of the molecule with the crystal. The stronger the bond is, the more probable the second case is. But only the case of strong bonding multiphonon transitions due to anharmonism can give the main contribution to vibrational deactivation of impurity of the molecule. Therefore, the application of these formulas[471-473] to the problem under discussion seems to be questionable.

For the adsorbed molecule this mechanism is of importance only in the case of chemisorption with a large binding energy compared with binding energies of atoms in a solid. But even in this case, if a molecular weight is substantially less than that of crystal atoms, the inner molecular vibration couples mainly with the local vibration of the molecule as a whole. The inner molecular vibrational quantum will transfer to the vibration of the molecule as a whole. As we have already shown above, the region with dominantly repulsive forces is of first importance.[474]

The deactivation probability per unit time w'_R can be represented as the transition probability in one collision $w_{1\text{-}0}$ multiplied by the number of collisions per second, i.e., the molecular vibrational frequency ν_o in the adsorption well. The density of states for translational molecular motion is assumed to be the same as for a quasicontinuous spectrum

$$w'_R = \nu_o w_{1-o} \tag{5.28}$$

$$w_{1-o} \approx \frac{8\pi^2}{h^2}\frac{m^2}{\mu(\alpha')^2}\hbar\omega_o \exp\left(-\pi\frac{2\sqrt{2m\hbar\omega_o}}{\alpha'\hbar}\right) \tag{5.29}$$

where m is the molecular mass, α' the coefficient in the exponential potential $U' = U_1\exp(-\alpha' r)$, and $\alpha' \approx 2\alpha$. The latter relationship is valid near the bottom of the well and differs from the value of α' at energies close to the dissociation energy as we discussed the vibrational excitation of the molecule in the adsorption process. In Equation 5.29 it is also supposed that the molecular mass is significantly less than the mass of a crystal atom, and that $\hbar\omega_o >> 2\pi h\nu_o$. Due to this fact, Equation 5.29 does not include the initial energy of the molecule in one of the low-lying energy levels. On substitution of ν_o by corresponding characteristics of the adsorption potential we have

$$w'_R = \pi\sqrt{D_\alpha}\,\frac{1}{2}\,\frac{m^{3/2}\omega_o}{\mu\hbar}\exp\left(\frac{-\pi\sqrt{2m\hbar\omega_o}}{2\hbar}\right) \tag{5.30}$$

It is obvious from Equation 5.30 that the exponent exhibits a weak isotopic dependence

$$\sqrt{\omega_o m} = \left[\sqrt{\frac{m_1 + m_2}{m_1 m_2}}(m_1 + m_2)\right]^{1/2} = \frac{(m_1 + m_2)^{3/4}}{(m_1 m_2)^{1/4}} \tag{5.31}$$

It is worth noting that large values for the excitation probability and the lifetime of a molecule in the vibrationally excited state, which we have derived above, are mainly connected with the neglect of other nonphonon deactivation mechanisms. For example, the electronic mechanism is shown to be more important for deactivation on metals;[101] hence, the probability of molecular vibrational excitation there must be appreciably lower.

5.2. EXPERIMENTAL LIFETIMES OF VIBRATIONALLY EXCITED MOLECULES ON SURFACES

We have already mentioned in Section 2.2.2 the difficulties encountered in determining the lifetimes of vibrationally excited molecules and surface vibrational excitations using the line widths of IR or HREELS spectra. The line width $\Delta\nu$ in the vibrational spectrum is inversely proportional to the excitation lifetime τ. Strictly speaking, the value of τ is defined by two terms

$$\Delta\nu \sim \frac{1}{\tau} = \frac{1}{2T_1} + \frac{1}{T_2} \quad (5.32)$$

where T_1 is the characteristic time for the reverse process and T_2 is the vibrational dephasing time, i.e., the characteristic time for elastic scattering of the molecular vibrational quantum by solid phonon which interrupts the vibrational phase without energy transfer. The factor of 2 is due to inconsistency in definitions: T_2 is defined relative to the amplitude of vibrations, and T_1 to the population, i.e., to the square of the amplitude. Thus, the line width measured experimentally is connected with the upper T_1 limit. If the dephasing dominates, τ can be greater than $1/\Delta\nu$. Sometimes the information about relative contributions of T_1 and T_2 can be taken from the temperature dependence: T_2 tends to infinity as the temperature approaches absolute zero.[475]

There are several theoretical works in which attempts to account for all possible broadening effects have been undertaken. Tobin[476] performed a detailed analysis of vibrational line widths $\Delta\nu$ for adsorbed molecules. As noted previously, the radiation decay of the vibrational excitation has a very small probability $\Delta\nu_{rad} >> \Delta\nu_{obs}$. Phonon relaxation has also a weak effect on the experimental line width. Persson[477] has shown that the phonon relaxation and vibrational dephasing can give rise to an IR line width value $\Delta\nu \sim 10\ cm^{-1}$ rather close to experimental. For example, for CO on Ni(100) and Ni(111) the line width $\Delta\nu$ is found experimentally to be $10\ cm^{-1}$, which is consistent with the theory of two-phonon quenching. For higher vibrational frequencies the lifetime τ grows exponentially with the growth of the phonon number. Electron-hole-pair excitation makes a significant contribution to $\Delta\nu$ and τ on metals and semiconductors. Dipole-dipole broadening plays an important part on metals, which arises on account of the change of positions of vibrationally excited atoms.

Sutsu et al.,[476] with the aid of IR diode laser, measured $\Delta\nu$ of the C-O valence vibration for monolayer CO coverage on Pt(111): $\Delta\nu = 2.3\ cm^{-1}$ at 100 K. The temperature dependence of the line width allows us to determine $\Delta\nu$ due to the excitation of the electron-hole pair, $2.0 \pm 0.4\ cm^{-1}$, which corresponds to the lifetime of 2.7 ± 0.6 ps. To determine the characteristic time of the excitation transfer from the adsorbed molecule to the electron-hole pair, Persson et al.[478] suggested measuring the resistance of a thin metal film. Their results were in agreement with experimental $\Delta\nu$ values for chemisorbed molecules: $1.4\ 10^{-11}$ s for CO/Ni; $4.6\ 10^{-11}$ s for N_2/Ni; and $6.9\ 10^{-12}$ s for O/Cu. For physisorbed molecules the lifetimes are somewhat greater: $3.6\ 10^{-10}$ s for CO/Ag; $3.2\ 10^{-10}$ s for C_2H_4/Ag; $3.6\ 10^{-9}$ s for C_2H_6/Ag.

New possibilities for determining the lifetimes of vibrational excitation on the surface using absorption line widths of vibrational spectra of adsorbed particles have appeared in connection with the development of SEW-spectroscopy (see Section 2.2.2). A striking result has been obtained in our work[479] in studies of SEW absorption with different radiation sources. Taking globar as a radiation source for the band 944 cm^{-1} which corresponds to Al-OH vibration in the film of Al_2O_3, we observed $\Delta\nu = 100$ cm^{-1}; for the laser source $\Delta\nu = 50$ cm^{-1} at power 0.1 W, and $\Delta\nu = 20$ cm^{-1} at 3 W. Special studies revealed that absorption vs. power dependence is linear in the band center, but it is substantially nonlinear for band edges at 948 and 960 cm^{-1}, the deviation from the linear law disappearing at lower powers. Similar results were obtained in experiments on laser reflection-absorption spectroscopy. Measurements showed that at high powers the bandwidth is inversely proportional to the source power. Because the observed nonlinearity could not be explained in the framework of transition saturation, experiments on ''hole burning'' in the band under investigation would be worthwhile. It was shown that at the band center such a burning is not observed, but at 948 and 940 cm^{-1} irregular changes of transmission were obtained. Using a probe, weak laser revealed decreasing absorption in the close vicinity of the frequency of the general laser, but further tuning away from it to the band center resulted in the reverse effect.

A model is suggested which explains all the observed effects. It is based on the nonlinear light absorption by large collectives of adsorbed molecules. The derived equations enable us to find T_1 and T_2, i.e., the times of vibrational and phase relaxation, respectively, using the width and the depth of the burned hole.

The number of direct measurements of the lifetimes of vibrational excitations on surfaces using a picosecond laser technique is rapidly increasing.

The lifetimes of vibrationally excited OH-groups on SiO_2[480,481] mica,[482] and zeolite[483] were studied. Vibrational transition in the OH-group was excited using an IR pulse of the neodymium laser and detected using the LIF method (see Section 2.2.1). Approximately 10% of hydroxyls are excited into the $v = 1$ state. The lifetime τ of an excited, isolated OH-group is 204 ps on SiO_2 at 293 K ($\nu \approx 3660$ cm^{-1}). The same lifetime in CCl_4 solution is 159 ps, and 87 ps in benzene. The value of τ for OH-groups in the bulk of fused silica is 109 ps, and in the bulk of zeolite ZSM-5 is 140 ps. For physically adsorbed water ($\nu = 3400$ cm^{-1}, $5H_2O$ molecules per 1 nm^2) the value of τ is 56 ns. This value of τ corresponds to the IR line width of 0.03 cm^{-1}. The observed line width of the OH group is 60 cm^{-1}, i.e., 2000 times greater than is attributed to the surface inhomogeneity and dephasing phenomena.

Close values 80 to 250 ns were obtained for excited states of OH groups on mica. OH-groups with higher frequencies have longer lifetimes. The most probable schemes of relaxation are converting of $\nu = 1$ quantum of O-H stretching vibration (~3600 cm^{-1}) into vibrations of adjacent SiO_4 tetrahedrons with simultaneous creation of 3 to 5 quanta of valence Si-OH vibrations (800 to 1060 cm^{-1}) or bending Si-O-Si vibrations (~400 cm^{-1}).[483]

Similar results were obtained for CO-group vibrations on surfaces.[484-487] The transition in the valence band (2096 cm^{-1}) was excited by a picosecond laser. The observed lifetime of the CO-group in the $Rh_2(CO)_4Cl_2$-complex supported by SiO_2 is 180 ns which corresponds to energy changing into 4 or 5 vibrations inside the complex itself or energy transfer into the support, for instance, into Si-O vibrations. The lifetimes of CO molecules adsorbed on supported Rh, Pt, or Pd were 6 to 8 times shorter. Another deactivation mechanism seems to be valid in this case, namely, the energy transfer into electronic excitation of metal. The line 2106 cm^{-1} of the CO linear form adsorbed on Pt(111) was pumped with a 0.7-ps saturating laser pulse at ν = 2107 cm^{-1}. It returned to the equilibrium shape in 3 ps. The latter value is dependent on Θ_{CO} and temperature at 150 to 300 K.[487]

Harris et al.[488,489] employed IR picosecond pulses for excitation and sum frequency spectroscopy for measuring τ (see Section 2.2.2). These authors found two characteristic relaxation times for excited C-H vibrations of $(CH_3)_2S$ adsorbed on a Ag(111) surface. The fast one had τ = 2.5 to 3 ps independent of temperature in the range 110 to 380 K; the longer time had τ = 55 ps at 380 K and 90 ps at 110 K. Both time constants were attributed to transfer to other vibrational modes of the molecule. The calculated τ value for electron-hole pair excitation is approximately 2 orders shorter.[488] For CO on Cu(100) it was measured that τ_1 = 3 ps and τ_2 = 0.9 ps.[489] Studies of hydrogen adsorption on Si(111) using the same method revealed that the lifetime of the Si-H vibration is 0.8 ns.[490]

Chang and Ewing[491] applied another direct method of observation, i.e., laser-inducted IR luminescence of adsorbed layers. An extremely long lifetime value of 4.3 ms has been found for CO physisorbed on NaCl(100) at 22 K which is indicative of a very weak energy exchange with the surface.

Vach[492] has measured the vibrational relaxation time of NO(v = 1) adsorbed on graphite using the molecular beam scattering technique. The excitation lifetime was found to be of the same order as the lifetime in the adsorbed state. On the basis of measured lifetimes of vibrational excitations on surface (>10^{-10} s), Misevich et al.[481] suggested that excited particles could participate in catalytic reactions.

5.3. EXPERIMENTAL LIFETIMES OF ELECTRONICALLY EXCITED MOLECULES ON SURFACES

Few data on the lifetimes of electronically excited molecules on surface are available as well as on vibrational excitations. Now such studies are only starting.

Avouris and Persson[483] reviewed the investigations concerning the lifetimes of electronic excitations on surfaces. Mainly physically adsorbed molecules were examined. For example, nitrogen molecules, physically adsorbed on Al(111), were excited by electron impact into the $C^3\Pi_g$ state. The lifetime was determined using HREELS to be $5 \cdot 10^{-15}$ s. A rapid exchange with the formation of the electron-hole pair in metal occurs. The measured lifetime τ is approximately 10^6 times shorter than the corresponding value for $N_2(C^3\Pi_g)$ in the gas phase.

A similar short lifetime was registered for the electronically excited state of pyrazine on Ag(111). In the first adsorbed layer the value of τ is equal to $5 \cdot 10^{-15}$ s, and in the second $3 \cdot 10^{-4}$ s. Pyrazine excitation happens through electron interband transition from the 5p zone of Ag into the low-lying vacant π^*-orbital of pyrazine (transition energy is 4 eV).

An elegant experimental technique was utilized to measure the dependence of the lifetime of electronically excited molecules on the distance d from the surface.[494] Fatty acids were preliminarily adsorbed on a Ni surface and pyrazine was adsorbed in the top third layer. When $d \rightarrow \infty$, $\tau \rightarrow \tau_o$, where τ_o is the lifetime of a free pyrazine molecule. At intermediate d, oscillations of τ were observed owing to the interference between incident and reflected light waves. Very short lifetimes were found at small distances due to direct energy exchange with metal electrons.

Campion et al.[495] improved this method and applied three-layer composition: Ni surface, a layer of physically adsorbed Ar atoms with varied thickness, and then the fluorescent dye (pyrazine). In the range of d from 0.8 to 10 nm the radiation intensity (the lifetime to the electronically excited state) increased proportionally to d^3. In another work[496] the dye molecules were separated from silver by a layer of SiO_2.

The lifetimes of various electronically excited aromatic molecules adsorbed physically on metal surfaces are of the order of 10^{-15} s.[497]. Electron transfer from the excited adsorbate to the metal conduction band seems to be the prevailing mechanism of their quenching.[493-496,498] Electron transfer between adsorbed aromatic molecules up to 50 nm is considered by Waldeck et al.[499]

In these examples the lifetimes of electronically excited molecules on the surface are extremely short, excluding their possible participation in catalytic processes. For dielectric surfaces, for instance, on glass surface, the lifetime is longer, as was shown by picosecond laser experiments. In dispersed CdS, TiO_2, Cd_3P_2, ZnS, and Fe_2O_3 excitation creates an electron-hole pair in the forbidden zone which rapidly migrates to the surface ($\tau << 10^{-9}$ s). Recombination of this pair on the surface results in luminescence. Electronically excited pyrene adsorbed on Al_2O_3 has two characteristic quenching times: 100 ns and a much shorter one. Excited aminopyrene adsorbed on isolated OH-groups of SiO_2 has a lifetime of the order of 100 ns at $\lambda \approx 420$ nm, and adsorbed heminal (coupled) OH-groups also 100 ns at $\lambda = 370$ to 400 nm.[500]

Substantially more long-lived electronically excited molecules are known as well. For example, the lifetime of the excited NO^*_2 molecule created in the reaction NO + O over a Ni catalyst was 79 μs, which is close to the radiational lifetime. Adsorbed singlet oxygen seems to be even more long-lived. The low-energy spectrum of electrons scattered at the Pt(321) surface has been attributed to the formation of $^1\Delta_g$ or $^1\Sigma_g^+$ states of the O_2 molecule as a result of electron bombardment.[501]

Chapter 6

PRECURSOR AND NONEQUILIBRIUM DIFFUSION

We have shown in previous chapters that there are two basic mechanisms of excited particle deactivation in adsorption and catalysis

1. Nonelastic scattering and direct adsorption with the impact, or Eley-Riedel mechanism in catalysis, corresponding to them
2. The interaction of a trapping-desorption type and indirect adsorption with the surface interaction, or Langmuir-Hinshelwood mechanism corresponding to them

Many researchers postulate for the processes of the second type the existence of the precursor — some intermediate preadsorbed state on the surface. This state can migrate over the surface. Therefore, the study of the role of surface diffusion in adsorption and catalysis is important.

6.1. THE NOTION OF A PRECURSOR IN ADSORPTION AND CATALYSIS

The notion about a precursor and its role in chemical adsorption goes back to the pioneer work of Langmuir, and in catalysis to the work of Roginsky and Zeldovitsch (see Chapter 1). However, until recently there were only a few works in which the existence of the precursor was proved directly. The last conference on this subject had the subtitle "Precursor: Myth or Reality?"[285]

Let us consider in more detail the preadsorbed state. For a long time the experimental dependence of the sticking coefficients during adsorption on coverage was the main argument in favor of existence of the precursor. If adsorption would proceed in accordance with the Langmuir isotherm via successive statistical population of adsorption centers, then the linear dependence of s on $1 - \Theta$ would be observed. Actually, as shown in Figure 6.1 and Table 6.1, s is practically independent of $1 - \Theta$ over a wide range of Θ values.[502] It is apparent that the precursor is formed, i.e., the preadsorbed state of the molecule, when it can move freely over the whole surface including the sites occupied by previously chemisorbed molecules, until strong adsorption or a catalytic reaction occurs. Later on, that precursor, which can exist over occupied adsorption sites, is called "extrinsic". A precursor that exists only on free sites of adsorbent is called the "intrinsic" precursor.

Note that the independence of s on Θ is typical for the pressure only in the case of fast chemical adsorption. Chambers and Ehrlich[503] indicated that in the case of slow activated adsorption, the dependence of s on Θ is observed neither in the presence nor in the absence of the precursor.

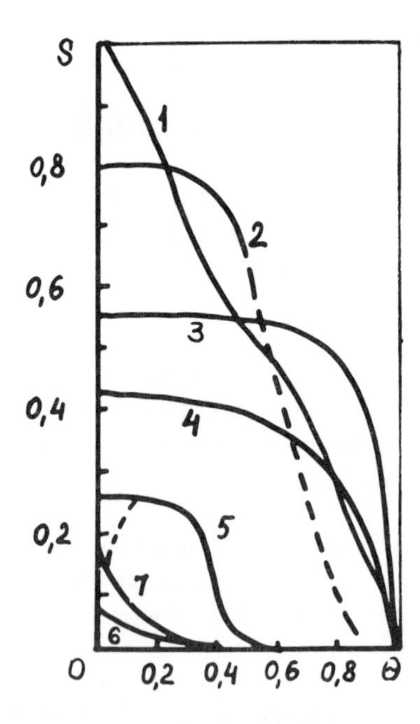

FIGURE 6.1. Sticking coefficient s as a function of coverage Θ for different gases on metals.[502] (1) O_2/W(100); (2) O_2/W(211); (3) CO/W(110); (4) N_2/W(100); (5) O_2/W(110); (6) N_2/W(110); (7) O_2/Pt(111).

The energy scheme of adsorption through the precursor is shown in Figure 6.2. The molecular transition from the precursor state to a strongly bound chemisorbed one can have the activation energy E_o.

There are two main reasons for the nature of the precursor: (1) the precursor is a physisorbed state of a molecule and (2) the precursor is in nonequilibrium, probably, in the vibrationally excited state.

Many authors assign the formation of a preadsorbed state to physical adsorption. There is strong experimental evidence for the precursor to be a physically adsorbed molecule. At low temperature it can be observed using the methods of work function or XPS. For example, Norton et al.,[504] using XPS, observed physically adsorbed CO on the Cu surface at 20 K. At 30 K approximately 40% of CO was desorbed, and 60% transformed into the chemisorbed state, with was stable above 100 K.

By means of photoelectron spectroscopy, analogous transformation of O_2 physisorbed on an Al(111) into an oxide layer was observed by Schmeisser et al.[505] Grunze[506] observed the existence of a physisorbed precursor in the process of chemisorption of Na on Ni(100), Ru(001), and W(100). In these cases physical adsorption is a precursor for which the activation energy of chemical adsorption is lower than Ea from the gas phase.

However, these studies are indicative of the nonequilibrium nature of the precursor. For example, in the temperature range 300 to 500 K the lifetime

TABLE 6.1
Initial Sticking Coefficients for Gases on Metal Surfaces

Adsorbate	Adsorbent	s_o	Coverage dependence
H_2	W(100)	0.18	p*
	W(100)	0.07	$1 - \Theta$
	W(111)	0.24	p
	Mo(110)	0.10	$1 - \Theta$
	Ta(110)	0.18	p
D_2	W(100)	0.25	p
	W(110)	0.7	$1 - \Theta$
N_2	W(100)	0.41	p
	W(110)	0.004	$1 - \Theta$
	Mo(110)	0.09	$1 - \Theta$
CO	W(100)	0.49	p
	W(110)	0.55	p
O_2	W(100)	1.0	$\sim(1 - \Theta)$
	W(110)	0.34	p
	W(211)	0.8	p
	Pt(111)	0.1	$\exp(-m\Theta)$
CO_2	W(100)	0.05	p

Note: * Indicates precursor.

From Steinbruchel, C. S. and Schmidt, L. D., *Phys. Rev. Lett.*, B10, 4209, 1974. With permission.

τ of physisorbed H_2, N_2, and other molecules calculated using the Frenkel formula $\tau = \tau_o \exp(Q/RT)$ is close to τ_o, whereas experimental lifetimes of the preadsorbed state are longer: it is known that a molecule captured by the surface can make tens and hundreds of elementary jumps until it is strongly chemisorbed. Furthermore, as seen in Table 6.1, a different dependence of $s(\Theta)$ is observed on different faces of the same monocrystal, which is unlikely for physisorption.

Certain experiments make us suppose that molecules in preadsorbed state are likely to be vibrationally or electronically excited. An adsorbed excited molecule successively loses its energy in jumps on a surface with the following chemical adsorption. According to the principle of microscopic reversibility, precursor formation during desorption is to be expected as well. As shown in experiments on metals,[502] the ratio of the rate constant of strong chemisorption k_α and that of desorption from the preadsorbed state k_d^* is given by the expression

$$K = \frac{k_\alpha}{k_d^*} = \frac{k_\alpha^{(o)}}{k_d^{*(o)}} \exp[(E_d^* - E_\alpha)/RT] \tag{6.1}$$

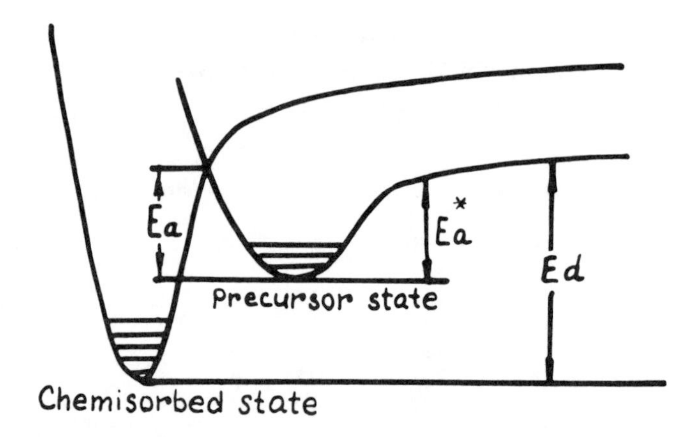

FIGURE 6.2. Potential curves for formation of preadsorbed state in chemisorption.

where K increases with the temperature growth, i.e., $E_d^* < E_\alpha$. In other words, the energy barrier between preadsorbed and chemisorbed states is greater than the adsorption heat in the precursor state (see Figure 6.2).

According to Tully,[507] in the case of Xe adsorption on Pt(111) at 779 K the molecular precursor with a short lifetime is formed not in the physisorbed but in the chemisorbed state. All Xe atoms with initial temperature T_i = 1950 K accommodate to the surface temperature T_s (for the motion perpendicular to the surface) during 20 ps (approximately 10 jumps on the surface). For the motion parallel to the surface it occurs approximately during 100 ps.

Molecular beam experiments elucidate the role of the precursor in adsorption (Figure 6.3). In most cases, s depends weakly on the angle of incidence φ and slightly increases with the increase of φ. A molecule is adsorbed if, on collision, its normal velocity is less than that required to overcome the potential barrier: $v_n < (2E_d^*/m)^{1/2}$, where m is the molecular mass and E_d^* is the activation energy for desorption from the precursor state (see Figure 6.2). Steinbruchel and Schmidt[508] proposed a theory that takes into account vibrational excitation of the solid. They obtained good agreement with the experiment[502] for s vs. the angle of incidence φ_i; s increases with the increase of φ_i and decrease of the normal velocity component. The decrease of s with increasing φ for H_2 and D_2 on W(110) (see Figure 6.3) was attributed to the activation barrier for chemisorption $E_\alpha > E_d^*$ (see Figure 6.2).

Molecular beam experiments give other evidence for precursor existence, i.e., the decrease of s with increasing translational energy of molecules E_d in the beam. It is observed for low translational energies $E_i < 10$ kJ/mol at low surface temperatures T_s.[285,287,326,509-513] We have already mentioned this fact in Chapter 4 (see, for example, Figure 4.4, where the data on O_2 adsorption on W are presented; for high E_i the dependence of s on T_i is typical for direct adsorption. For adsorption with a precursor s depends on T_s. It is

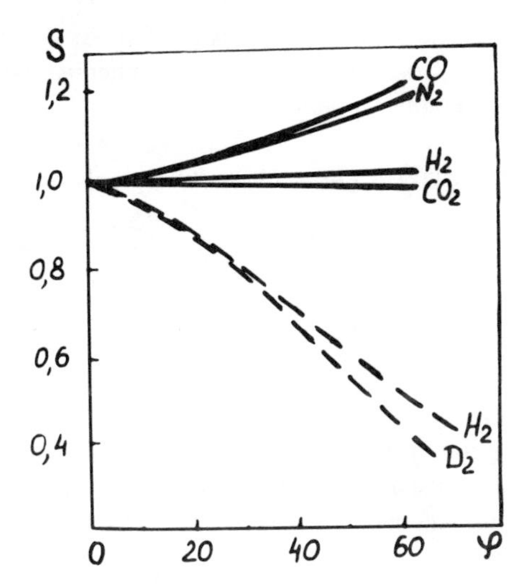

FIGURE 6.3. The sticking coefficient s (normalized values) as a function of the incidence angle φ_i for molecular beams of different gases on W(110) (solid line) and W(100) (dashed line). (From Steinbruchel, C. S. and Schmidt, L. D., *Phys. Rev. Lett.*, B10, 4209, 1974. With permission.)

also worth mentioning that low initial sticking coefficients $s_o << 1$ are characteristic of the kinetics of adsorption through the precursor.

A large number of studies have been devoted to the theory of the precursor. Doren and Tully[513] carried out trajectory calculations on the basis of a complicated model for nondissociative adsorption involving multidimensional potential surfaces, surface vibrations, interaction between different molecular and solid modes, etc. As a result, a strong interaction between translational and rotational molecular degrees of freedom has been observed. The model revealed a rather high diffusion rate of the precursor over the surface. Strong, rotational polarization of the molecule desorbing from the precursor state is one more characteristic property of the precursor.

According to these authors,[513] a precursor state is a secondary local minimum in the potential of mean force located on the reaction coordinate between the chemisorbed state and the asymptotic state of infinite separation between the molecule and the surface in the framework of the multidimensional model.

One of the first kinetic adsorption models for a molecule A with a precursor A_p was proposed by Ehrlich[514]

$$A_{gas} \underset{k_d^*}{\overset{s^*}{\rightleftarrows}} A_p \xrightarrow{k_\alpha W(\Theta)} A_{chem}$$

where A_{chem} is the chemisorbed state and $k_\alpha W(\Theta)$ is the rate constant of transition from the precursor to the chemisorbed state, which is proportional to the probability $\omega(\Theta)$ to find a vacancy in the chemisorbed layer.

In the steady-state approximation,

$$s = k_\alpha w(\Theta)n^* = s_o (1 + K)/[1 + K/\omega(\Theta)] \quad (6.2)$$

where

$$K = k_d^*/k_\alpha$$

$$s_o = s^*/(1 + K)$$

Kisliuk suggested a similar model involving diffusion of molecules above the surface with the rate constant k_{dif} for transition from one center to another.

According to his scheme,

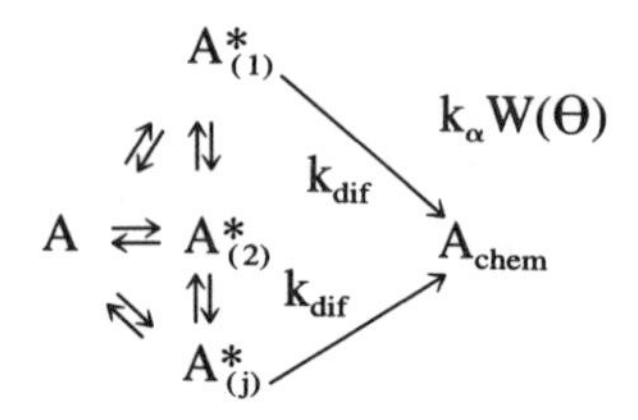

where A_1^*, A_2 . . . A_j^* are precursors, absorbed on different sites, and k_{dif} are diffusion rate constants, assuming that a concentration of precursor is negligible in comparison with a concentration of chemisorbed and vacant sites, and is given by the expression

$$s = \sum_{i=1}^{\omega} w_{\alpha i} = s_o/[1 + k'(\Theta)/w(\Theta)] \quad (6.3)$$

where k' is the ratio of the probabilities of a precursor conversion and desorption.

Equations 6.2 and 6.3 become identical for $W(\Theta) = 1 - \Theta$. As is shown in a number of studies concerning adsorption and desorption kinetics with a precursor, the precursor existence gives rise to underestimated k_o values for desorption and the so-called compensation effect.[516,517] Abnormally high reaction rates in excess of the number of collisions between incident and adsorbed reagents has been revealed in some cases on pure metals. For example, Monroe and Merrill[518] observed the reaction of incident hydrogen with adsorbed oxygen on Pt(111) with a cross-section of ~4 nm^2, i.e., each H_2 molecule that falls on this area reacts with an adsorbed oxygen atom. The corresponding value for CO is 0.8 nm^2. These phenomena were attributed to diffusion of a mobile H_2 and CO precursor.

Gland and Korchak[519] have shown that NH_3 oxidation occurs in atoms at the edge of a stepped Pt surface. But the rate of NH_3 oxidation on these atoms was approximately an order of magnitude higher than that of a number of molecular collisions with these atoms at the edge. It was concluded that oxygen is in the state of a mobile precursor on the Pt surface, with diffusion to atoms at the edge.

According to the scheme of Harris et al.,[510] hydrogen oxidation on Pt occurs via an excited precursor H*

$$H_2 \longrightarrow H^*(ads.) + H(ads.)$$

$$H^*(ads.) + O(ads.) \longrightarrow OH(ads.)$$

$$OH(ads.) + H^*(ads.) \longrightarrow H_2O$$

This conclusion is apparent from the analysis of kinetic data. The following experiments give experimental evidence of the existence of a precursor in this reaction

1. The presence of the "hot" reaction products (estimations for H_2O give the energy up to 1 eV)
2. The absence of a time delay between the H_2 impact and H_2O desorption ($\tau \leqslant 10^{-9}$ s)
3. Isotope effect for H_2/D_2
4. The presence of OH species on the surface
5. The absence of temperature dependence for this reaction (especially at low temperatures)

Of most importance is the observation of the precursor on the surface using direct physical methods.

Finally, let us answer the question "Precursor: myth or reality?" We can state with assurance that the precursor is a reality. As to its nature, there are apparently two opinions: in some cases it might be a physically absorbed molecule; in others, a short-lived excited state.

6.2. MIGRATION OF ADSORBED PARTICLES

The role of energy activation and relaxation of adsorbed particles becomes clearer when considering the mechanism of the surface migration and kinetics of surface processes. Indeed, an adsorbed atom is localized in the potential well. To move over the surface it requires an activation energy from lattice phonons. On gaining sufficient energy, the adsorbed atom can migrate over the surface, its displacement distance depending on the relaxation time. As will be shown, the latter defines the kinetics of the process (for example, atomic recombination).

We have analyzed on the microscopic level the kinetics of the simplest catalytic reaction on the ideal monocrystal surface at small coverages. Such an approach significantly simplifies all calculations because in this case the surface can be treated as homogeneous, and adsorbed particles as noninteracting.[520-522]

It is reasonable to suppose that the simplest surface reaction involves at least two stages: the stage of reagent transport and the stage of their interaction. Here the term "transport" refers to the process of surface migration of one reagent over the surface either to the second one or to the surface reaction site, but not to the transport of one reagent to the surface from the gas phase.

Let us discuss one more problem that is of importance for the kinetics of surface reactions, namely, the applicability of the mass action law (or surface action). It is well known that this law refers to local concentrations, which are commonly substituted by their mean values, i.e., the total number of adsorbed species is divided by the total number of sites. Thus, the mean concentration of adsorbed particles is calculated. In gas and liquid phases, when the role of fluctuation is negligible, all kinetic characteristics in the gas or liquid phase are usually averaged due to chaotic molecular motion. In this case, the local concentrations can be substituted by their mean values. But the validity of that assumption relative to the surface process remains to be proved because it is well known that surface migration, as a rule, requires appreciable activation energies, and particles cannot move freely on the surface.

It should be noted, that most of the adsorption isotherms do not take into account the surface migration. Even the Langmuir isotherm "permits" the particles to move on the surface only via the gas phase, in accordance with the adsorption equilibrium rule. However, experiments show that the activation energy of the surface migration is substantially lower than that of desorption. Therefore, let us consider the surface diffusion directly rather than in terms of adsorption-desorption stages.

6.2.1. EXPERIMENTAL COEFFICIENTS OF SURFACE DIFFUSION

The importance of the surface has long been understood and wide experimental information on this subject is available. Systematic investigation of the diffusion of chemisorbed atoms has been started in the works of Taylor and Langmuir.[2] They examined the surface diffusion of Cs atoms on a vanadium surface back in 1932. These authors reported that for the coverage of $\Theta = 0.03$ the preexponential factor of the diffusion coefficients is equal to 0.2 cm^3/s, and the activation energy of the migration $E_m = 60$ kJ/mol which is about 1/5 of the value required for desorption. It was this work that suggested the idea of adatom migration over the surface in the adsorbed state due to substantial difference in activation energies for diffusion and adsorption.

An impressive experimental progress in this area has been achieved with the appearance of the field emission microscope (FEM) for studying surface

TABLE 6.2
The Activation Energies for Diffusion of Chemisorbed Atoms[528]

System	E_m (kJ/mol)	E_d/E_m	System	E_m (kJ/mol)	E_d/E_m
O/W	126 ± 6	4.3	H/Ni	29 ± 4	9.2
N/W	147	4.4	H/Pt	19	13.4
H/W	67 ± 13	4.3	O/Pt	143	2.5

diffusion.[523] Gomer et al.[524-527] investigated a variety of the adsorbent-adsorbate systems, in particular, the diffusion of chemically active H, O, and N atoms (Table 6.2). The values of E_d/E_m are close to those obtained by Taylor and Langmuir.[2]

Gomer[529] suggested the technique for measuring the diffusion coefficient using FEM. It is based on the correlation between the electron current fluctuations, i.e., FEM noise, and the density fluctuations of the layer adsorbed on the emitter surface under investigation. The density fluctuations of the layer in equilibrium conditions, in turn, correlate with the diffusion coefficient.[530,531] Using this method, Chen and Gomer determined the diffusion coefficient of oxygen atoms[532] and CO molecules[533] on W(110) as a function of coverage. In the case of oxygen atoms they found that at $\Theta \approx 0.25$ the adsorption energy was sharply increased from 60 to 90 kJ/mol, which was attributed to lateral interaction in the adsorbate layers and, possibly, to a formation of the p (2×1) superstructure. At this point, the preexponential factor of the diffusion coefficient also dropped sharply when decreasing the coverage from 10^{-4} cm²/s for $\Theta = 0.3$ to 10^{-6} cm²/s for $\Theta = 0.2$. On further Θ decrease, the preexponential factor, in contrast to the activation energy, continued to diminish rapidly, and in the limit of zero coverage had an extremely low value of the order of $10^{-10} \approx 10^{-11}$ cm²/s. This fact has not been explained. For CO diffusion such behavior of the diffusion coefficient as a function of Θ was not observed.

The diffusion of hydrogen and deuterium atoms on W(110) was studied using this method by Di Foggoli and Gomer.[534] These authors found that at T below 150 K the diffusion coefficient is independent of temperature. It was interpreted in terms of the transitions from the activated diffusion to tunneling. Gonchar et al.[535] reported that they failed to observe such a transition under similar conditions. Braun and Pashitsky[536] proposed an alternative interpretation of this phenomenon. These authors assumed that at low temperatures diffusion is also activated; however, an adatom is excited into the activated state by FEM electrons rather than by lattice phonons, which results in the absence of temperature dependence.

In 1966, Ehrlich and Hudda[537] employed for the first time the field ion microscope (FIM) for investigation of W self-diffusion on different crystal faces. Unfortunately, the variety of metals and adatoms was limited in this technique by the possibility of obtaining an image in microscopy. The diffusion coefficient was calculated using the mean-square displacement of the atom under investigation.

TABLE 6.3
The Activation Energy E_m (kJ/mol) and the Preexponential Factor D_o(cm^2/s) for Atomic Migration of Different Metals on Oriented Planes of Tungsten

Atom/value		(110)	(211)	(321)
W	E_m	87.4 ± 6.7	82 ± 8	79.8
	D_o	(6.2 ± 1.1) 10^{-3}	2 10^{-2}	10^{-4}
Ta	E_m	75.6	46.2	63
	D_o	4 10^{-2}	0.9 10^{-7}	—
Mo	E_m	—	55.4	52.5
	D_o	—	9 10^{-7}	1.2 10^{-7}
Ir	E_m	68	51.2 ± 5	—
	D_o	10^{-5}	5 10^{-7}	—
Pt	E_m	66	—	—
	D_o	3 10^{-3}	—	—
Si	E_m	67.6 ± 6.7	—	—
	D_o	3.1 $10^{-4 \pm 1.3}$	—	—
Pd(Ni)	E_m	49	—	—
	D_o	—	—	—

Some modern devices enable the migrated particle to be identified by means of its evaporation by the field and subsequent detection with the aid of a mass spectrometer. Experimental data on the atomic diffusion on oriented crystal faces of different metals are given in Table 6.3.

Kisljuk, Tretjakov, and Nartikoev[538] developed a macroscopic method for determining diffusion coefficients with the help of fast thermodesorption heating of a small catalyst. When heating was stopped and the temperature became equalized, the system was kept for a certain interval in a stationary state, and then the thermodesorption cycle from the same area was repeated. The surface diffusion of O_2 on polycrystalline platinum was studied by this method.

The main conclusion from the data is that surface diffusion can be described in terms of the diffusion coefficient in the usual Arrhenius form

$$D = D_o \exp(-E/k_b T) \tag{6.4}$$

Activation energy of the surface migration lies in the range 1/3 to 1/10 of the value required for desorption, which is usually from 20 to 60 kJ/mol, and the preexponential factor varies from 10^{-2} to 10^{-4} cm^2/s.

The simplest estimation of the preexponential factor is $D_o \approx vl$ (here v is a characteristic velocity of a particle and l is the mean length of a jump) for $v \approx 10^4$ to 10^5 cm/s; $l = 10^{-8}$ to 10^{-7} cm gives $D_o \approx 10^{-4}$ to 10^{-2} cm^2/s. As seen from Table 6.3, experimental data, as a rule, lie in this range. Sometimes considerable deviation to much lower D_o values are reported, but as noted in Bassett's review[539] testing the results usually gives "normal" D_o values.

6.2.2. THEORY OF MIGRATION OF ADSORBED ATOMS

As a rule, the authors of experimental works do not give the accuracy of their values. Nevertheless, according to Tsong[540] the accuracy of experiments on determining the activation energy is 10%, which results in an error of about 1 or 2 orders in the preexponential factor. This fact makes the creation of an adequate theory extremely difficult for such a low accuracy the agreement between the calculated and the experimental diffusion coefficients cannot prove the validity of theoretical presumptions.

This is the reason why the simple model of jumps to the nearest adsorption site is widely practiced for interpretation of experimental results and in various empirical estimations of the diffusion coefficient. In accordance with this model, migration proceeds through successive jumps of an adsorbed particle from the initial site to an adjacent one. The diffusion coefficient is written as a product of the square of jump length, i.e., the lattice period, and a frequency of jumps, which can be evaluated within the framework of the theory of transition state. The simplest estimations of the diffusion coefficient according to this model give reasonable values which lie within the experimentally observed region.

An alternative situation arises when an adsorbed particle "flies" through an arbitrary distance along a certain direction. In this case, the diffusion coefficient is defined by the mean length of the jump that may be much longer than the lattice period. Only exact quantum mechanical calculations, extremely complicated for any system with strong chemical bonding, could elucidate the situation. Thus, the matter is how in reality adsorbed particles move over the surface rather than how to choose the model of migration.

It should be noted that the much-used theory of transition state cannot answer which of the concepts is true because the theory does not take account of the motion of reaction products after decomposition of the activation complex. Thus, the theory of transition state fails to predict the characteristic length of elementary jumps.

Theoretical estimations of diffusion coefficients are very few in number[541-544] and contain a large number of fitting parameters, which themselves must be determined from experiments. As a result, it is not possible to draw any definite conclusions as to the mechanism of migration.

In a number of works[545,546] the surface diffusion is treated in terms of the well-known Kramers model[547] for chemical reactions. According to this model the motion of the particle along the reaction coordinate is governed by random force, which simulates quite well the interaction of a particle with a heat bath. It should be noted that the Kramers theory, as well as the theory of transition state, gives only the flux value through the saddle point along the reaction coordinate. It works well for chemical reactions which have two

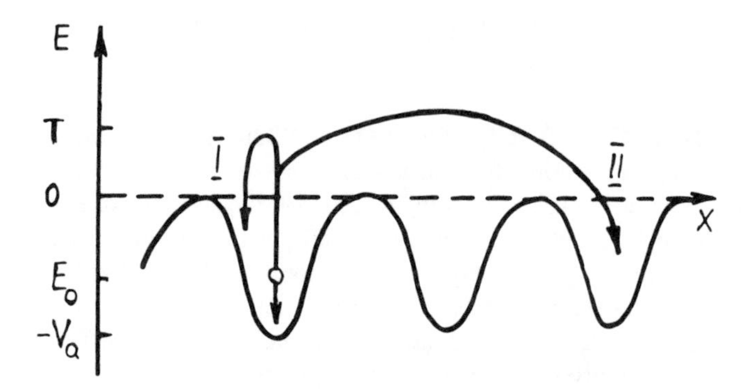

FIGURE 6.4. Adatom motion in the surface periodic potential. (I) Fast relaxation, $\lambda << 1$; (II) slow relaxation $\lambda >> 1$.

adjacent potential wells. But for surface migration in a periodic potential, when the number of wells behind the saddle point, generally speaking, is unlimited, this question requires special consideration. In the papers cited such an analysis has not been carried out; instead it was assumed that an adatom can jump only to one of the adjacent sites. The methods of molecular dynamics[548-551] along with analytical studies are rapidly progressing in this field.

We have studied in detail the model of surface migration in which energy relaxation was described in terms of the Fokker-Plank equation, similar to the Kramers theory.[520-522] The periodicity of the surface potential was explicitly taken into account in the convection term of the kinetic equation. This enabled us to calculate the transition probability not only into the nearest adsorption site, but into more distant ones as well. Much attention has been paid to an estimation of the energy diffusion coefficient due to its relation to the friction factor in the Kramers problem. When the velocity of a moving particle is much less than the sound velocity in a solid, an analytical expression containing only highly reliable parameters has been obtained.

A picture of migration of an adsorbed particle can be represented as follows. Suppose that at the initial moment a particle is localized in a certain potential well of the periodic surface potential (Figure 6.4). When its energy $E_o < 0$, a heavy adatom cannot move over the surface since its tunneling probability is negligible. Tunneling can be neglected even for the hydrogen atom, if the temperature is not too low, and, consequently, the activated diffusion dominates. The interaction with phonons of a solid results in the equilibrium energy distribution; therefore, a particle has a finite probability to transit into a state with positive energy. In such a state a particle, while being bound to the surface, can move over it. The mean velocity v_α of the particle in such a state is defined by the well depth V_α, and can be described by the expression of a type of $v_\alpha = (2V_\alpha/m_\alpha)^{1/2}$, where m_α is the mass of the adatom. The value of V_α far exceeds the characteristic thermal velocity for a given temperature, being practically independent of it.

The Boltzmann distribution must be established in the migration state as well, which one can conceive as the excitation up to energy ~T (in the system of units, where $k_B = \hbar = 1$) followed by deactivation into a more probable low-lying energy state. Hence, adatoms have a certain lifetime in the migration state, which has the same order of magnitude as the energy relaxation time τ_r for $E > 0$. Suppose that the time of flight for individual well is defined as $\tau_t = a/v_\alpha$, where a is the period of the lattice potential, and the mobility parameter $\lambda = \tau_r/\tau_t$. Then the trajectory of a particle depends on λ: for $\lambda << 1$ an adatom will return with maximum probability back to the starting site and with much less probability will transit into one of the adjacent wells. Therefore, the model of jumps to the nearest well holds at $\lambda << 1$. For $\lambda << 1$ long jumps become also possible, the mean jump length being of the order of $\tilde{L} \approx \lambda_\alpha$. Therefore, it is clear that in order for the diffusion characteristic to be determined, it is necessary to calculate not only the lifetime of the particle in a potential well $\tau_l \approx \tau_r \exp(V_\alpha/T)$ but also for the mobility parameter λ.

For $\lambda \approx 1$, when the region of particle motion is of the order of a, the jump transition must be treated using quantum mechanics. At the same time for $\lambda >> 1$, there are two considerably different times, i.e., τ_t and τ_τ, so the problem is substantially simplified.

Microscopic processes with the characteristic time τ_t (which is of the order of an oscillation period in the potential well) should be treated on a quantum mechanical basis, as previously but much slower processes can be described in terms of some kinetic equation that takes into account the motion of an adatom over the surface and energy relaxation. Thus, for $\lambda >> 1$ the problem can be divided into two stages: the first is to consider the quantum mechanical motion in the periodic surface potential and to calculate the transition probabilities of the particles involving the energy transfer to the lattice phonons; the second is to derive and to investigate a kinetic equation, to obtain necessary migration parameters using this equation and calculated transition probabilities.

This program has been successfully completed, and the following expression for the coefficient of the surface diffusion has been obtained

$$D = (a\, v_\alpha/2)[1 + (\lambda/2)] \exp(-E_m/k_B T) \tag{6.5}$$

There are no fitting parameters in this expression, excluding experimentally determined energy E_m, which can be found with good accuracy. In the case of heavy particles the expression for the parameter λ can be written in the form

$$\lambda \sim \sqrt{m_\alpha}\, T V_\alpha^{3/2} \tag{6.6}$$

As is seen, a characteristic length is shorter for more light atoms and increases considerably with decreasing activation energy.

The derived theoretical expressions were applied to the diffusion of W atoms on W: for $V_\alpha = 1$ eV, $m_e = 3\ 10^{-22}$ g, E = kT, T = 300 K, a = $3\ 10^{-8}$ cm, $v_\alpha = 0.5\ 10^5$ cm/s, and $\tau = 6\ 10^{-13}$ s, we obtained $\lambda = 3$ to 5, and consequently, the characteristic length is considerably longer than the lattice period. The preexponential D_o is about $5.8\ 10^{-3}$ cm^2/s. On decreasing the activation energy of migration, λ rapidly rises and may reach a value of several tens.

One more essential feature of the developed theory should be noted. The Einstein relationship for diffusion $\langle x^2 \rangle = 2Dt$ holds, naturally for any migration mechanism. Using the mean-square value $\langle x^2 \rangle$, one can determine the diffusion coefficient D, but it is impossible to find such an important migration characteristic as the jump frequency. However, if experimental data enable us to calculate $\langle x^4 \rangle$, one can determine the mobility parameter λ, and, consequently, elucidate the mechanism of surface migration. It should be noted that in order for the reliable $\langle x^2 \rangle$ value to be determined, more statistical data are needed compared with those commonly used. But there is no possibility to increase the observation time because the latter is limited by adatom desorption.

This picture is in a good agreement with the available experimental data on diffusion coefficients. However, what is more important is that our results closely coincide with those obtained by more elaborate dynamic calculations made by Tully et al.[548] The results of this work also support the possibility of the adatom migration through a large distance during one jump.

6.2.3. ROLE OF MIGRATION FOR THE KINETICS OF SURFACE REACTIONS

As was mentioned previously one can conceive a simple surface reaction involving two stages: adsorbate migration and interaction itself. Theoretical investigations of the diffusion-controlled reactions have shown that in contrast to the three-dimensional case, in one- and two-dimensional cases, which are of significance for surface reactions, it is not possible to introduce a certain rate constant, i.e., to use formal kinetic equations.

The main reason for the qualitative difference in kinetics of the bulk- and surface-diffusion-controlled reactions is due to the diffusion peculiarities. In the three-dimensional case, the collision probability is low and equal to the ratio of characteristic size of a particle to the interparticle spacing. In one- or two-dimensional cases the collision probability is always equal to unity however small the coverage may be.

As we have shown, it is for this reason that the type of kinetics of the surface reactions is determined by the relationship between the diffusion rate and the collision efficiency, i.e., between diffusion and kinetic constants. The temporal dependence of the process coincides with that following from the formal kinetic equations in the kinetic regimen, when the limiting stage is interaction itself. Therefore, it is of importance to describe the collisional kinetics correctly.

It should be emphasized that a formal kinetic description is applicable for surface reactions either in the case of the high mobility of particles, i.e., $\lambda >> 1$, or low reaction probability. Both these conditions allow us to substitute the local concentrations, involved in the law of surface action, by the mean concentration.

Consider in more detail[552] the A + B → C type of reaction between adsorbed particles A and B for low coverages, when $\Theta_A << \Theta_B << 1$. The reagents have low mobilities since their lifetimes in adsorption sites are rather long. Mobility values vary over a wide range for different species. Therefore, only one of the reagents is usually assumed to be active and responsible for approaching closely enough for the reaction to occur. Let it be reagent A; then its partner B is regarded to be practically immobile during the characteristic reaction time and occupies a certain fixed site on the surface.

Suppose the distribution of adsorption sites on the surface to be random; then the lifetime of an active particle on the surface depends on a given configuration. The reaction rate is maximal for the regions with the most dense population of B particles, next the regions with some mean density, and at last, the most rarely populated configurations will "burn up". Even though for each configuration the reaction could be described in terms of a certain rate constant k, as a consequence of statistical averaging, the total product yield will have a very complicated temporal dependence because k depends on the local density.

The reaction is simulated in the following model. Being strongly bounded to the lattice, B molecules are assumed to have sharp perturbation of the surface potential for the motion of A particles. Therefore, when colliding with B, an active adatom A can easily lose its energy of the order of kT_B by means of multiphonon processes or through excitation of high-frequency local vibrations. As a result of collision, the A particle will occupy with a high probability an adjacent site to the B adsorption site. Suppose that the reaction probability for each collision of A with B is w, then $(1 - w)$ is the probability for A to occupy an adjacent site.

An analysis of this model has shown that for two-dimensional migration of active species, the surface chemical reaction can be described by the following kinetic equation in the whole range of the B coverage

$$\dot{\Theta}_A = -\tau_{chem}^{-1}\,\Theta_A = -k_{eff}\Theta_A \tag{6.7}$$

but the rate constant τ_{chem}^{-1} or k_{eff} depends nonlinearly on the density of B particles. A qualitative curve for τ_{chem}^{-1} as a function of Θ_B is shown in Figure 6.5. For $\Theta_B \approx \lambda^{-1}$ the reaction order varies from the second, which is observed in the limit of $\lambda\Theta_B << 1$, to the first one in the limit of $\lambda\Theta_B >> 1$.

An expression for the rate constant at $\Theta_B << \lambda^{-1}$ has the form

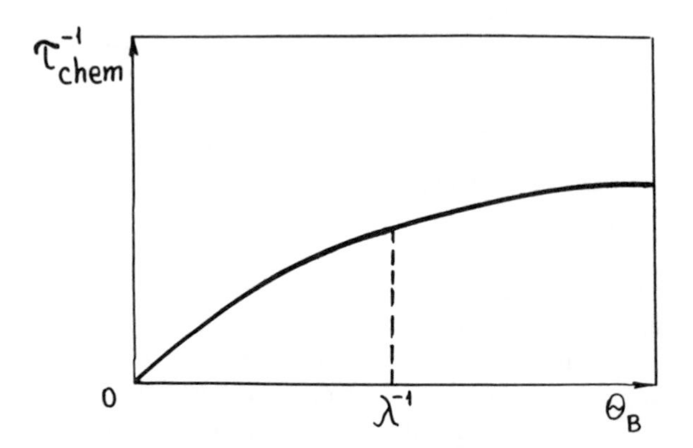

FIGURE 6.5. The rate of reaction as a function of density of B particles.

$$k = \frac{4wv_\alpha a}{1 + (1 + w)/\sqrt{2}} \exp(-E_m/k_BT) \tag{6.8}$$

According to Equation 6.8, k is independent of the mobility parameter λ; however, a high mobility of A is incorporated into the model from the very beginning. An estimation of the preexponential factor using Equation 6.7 for $w = 1$ gives a value of $2\ 10^{-3}$ cm^2/s.

Note once again a difference between this approach and the traditional one. The latter leads to the time-dependent rate constant and, therefore, fails to describe the kinetics of the process. First of all, it should be noted that only taking into account long jumps, which far exceed the lattice constant, enables us to describe the formal kinetics of the process, i.e., in terms of rate constants. On the other hand, within the framework of formal kinetics, the reaction of the A + B → C type is usually described by Equation 6.8 for a constant Θ_B, where k_{eff} is not a linear function of Θ_B; for small Θ_B, k_{eff} is proportional to Θ_B; for large Θ_B, k_{eff} is independent of Θ_B, i.e., on first reaction is second order, and then it becomes first order.

The sketchy model shown in Figure 6.6 could formally describe the above results. Suppose that particles A and B are in adjacent adsorption wells. Their interaction is possible only in the case when A particle transits into the migration state A*. The scheme for the process is given by

$$A \underset{k_2}{\overset{k_1}{\rightleftarrows}} A^*$$

$$A^* + B \xrightarrow{k_3} C$$

Assuming [A*] to be quasistationary, we arrive at the following equation for Θ_A

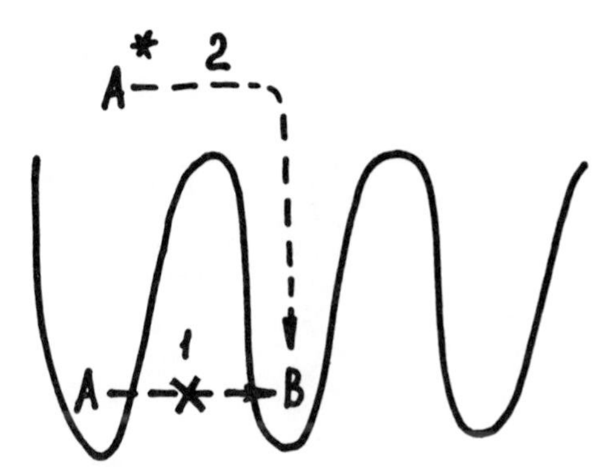

FIGURE 6.6. The model of atomic recombination. (1) Recombination is impossible; (2) recombination path.

$$\dot{\Theta}_A = k_1k_3\Theta_A\Theta_B/(k_2 + k_3\Theta_B) \qquad (6.9)$$

Then for small Θ_B the reaction is second order, $\sim\Theta_A\Theta_B$, and for large it is first order, $\sim\Theta_A$. The similarity of these results with the above one is rather formal: rate constants involved in this equation have no physical meaning and are not connected with microscopic parameters of the system.

Later, we studied the kinetics of the surface recombination of mobile atoms.[553] Associative desorption is the simplest example of such a process. Formally this reaction can be written in the form of $A + A \rightarrow 0$, implying that the molecule created quickly desorbs and does not participate in the further process. For this case, taking into account the jump migration into the nearest complex site, the task has been solved exactly as suggested by the unity probability for the reaction. The asymptotic limits of the reaction course have been analyzed as well.

Chapter 7

CHEMOENERGETIC STIMULATION

In the previous section we considered nonequilibrium phenomena in catalysis which arise at microlevel due to translational, rotational, vibrational, and electronic excitations of a molecule, interacting with a catalyst. On interaction of molecules with the surface, the population of levels may alter those results in the violation of the Boltzmann energy distribution. On account of the energy released in surface exothermal reactions, the formation of "hot" molecules with high translational, rotational, and vibrational energy is possible. This, in its turn, may result in changes of the adsorption, desorption, and catalytic reaction rates.

The origination of nonequilibrium distribution is of general importance in chemical kinetics, and not just in the kinetics of heterogeneous reactions. Interest in the possibility of using these phenomena for nonthermal stimulation of other chemical reactions has quickened in the past few years, especially in connection with the experimental observations of energetic acceleration in some branched — chain reactions.

Zhdanov[554] undertook a theoretical investigation in order to elucidate the possibility of the use of the energy released in one reaction for direct acceleration of another one. Among conditions necessary for the efficient use of this energy are the following: (1) the second reaction must be in nonequilibrium (here, by nonequilibrium is meant the depletion of the excited state in the course of reaction, when the reaction probability of excited reagents transformation far exceeds their relaxation probability); and (2) the energy released in the first must exceed the activation energy required for the second one. Finally, it was concluded that there is little likelihood of such processes.[554]

In our view, this conclusion is in error. In collaboration with M. Rozovsky, we have analyzed some of the simplest kinetic models for nonthermal stimulation of chemical reactions and found conditions at which the efficiency of stimulation is sufficiently high.

7.1. CONDITIONS REQUIRED TO REALIZE CHEMOENERGETIC STIMULATION

7.1.1. BASIC DEFINITIONS AND STIMULATION MECHANISMS

By chemoenergetic stimulation (CES) is meant the stimulation of one chemical reaction by products of another one via nonthermal use of their overequilibrium energy. The main features of this effect are the nonthermal origin of stimulation and the chemical source of nonequilibrium stimulation.

The term "stimulation of chemical reaction" in our opinion, needs some clarification. By stimulation efficiency, we imply the production of maximum

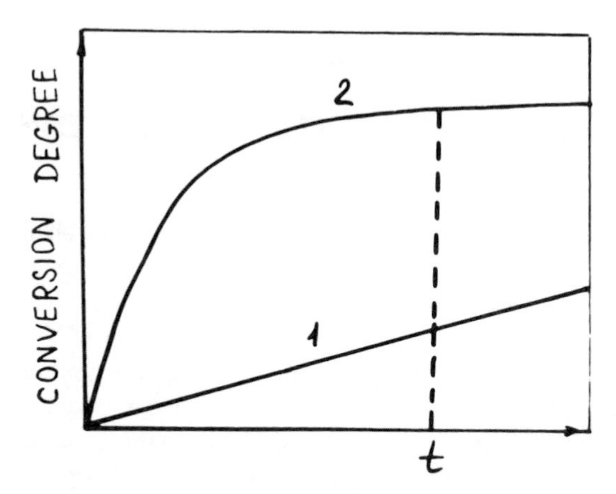

FIGURE 7.1. The scheme of reagent conversion in the absence of stimulator (1) and in the presence of stimulator (2), at the moment t the reaction rate $r_1 > r_2$.

reaction outputs in the same time interval t. It should be noted that using differential characteristics, such as the rate or effective rate constant, can lead to incorrect conclusions about the stimulation efficiency. For example, one can conceive the stimulation (Figure 7.1) when by the time t the greater conversion degree is achieved in the presence of a stimulation factor. Because an appreciable decrease of the reagent concentration has occurred, the reaction rate r_2 by the time t may be less than the rate r_1 in the absence of the stimulator. In contrast, the reaction output by this time is greater in the first case than in the second one. A similar situation takes place in the case of the effective rate constant.

The investigation of the stimulation effect should be carried out when all other factors are equal. As to CES, it implies that the yields of reaction products must be compared when all thermodynamics parameters (volume, pressure, temperature, etc.) as well as the inputs of each reagent are the same.

According to that speculation, the CES problem can be formulated as follows. Suppose that without stimulation the input reagents of a chemical reaction are supplied to the reactor in accordance with a certain law N(t), the output kinetics being described by the function n(t). In the presence of some chemoenergetic stimulation α, the output function is $n_\alpha(t)$ under the same conditions. Then the efficiency of the chemoenergetic stimulation is defined by the CES coefficient

$$\zeta = n_\alpha(t)/n(t) \tag{7.1}$$

CES mechanisms are reduced to two main cases

1. Nonequilibrium outputs of one reaction are inputs or intermediates for another reaction. This situation will be called the stimulation by gen-

eration because the first (stimulating) reaction is a generator of reagents for the second (main) reaction.

2. Nonequilibrium products of the stimulation reaction can exchange their energy with the inputs or intermediates of the main reaction. This case will be spoken of as exchange stimulation.

Here we restrict the analysis to the first case.

7.1.2. STIMULATION BY GENERATION

Consider the simplest two-level model of the main reaction and the following kinetic scheme of the process:

$$\xrightarrow{\alpha r(t)} A^* \xrightarrow{w} B$$

$$k_{o1} \uparrow\downarrow k_{1o}$$

$$\xrightarrow{(1-\alpha)\, r(t)} A_o$$

Here A_o and A^* are reagent molecules of the stimulating reaction in the ground and excited state, respectively; B is a product of the main reaction; k_{01} and k_{10} are interlevel rate constants related to each other by detailed balancing; w is the reaction probability from the excited state; r(t) is the total generation rate of a reagent, and α is the portion generated in the excited state via the stimulating reaction.

The kinetic scheme of this process is described by the following system of equations

$$\begin{aligned} dn_0/dt &= -k_{01}n_0 + k_{10}n_1 + (1-\alpha)r(t) \\ dn_1/dt &= k_{01}n_0 - k_{10}n_1 - rn_1 + \alpha r(t) \end{aligned} \qquad (7.2)$$

where n_0 and n_1 are the concentration of A molecules in the corresponding states.

The output of the main reaction is given by

$$n_\alpha(t) = \int_o^t wn_1(t)dt \qquad (7.3)$$

To calculate the CES coefficient defined by Equation 7.1, we should compare $n_\alpha(t)$ with that obtained in equilibrium conditions, i.e., when

$$\alpha = \alpha_{eq} = k_{o1}/(k_{o1} + k_{1o}) \qquad (7.4)$$

The system (Equation 7.2) can be solved by quadratures at an arbitrary generation rate r(t). In order for the analysis to be simplified, consider only two limiting cases: (1) the pulse generation when some amount of the input reagent of the stimulating reaction converts very rapidly (in a time much less than the characteristic reaction time) into the input for the main reaction; or (2) the stationary generation when the stimulating reaction is a steady source of reagents for the main reaction.

Pulse Generation

In this case the rate of A formation is simulated by the δ-function

$$r(t) = N\,\delta(t) \qquad (7.5)$$

where N is the total amount of A component resulting from the stimulating reaction. The solution of the system (Equation 7.2) for the excited state with a source function (Equation 7.5) has the form

$$\frac{n_1(t)}{N} = \frac{\alpha\lambda_1 - k_{o1}}{\lambda_1 - \lambda_2}\exp(-\lambda_1 t) - \frac{\alpha\lambda_2 - k_{o1}}{\lambda_1 - \lambda_2}\exp(-\lambda_2 t) \qquad (7.6)$$

where characteristic constants are defined by the formula

$$\lambda_{1,2} = \frac{k_{o1} + k_{1o} + w \pm \sqrt{(k_{o1} + k_{1o} + w)^2 - 4k_{o1}w}}{2} \qquad (7.7)$$

Substituting Equation 7.6 into the integral Equation 7.3, we find the kinetic function for the product formation

$$\frac{n_\alpha(t)}{N} = 1 - \frac{\alpha w - \lambda_2}{\lambda_1 - \lambda_2}\exp(-\lambda_1 t) + \frac{\alpha w - \lambda_1}{\lambda_1 - \lambda_2}\exp(-\lambda_2 t) \qquad (7.8)$$

The concentration of the ground-state molecules can be obtained, for example, from the balancing condition

$$n_o(t) + n_1(t) + n_\alpha(t) = N \qquad (7.9)$$

If the characteristic constants λ_1 and λ_2 are sufficiently distinct from each other, the characteristic stages can appear in the reaction kinetics, and the values λ_1 and λ_2 may have a clear physical meaning. Thus, if the excited-stage energy is sufficiently high compared with that of the thermal one, the equilibrium interlevel constant is given by

$$K_{eq} = \frac{k_{o1}}{k_{1o}} = \exp\left(-\frac{\Delta E}{RT}\right) << 1 \qquad (7.10)$$

where $\Delta E = E_1 - E_0$ is the excitation energy. In this case Equation 7.7 has the form

$$\lambda_1 = k_{1o} + w$$

$$\lambda_2 = \frac{k_{o1}w}{k_{1o} + w}$$

$$\lambda_1 >> \lambda_2 \tag{7.11}$$

Now λ_1 has a physical meaning of the relaxation constant in reaction conditions, and λ_2 may be referred to as the effective reaction rate constant when the relaxation process is already completed ($\lambda_1 t >> 1$).

Reaction kinetics in the absence of stimulation is defined by Equation 7.8 where α is given by Equation 7.4. The resulting coefficient for the pulse stimulation can be found on corresponding substitutions into Equation 7.1.

The CES coefficient ζ is defined by four independent dimensionless parameters: (1) nonequilibrium degree created by the stimulator, $\beta = \alpha/\alpha_{eq}$; (2) the equilibrium interlevel constant K_{eq}; (3) reduced reactivity of the excited stage $\rho = w/k_{10}$; and (4) the time t expressed in terms of $\tau = k_{01}$ t.

It can be shown that there are three characteristic stages of the reaction for which the expression for ζ can be simplified, and the results can be easily interpreted.

In the initial stage of the reaction (λ_1 t << 1) when the excited stage remains unrelaxed, we have an obvious result $\zeta = \beta$, i.e., the stimulation efficiency coincides with the efficiency of the generator.

The intermediate stage ($\lambda_1^{-1} << t << \lambda_2^{-1}$) can be separated only when λ_1 and λ_2 are sufficiently distinct from each other, for example, when the condition in Equation 7.10 is fulfilled. In this stage the CES coefficient decreases according to the law

$$\zeta = (\beta + \tau)/(1 + \tau) \tag{7.12}$$

and as long as $\tau < \beta$, the stimulation efficiency can be sufficiently high. In this stage the stimulation has the following meaning. By the beginning of the intermediate stage, relaxation is already completed. Product accumulation in the case of the stimulated reaction will occur at an even lesser rate than in its absence because of the decrease of the reagent concentration in the relaxation stage. But the total yield will be greater than in the absence of a stimulator (see Figure 7.1) due to the contribution of the relaxation stage after a time. The latter becomes comparable with the product yield in the case of nonstimulated reaction, which is indicative of the decrease of CES efficiency. The time taken for the CES coefficient to be decreased to the value of $\zeta = 2$ may be accepted as the characteristic time of the process. From Equation 7.12 it is seen that this time is $\tau_2 = \beta$; thus, the CES efficiency is sufficiently high even for the time intervals longer than the relaxation time, providing the generation is sufficiently power.

Finally, when the reaction is almost completed (λ_2 t << 1), the CES coefficient tends to unity, i.e., the reaction "forgets" about nonequilibrium initial conditions.

Stationary Generation

This model is applicable to stimulators operating throughout the main reaction. Now the stimulating reaction is a steady source of reagents for the main reaction, i.e., the generation rate

$$r(t) = r = \text{const} \tag{7.13}$$

The concentration of molecules in the excited stage can be derived from the system (Equation 7.2) with the source function (Equation 7.13)

$$n_1(t) = \frac{r}{w}\left[1 - \frac{\alpha w - \lambda_2}{\lambda_1 - \lambda_2}\exp(-\lambda_1 t) + \frac{\alpha w - \lambda_1}{\lambda_1 - \lambda_2}\exp(-\lambda_2 t)\right] \tag{7.14}$$

As in the case with pulse generation, the characteristic constants λ_1 and λ_2 are defined by the formula in Equation 7.7 because the steady source does not introduce an additional characteristic time constant in the reaction process. Substituting the function (Equation 7.14 into Equation 7.3), we find an equation the product formation

$$\frac{n_\alpha(t)}{r} = t + \frac{\alpha}{k_{o1}} - \frac{1}{\lambda_1} - \frac{1}{\lambda_2} + \frac{\alpha w - \lambda_2}{\lambda_1(\lambda_1 - \lambda_2)}\exp(-\lambda_1 t) - \frac{\alpha w - \lambda_1}{\lambda_2(\lambda_1 - \lambda_2)}\exp(-\lambda_2 t) \tag{7.15}$$

The concentration of molecules in the ground state can be found from the condition of the material balance that is written for the stationary generation in the form

$$n_o(t) + n_1(t) + n_\alpha(t) = rt \tag{7.16}$$

The equilibrium product concentration n(t) is defined by Equation 7.15 with $\alpha = \alpha_{eq}$. On substitution of the corresponding functions into Equation 7.1, the CES coefficient can be obtained. From Equations 7.7 and 7.15 it follows that for stationary generation the ζ value depends on the same dimensionless parameters and, as expected, is independent of the total reagent amount.

The characteristic time required for the steady-state regime is of the order of λ_2^{-1}. The corresponding stationary concentrations can be derived, for example, by equalizing the right-hand sides of Equation 7.2 to zero

$$n_1^{(o)} = \frac{r}{w}$$

$$\frac{n_1^{(o)}}{n_o^{(o)}} = \frac{k_{o1}}{k_{1o} + (1 - \alpha)w} \tag{7.17}$$

When $t >> \lambda_2^{-1}$, the steady-state regime of the product accumulation with a constant rate is reached. The term ''stage'' in the case of stationary generation has to some extent another meaning than that in the case of pulse generation. Of course, the λ_1 value can be also treated as the relaxation rate constant since in the time of the order of λ_1^{-1} a portion of particles in the excited state drops sharply compared with α value induced by the stimulator. In contrast, the λ_2 value cannot be regarded as the effective rate constant of the process; it is only a characteristic value of the transition to the regime of stationary concentration transition.

For different reaction stages, the CES coefficient behaves as follows. In the initial stage of the reaction ($\lambda_1 t << 1$), $\zeta = \beta$ as in the case of pulse generation. In the intermediate stage ($\lambda_1^{-1} << t << \lambda_2^{-1}$) the CES coefficient drops with time increase according to the law

$$\zeta = (2\beta + \tau)/(2 + \tau) \tag{7.18}$$

Equation 7.18 differs from that for the pulse generation Equation 7.12 only by the factor of 2, which defines the total effect of the continuous operation. The stimulation effect in this stage is also a consequence of the relaxation-stage contribution to the product accumulation. The time it takes for the CES coefficient to be decreased to the value of $\zeta = 2$, is $\tau = 2\beta$, i.e., it is only twice as long as that for the pulse stimulation.

At the final stage ($\lambda_2 t >> 1$), when the steady-state regime is established, the CES coefficient tends to unity, i.e., despite its continuous action, the stimulator ''fails'' to affect the reaction yield. This situation can be easily explained by the fact that in the stationary regime the rate of product accumulation equals the generation rate of the total reagent amount, irreversible of their distribution over energy states. As it follows from Equation 7.15, the conversion degree is little different from 100% both with and without stimulation, provided the stationary regime is achieved.

7.1.3. NUMERICAL ESTIMATIONS OF STIMULATION EFFECTS

The previous analysis allows us to make the following conclusions about the characteristic features and conditions necessary for CES

1. CES efficiency depends weakly on whether the kind of stimulating action is pulse or stationary.
2. As expected, most CES efficiency is reached in the relaxation stage of the reaction, the maximum value of the CES coefficient being equal to β, the nonequilibrium degree of stimulating reaction products. The only necessary and sufficient condition for high CES efficiency has an obvious meaning

$$\beta = \alpha \exp(\Delta E/RT) >> 1 \tag{7.19}$$

3. When the relaxation stage is completed, stimulation reflects the influence of the relaxation stage on the total accumulation of the product, and CES efficiency decreases as time passes. This effect can be characterized by the interval in which the CES coefficient drops to $\zeta = 2$. The time interval of effective CES $\tau_2 \sim \beta$ is appreciably longer than the relaxation time, provided the condition (Equation 7.19) is fulfilled. For $\tau > \tau_2$ the CES coefficient tends to unity even for the stationary operation of a highly efficient stimulator: the product yield of the main reaction becomes insensitive to the existence of the stimulating reaction.

These conclusions are illustrated by numerical data in Table 7.1.

For example, if the stimulating reaction generates 10% of reagents in the excited state at T = 300 K, and the excitation energy is 35 kJ/mol, the maximum CES coefficient is $\zeta_{max} = 10^5$, and the period of effective CES is 10^5 times as long as the relaxation time. If the excitation energy under the same conditions increases up to 45 kJ/mol, ζ_{max} grows to 10^7, and the period of effective CES is 10^7 as long as the relaxation time.

So, a high CES efficiency can be achieved for sufficiently powerful generators independent of whether the main reaction is in equilibrium or not and irrespective of the location of the excited state (above or below the activation barrier). Our statement contradicts that made by Zhdanov.[554]

Let us return to the above numerical estimations. If the relaxation time of the vibrational excitation is 10^{-5} s, the period of effective CES will be 100 s for $\Delta E/RT = 18.4$. If $\tau \approx 10^{-2}$ s, all other conditions being the same, τ_2 becomes 10^5 s. This estimation is of general importance and may be applied, for example, to excitation of active sites in the catalyst itself. Here the characteristic reconstruction time of the active site is very long, much longer than the lifetime of the vibrational excited molecule. Hence, in heterogeneous catalysis τ_2 is extremely long; therefore, the chemoenergetic stimulation effect is of widespread occurrence. It is this reason why the kinetics of heterogeneous catalysis is similar to that in chain reactions.

7.2. CHAIN MECHANISMS IN CATALYSIS AND CHEMOENERGETIC STIMULATION

One of the characteristic properties of chain reactions is the possibility of partial utilization of the energy released in the process to accelerate the reaction. Owing to this fact, the over-equilibrium concentrations of intermediates appear. Active species arise in the heterogeneous catalysis as well. They may be free atoms and radicals desorbed from the surface into the gas phase, quasifree particles reacting in the 2-D adsorbed layer, and excited molecules as well. However, the most commonly encountered case is the

TABLE 7.1
Numerical Estimation of the Stimulation Effect

$\alpha\%$	$\Delta E/RT$	α_{eq}	β	ζ_{max}	τ_2
10	6.9	10^{-3}	10^2	10^2	10^2
10	9.2	10^{-4}	10^3	10^3	10^3
10	11.5	10^{-5}	10^4	10^4	10^4
10	13.8	10^{-6}	10^5	10^5	10^5
10	16.1	10^{-7}	10^6	10^6	10^6
10	18.4	10^{-8}	10^7	10^7	10^7

origination of nonequilibrium active site concentration in the catalyst itself, similar to active particles in the gas-phase reaction. Consider the simplest two-site schemes for the reaction A → B

$$A + Z \xrightarrow{k_1} A^1 + Z^*$$

$$A + Z^* \xrightarrow{k_2} B + Z^*$$

$$Z^* \underset{k_{-3}}{\overset{k_3}{\rightleftarrows}} Z$$

where Z is an active site, k_1 is the activation rate constant of the catalyst by A reagent, k_2 is the rate constant of catalysis, k_3 is the deactivation rate constant of the active site, and k_{-3} is the thermal excitation rate constant.

The first of these stages is similar to initiation in the chain reaction, the second to the stage of chain propagation, and the third to linear termination. The activation effect is of importance when $k_1 [A] >> k_{-3}$, i.e., when thermal activation is less probable than activation through the chemical reaction.

If n is a concentration, its kinetics is described by the equation:

$$dn/dt = -k_2n[Z^*] \tag{7.20}$$

and the kinetics for a number of active sites by

$$d[Z^*]/dt = k_1n[Z] - k_3[Z^*] \tag{7.21}$$

Taking into account that the total number of active sites is constant, i.e.,

$$[Z] + [Z^*] = N_o \tag{7.22}$$

the substitution of [Z] into Equation 7.21 by its value from the last expression gives

$$d[Z^*]/dt = k_1N_on - (k_1n + k_3)[Z^*] \quad (7.23)$$

Equation 7.23 described the transition process for quasistationary [Z*] concentration. Its characteristic time is

$$\tau_1 \approx 1/(k_1n + k_3) \quad (7.24)$$

The quasistationary [Z*] concentration is substantially higher than the equilibrium concentration*

$$[Z^*]_{eq} = [Z]k_3/k_{-3}$$
$$[Z^*]_{stat} = k_1N_on/(k_1n + k_3) >> [Z^*]_{eq} \quad (7.25)$$

Under the quasistationary conditions, the reaction rate is given by

$$r_{stat} = k_1k_2n^2N_o/(k_1n + k_3) \quad (7.26)$$

Equation 7.26 is at least superficially similar to the equations of the Langmuir type, though the nature of rate constants is quite different in this case. For $k_1n << k_3$ we have

$$r_{stat} = (k_1k_2/k_3)N_on^2 \quad (7.27)$$

and the activation energy is given by

$$E = E_1 + E_2 - E_3 \quad (7.28)$$

as in the case of simple (unbranched) chain reaction with linear termination. Activation in this scheme (the first stage) proceeds immediately, independently on catalysis when the reagents are passed over the surface.

The mechanisms, involving different active sites and with the activation of species in the course of reaction, seem to be more widespread. Consider the kinetic features of the catalytic reaction proceeding over the sites Z and Z*, where the latter arise in the course of reaction. The reaction rate is high for Z* sites and low for Z sites

$$A + Z \xrightarrow{(1 - \alpha)k_1} B + Z$$

$$A + Z \xrightarrow{\alpha k_1} B + Z^*$$

* In the following, the equilibrium concentration of active sites or intermediates means the concentration in thermal equilibrium in the absence of catalysis.

$$A + Z^* \xrightarrow{k_2} B + Z$$

$$Z^* \underset{k_{-3}}{\overset{k_3}{\rightleftarrows}} Z$$

where α is the excitation probability of the active site in a slow reaction $Z \rightarrow Z^*$.

The transformation from the slow regime to the fast one is possible if $k_2 >> k_1$ and $\alpha k_1 n >> k_3$. Analysis similar to that for Equations 7.20 through 7.28 results in the following kinetic equations

$$dn/dt = k_1 N_o n - k_2 n[Z^*] \quad (7.29)$$

$$d[Z^*]/dt = \alpha k_1 N_o n - (\alpha k_1 n + k_3)[Z^*] \quad (7.30)$$

The characteristic time required for the quasistationary regime to be established, is given by

$$t_1 \sim 1/(\alpha k_1 n + k_3) \quad (7.31)$$

The stationary concentration of the active sites can be expressed as follows

$$[Z^*]_{stat} = \frac{\alpha k_1 N_o n}{\alpha k_1 n + k_3} >> [Z^*]_{eq} \quad (7.32)$$

Consider several limiting cases.

When $k_2/k_1 >> k_3/k_{-3}$, even the initial concentration of excited sites is sufficient to neglect the slow reaction. Then, for $t << t_1$

$$dn/dt = -\frac{k_2 k_{-3}}{k} N_o n = -k_o n \quad (7.33)$$

In the case of rapid deactivation of the active site, under conditions of a catalytic reaction ($\alpha k_1 n << k_3$), we obtain

$$[Z^*]_{stat} = (\alpha k_1 n/k_3) N_o << N_o \quad (7.34)$$

and

$$dn/dt = -\frac{\alpha k_1 k_2 N_o}{k_3} n^2 = -k_{stat} n^2 \quad (7.35)$$

In this case the order of the reaction alters from first to second, the quasimonomolecular rate constant being $k_{stat}n > k_0$. In the opposite case, i.e., deactivation is slow and activation is fast, $\alpha k_1 n >> k_3$, we have

$$[Z^*]_{stat} \approx N_o \tag{7.36}$$

$$dn/dt = -k_2 N_o n = -k_{stat} n \tag{7.37}$$

The transition occurs from the regime characterized by the rate constant k_0 into the stationary one with k_{stat}, the following condition being fulfilled $k_{stat}/k_0 = k_3/k_{-3} >> 1$.

Consider another case when $k_2/k_1 << k_3/k_{-3}$, i.e., the initial contribution is insufficient to exclude the slow reaction. Then for $t << t_1$

$$dn/dt = -k_1 N_o n = -k_0 n \tag{7.38}$$

In the limiting case when $\alpha k_1 n << k_3$, we have

$$[Z^*]_{stat} = (\alpha k_1 n/k_3) N_o << N_o \tag{7.39}$$

$$dn/dt = -k_1 N_o n - (\alpha k_1 k_2/k_3) N_o n^2 \tag{7.40}$$

Hence, it follows that the reaction is first order in n with the rate constant $k_3 >> k_1$ when even for $k_2 >> k_1$ the stationary concentration of excited sites is too low to exclude the slow reaction, and it is second order when $\alpha k_2 n >> k_3$ and $k_1 n >> k_0$.

The scheme under consideration implies some probability for the use of the energy of the catalytic step for the acceleration of the catalytic process as a whole. At the same time the presence of less active sites with concentration [Z] is admitted.

Now consider another model when from the beginning the reaction proceeds only over excitation sites Z*, but in the course of the reaction there is a probability α to excite one more site Z*

$$A + Z^* \xrightarrow{(1-\alpha)k_1} B + Z^*$$

$$A + Z^* \xrightarrow{\alpha k_1} B + 2Z^*$$

$$Z^* \underset{k_{-2}}{\overset{k_2}{\rightleftarrows}} Z (k_2 >> k_{-2})$$

This scheme is analogous to chain branching with linear chain termination

$$dn/dt = k_1[Z^*]n \qquad (7.41)$$

$$d[Z^*]/dt = \alpha k_1 n[Z^*] - k_2[Z^*] + k_{-2}[Z]$$

$$= (\alpha k_1 n - k_2)[Z^*] + k_{-2}N_o \qquad (7.42)$$

$$[Z] + [Z^*] = N_o \qquad (7.43)$$

For $\alpha k_1 n > k_2$ the chain inflammation is possible with the characteristic time

$$\tau = (\alpha k_1 n - k_2)^{-1} \qquad (7.44)$$

There exists a lower pressure limit defined by the condition $\alpha k_1 n = k_2$. Unlike in the gas-phase process, the upper limit is restricted, not by the second-order annihilation, but by the limited maximum number of active sites ($[Z^*]_{max} = N_0$). However, if the active sites are available in the catalyst bulk, and not just on the surface, this limit can be very large.

It is often assumed that the difference between catalysis and chain reactions lies in the fact that active sites in catalysis exist before starting the reaction, whereby in chain processes they arise only on the initiation of the reaction.[555] In the above schemes of catalytic reactions the less-active sites Z exist from the very beginning, and more-active sites Z* (unrelaxed, or "excited" ones) appear in the course of the reaction which proceeds further over new sites. The difference between catalysis and chain reactions becomes of less importance in this case.

There are a lot of proposals in the literature about heterogeneous reactions proceeding via chain mechanisms. In different schemes the role of active particles in the chain process was attributed to surface free radicals, atoms, ions, excited particles, and sites on the catalyst surface. Nevertheless, there are a few cases where the chain mechanism has been proved with assurance. We shall give examples of some work on this subject.

Kazansky et al.[11] have shown that CO oxidation over supported V_2O_5/SiO_2 catalyst proceeds according to the scheme

$$2V^{5+} + O^{2-} + CO \longrightarrow CO_2 + 2V^{4+} \qquad (7.45)$$

$$V^{4+} + O_2 \longrightarrow V^{5+}O_2^- \qquad (7.46)$$

$$V^{5+}O_2^- + V^{4+} \longrightarrow 2V^{5+}O^- \qquad (7.47)$$

$$V^{5+}O^- + CO \longrightarrow CO_2 + V^{4+} \qquad (7.48)$$

$$V^{4+}O^- + V^{4+} \longrightarrow O^{2-} + 2V^{5+} \tag{7.49}$$

Here Reaction 7.45 is analogous to the chain initiation, it leads to the formation of active V^{5+} sites; Reactions 7.46 through 7.48 are similar to those of chain propagation, where the role of active intermediates is played by both V^{4+} sites and oxygen O^- and O_2^- species (or $V^{5+}O^-$ and $V^{5+}O_2^-$); Reaction 7.49 corresponds to the surface regeneration and chain termination. The stationary concentration of adsorbed oxygen species far exceeds the equilibrium one. CO oxidation occurs due to the conjugation with the reductive reaction of the catalyst surface. The chain length, which equals the ratio between the rates of Reactions 7.48 and 7.49, is of the order of several hundred links.

In the above example the catalyst was rather diluted: about 1% of V_2O_5 in SiO_2. The question arises about the role of electron and oxygen diffusion between sites in the course of this reaction. Roginsky[14] suggested the chain mechanism for CO oxidation on MnO_2 back in the 1930s

$$MnO_2 + CO = MnO_2(CO) \tag{7.50}$$

$$MnO_2(CO) = MnO_2 + CO \tag{7.51}$$

$$MnO_2(CO) + O_2 = MnO_2(O) + CO_2 \tag{7.52}$$

$$MnO_2(O) + CO = MnO_2 + CO_2 \tag{7.53}$$

$$MnO_2(CO) = MnO + CO_2 \tag{7.54}$$

$$MnO + \frac{1}{2}O_2 = MnO_2 \tag{7.55}$$

In his opinion, the mechanism of appearance of the active surface $MnO_2(CO)$ complex is similar in physical meaning to the chain one, if the catalytic sites with active oxygen are considered as initiating sites. This is possible for MnO_2 catalyst with a low number of active sites, but it is impossible for highly active catalysts. Reactions 7.50, 7.54, and 7.55 give rise in the latter case to the simple redox mechanism without chains. Note that at that time, there was no opportunity to observe the active sites under catalytic conditions.

In our work in collaboration with Spiridonov and Gati[556] the mechanism of olefins isomerization $[R_{(1)}H \rightarrow R_{(2)}H]$ has been studied over MoO_3 catalyst. The study of catalyst ESR spectra *in situ* has led to the conclusion about the formation of the active sites by the redox mechanism

$$Mo^{6+}_{s_s}O_s + R_{(1)}H \longrightarrow [Mo^{5+}_sO_sH]R^+_{(1)} \tag{7.56}$$

$$[Mo^{5+}O_sH]R^+_{(1)} + Mo^{6+}_BO_B \longrightarrow [Mo^{6+}_sO_s]R_{(1)} + Mo^{5+}_BO_B + H^+ \tag{7.57}$$

where S and B subscripts denote the catalyst surface and bulk, respectively.

The reduction reaction of the surface

$$[Mo^{6+}_sO_s]R_{(1)} \longrightarrow Mo^{6+}_s[\ ]_s[R_1]O \tag{7.58}$$

results in the vacancy o_s formation which diffuses slowly into the bulk with the stabilization near Mo^{5+} ion. This process is equivalent to the oxygen diffusion to the surface

$$Mo^{6+}_s[\ ]_sR_{(1)}O + Mo^{5+}_BO_B \longrightarrow Mo^{6+}_sO_s + Mo^{5+}_B[\ ]_B + R_{(1)}O \tag{7.59}$$

The olefin isomerization itself (Steps 7.60 and 7.61) likely proceeds through the acid (of the Lewis type) mechanism on Mo^{6+}_s $[\ ]_s$ site during the period of vacancy $[\ ]_s$:

$$Mo^{6+}[\ ]_sR_{(1)}O + R_{(1)}H \longrightarrow Mo^{6+}\begin{matrix} \diagup R_{(1)}H \\ \diagdown R_{(1)}O \end{matrix} \tag{7.60}$$

$$Mo^{6+}_s\begin{matrix} \diagup R_{(1)}H \\ \diagdown R_{(1)}O \end{matrix} \longrightarrow Mo^{6+}_s\begin{matrix} \diagup R_{(1)}H \\ \diagdown R_{(1)}O \end{matrix} \longrightarrow Mo^{6+}_s[\ ]_sR_{(i)}O + R_2H \tag{7.61}$$

The Stages 7.60 and 7.61 describe the chain process itself. Stages 7.56 to 7.58 form the active site and initiate the chain (catalytic) reaction. Activation proceeds irrespectively to the catalytic process immediately when the reagents are passed over the surface. The reaction (and chain) termination occurs in Stage 7.61 due to the occupation of the vacancy.

For each act of olefin oxidation ($R_{(1)}H \rightarrow R_{(1)}O$) and surface reduction ($Mo^{6+} \rightarrow Mo^{5+}$) there are hundreds of acts of catalytic isomerization. The process is not stationary unless a small amount of oxygen is doped to olefin.

Surface chains are observed in a lot of polymerization reactions. The most extensive study was carried out for the Fischer-Tropsch synthesis of hydrocarbons over iron and cobalt catalysts.[25,557-559] The chain mechanisms are accepted for description of hydrocarbon synthesis from methanol on zeolites (the so-called Mobile-process)[560] as well. In the chain mechanism of Nijs and Jacobs,[561] the hydrocarbon decomposition is in the branching stage

$$C_mM_{2m} + nCH_3OH \longrightarrow C_{m+n}H_{2m+2n} + nH_2O \longrightarrow C_mH_{2m} + C_nH_{2n}$$

Nevertheless direct evidence in favor of this mechanism is absent.

Barelko et al.[12,562,563] studied NH_3 oxidation on Pt and revealed a set of nonstationary phenomena such as abrupt increase of reaction rate at a definite temperature or concentration, effects of decay, ''memory'', and others. For explanation these authors suggested the hypothesis of the branched-chain reaction: an elementary reaction passing over one active site of the catalyst is accompanied by creation of at least one more active site. Such sites were adsorbed atoms of the catalyst itself that were supposed to form a two-dimensional quasigas in equilibrium with its own lattice. If a surface reaction is accompanied by the concentration increase of such Pt atoms over the equilibrium value, it would cause a catalyst reconstruction in the course of the reaction. In support of this conjecture these authors refer to the results of Roginsky, Tretjakov, and Shekhter[22] on catalytic corrosion.

Indirect evidence in favor of a chain mechanism may serve also the phenomenon of heterogeneous-homogeneous catalysis when active species (as a rule, free radicals) desorb from the surface into the gas phase and initiate a chain reaction there.

Thus, contrary to widespread views, active sites in catalysis are created, as a rule, in the course of the reaction rather than existing from the very beginning. This allows us to derive formal kinetic equations which are similar in a number of cases to those for chain reactions. Certain of the above schemes are supported by experiment. Naturally, in the formal kinetics of the catalytic reaction the characteristic features of chain reactions may not appear. This would be the case, for example, when the rates of formation and deactivation of active sites are high under catalytic conditions.

Both in chain and catalytic reactions nonequilibrium concentrations of intermediates may occur for certain relations between the rates of individual stages. A possibility of kinetic conjugation arises when thermodynamically hindered stages are allowed. An apparent shift of equilibrium in some individual stages is observed. We shall discuss these problems in Chapter 8.

Measuring the rates of direct and reverse catalytic reactions close to equilibrium is a possible check for the chain mechanism. For common (non-chain) chemical reactions the equilibrium constant is equal to $K = k_1/k_{-1}$, where k_1 and k_{-1} are the rate constants of the direct and reverse reactions, respectively, in accordance with the detailed balancing principle. For chain reactions (even for nonbranched-chain reactions in stationary conditions) this is usually not the case, i.e., the ratio of the direct and reverse rate constants is not equal to the equilibrium constant. Following Laidler and co-workers[564] consider, for example, a reversible reaction $A \rightleftarrows B + C$ passing in accordance with the chain mechanism through the formation of active intermediates (radicals) R and R′. The direct reaction is first order, and the reverse one is second order. For the direct reaction we have

$$2A \xrightarrow{k_1} 2R$$

$$A + R \xrightarrow{k_2} B + R'$$

$$R' \xrightarrow{k_3} R + C$$

$$2R \xrightarrow{k_4} RR$$

where the first stage is the initiation, the second and third ones correspond to chain propagation, and the fourth refers to the chain termination; RR is a minor product. The rate of direct reaction is defined by

$$-d[A]/dt = k_3(k_1/k_4)^{1/2}[A] \tag{7.62}$$

For the reverse reaction we have

$$2C \xrightarrow{k_1'} R' + R''$$

$$R' + B \xrightarrow{k_{-2}} R + A$$

$$C + R \xrightarrow{k_{-3}} R'$$

$$2R \xrightarrow{k_4} RR$$

where R″ is a radical not involved in the chain propagation. The rate of the reverse reaction is given by

$$-d[B]/dt = k_{-2}(k_1'/k_4)^{1/2}[B][C] \tag{7.63}$$

Note that the product $(k_3k_{-2})\ (k_1/k'_1)^{1/2}$ is not equal to the equilibrium constant K.

A question arises whether the chain mechanisms can be disregarded, if $K = k_1/k_{-1}$. For example, in the elementary reaction $O_2 + M \rightleftarrows O + O + M$ on metals, this relationship has been checked in the rate interval ranging over 40 orders and turned out to be valid.[565] The cases when the rate of initiation (the rate of creation of active sites) is high, but the rate of chain termination (disappearance of active sites) is low, seem to be of widespread occurrence under catalytic conditions. Then the rate of initiation can be neglected in the stationary approximation. The relaxation time of the catalyst active structure (deactivation or "chain termination") is extremely long, and the catalyst operates in the nonequilibrium and quasistationary (with respect to the reaction) state. Naturally, in this case the relationship between the equilibrium constant and the rate constant of the catalytic reaction remains valid.

In summary, let us enumerate all properties characteristic of the chain reaction that are often observed in catalysis: (1) the alternating creation and annihilation of active sites (occupation and vacation of surface sites); (2) the creation of overequilibrium concentrations of intermediates proving to overcome the endothermicity of unfavorable stages; (3) the attainment of a maximum rate in due course of the reaction.

Chapter 8

THERMODYNAMICS OF IRREVERSIBLE PROCESSES AND DISSIPATIVE STRUCTURES IN CATALYSIS

The generation of nonequilibrium distribution of particles over different degrees of freedom in heterogeneous processes has been illustrated in Chapter 3 by molecular deactivation and in Chapter 4 by molecular beam experiments. Hence, various extrapolations of the processes of adsorption, desorption, and catalysis have been made. Both deactivation and beam experiments are performed in nonequilibrium conditions by themselves. Only a few catalytic reactions give rise to the evident formation of excited particles in the course of reaction. In the previous chapter we demonstrated that on the macroscopic level, deviations from the equilibrium should also be expected. This makes catalytic reactions similar to chain ones.

Menzel[566] examined different causes of nonequilibrium effects in catalysis. Along with microscopic nonequilibrium, which he called nonequilibrium energy distribution over different degrees of freedom, are the following effects: the irreversibility of experimental conditions (for example, adsorption and desorption are measured under different conditions); the deviation from the equilibrium inside the adsorbed layer; and kinetic conditions which result in several alternative reaction paths.

Let us consider the nonequilibrium catalytic processes on the macroscopic level taking into account the interaction of reaction media with a catalyst. Such processes are the subject of thermodynamics of nonequilibrium processes[26] and synergetics.[27]

8.1. THERMODYNAMICS OF NONEQUILIBRIUM PROCESSES

8.1.1. PRINCIPAL CONCEPTS AND DEFINITIONS

In thermodynamics it is conventional to divide all material objects into two constituent parts: "the system" under investigation and "the environment" which supplies energy and matter to the system. The system is closed if it does not interact with the environment. Open systems interact with the environment.

The systems can be in an equilibrium and nonequilibrium state. The first ones do not change their properties without interaction with the environment. In contrast, the latter undergo changes without any effect on the environment.

Equilibrium (reversible) processes are described quantitatively in classic thermodynamics. Unlike the latter, thermodynamics of nonequilibrium (irreversible) processes establishes only nonequalities directing possible ways of the process.

The systems can be in stationary and nonstationary states as well. In the stationary state the properties of the system do not vary with time. For the closed system stationary states are always in equilibrium, and nonstationary ones are in nonequilibrium. The open system permanently exchanging with the environment comes to the stationary state at the end. The latter may be in equilibrium if the system does not alter its properties on isolation, or in a nonequilibrium stationary state in the opposite case.

In heterogeneous catalysis stationary states arise on passing over a catalyst reagent flow (for instance, in a flow reactor) with a constant velocity, temperature, and composition. Upon entering the stationary state, the system (in this case it is the catalyst plus reagents in the catalytic reactor) may be in equilibrium or nonequilibrium. The catalyst is often described as being in equilibrium with the environment if upon turning off the reagent flow the catalyst does not change its composition. Clearly, the reagent composition over catalyst is also taken into account.

The thermodynamics of irreversible processes is based on linear relations between fluxes J_i and thermodynamic forces X_i for small deviations from the equilibrium, so-called phenomenological relations

$$J_i = \sum_j L_{i,j} X_j \tag{8.1}$$

where L_{ij} are kinetic (phenomenological) coefficients, which are known from experiment. A flux is the quantity of matter, energy, or momentum passing through a unit area in a unit time; the thermodynamic force is defined as the gradient of concentration, temperature, potential, etc., which gives rise to the appearance of irreversible fluxes, e.g., diffusion, heat conduction, electric current, etc.

For the simplest cases of direct processes, the law $J_i = L_{ij}X_j$ has long been known from the experiment; for instance, the temperature gradient causes the flux of matter (diffusion — Fick law); the potential gradient, electric current (Ohm's law). In general, the fluxes are interplayed, and thermodynamic forces X_i can induce the flow I_i for $i \neq j$; for example, the temperature gradient results in a flux of matter in a multicomponent system (thermodiffusion), and the concentration gradient gives rise to the energy flow (Dufour effect). For such cross-coupled processes in the thermodynamics of irreversible processes, coefficients L_{ij} are usually assumed to obey Onsager's relation

$$L_{ij} = L_{ji} \tag{8.2}$$

which follows from the principle of microscopic reversibility. This relation is strictly valid only for small deviations from the equilibrium when fluctuations are linearly connected with acting forces.

Entropy production is another important quantity in the thermodynamics of irreversible processes. It is defined as the quantity of entropy produced in a unit time σ = dS/dt. The value of δ is related to the forces X_j and flows J_i by the relation

$$\delta = dS/dt = \sum_i J_i X_i = \sum_{i,j} L_{ij} X_i X_j \geq 0 \tag{8.3}$$

For equilibrium systems $\sigma = 0$ (the second law of thermodynamics), all flows J_i and forces X_i disappear; in the opposite case $\sigma > 0$.

Suppose an open system is described by n independent thermodynamic forces $X_1, X_2 \ldots X_m \ldots X_n$. If m is held constant (for example, the temperature gradient is kept constant supplying the heat to the boundaries of the system), then we can find the state with the minimal σ by the differentiation of Equation 8.3 over X_i:

$$\frac{\partial \sigma}{\partial X_i} = \sum (L_{ij} + L_{ji}) X_i = 0 \tag{8.4}$$

for i = m + 1 . . . n. Taking into account Onsager's relation (8.2) we arrive at

$$2 \sum_{j=1}^{n} L_{ij} X_j = 0 \tag{8.5}$$

where i = m + 1 . . . n. We have just proved the Prigogine theorem that in the stationary state with minimal entropy production, all flows for i = m + 1 . . . n disappear.

Variation of α over perturbations δX_i gives

$$\sigma = \sigma_o + \delta\sigma + \frac{1}{2} \delta^2 \sigma \tag{8.6}$$

Taking into account of Equation 8.5 and substituting Equation 8.3 for the entropy production in Equation 8.6, we obtain that the stability of the system is defined by the second-order term

$$\delta^2 \sigma = L_{ij} (\delta X_{ji})^2 > 0 \tag{8.7}$$

Hence

$$J_i \delta X_i > 0 \tag{8.8}$$

So, a variation δX_i induces in the system a flux J_i with the opposite sign, and the system returns to the state with constant entropy production.

8.1.2. APPLICATION TO CHEMICAL REACTIONS

In the thermodynamics of irreversible processes Equations 8.1 to 8.8 are only valid for deviations from equilibrium that are not very large. This is also true for chemical reactions. The application of the thermodynamics of irreversible processes often allows us to connect the thermodynamic and kinetic parameters, in particular, to determine the rates of irreversible processes vs. outside conditions.

When a chemical reaction is the only process in the system, the variation of its entropy is related to the change of the numbers of moles of components

$$dS = -\frac{1}{T}\sum_i \mu_i dn_i = -\frac{1}{T}\sum \nu_i \mu_i d\zeta > 0 \quad (8.9)$$

where μ_i is the chemical potential of the ith component; dn_i is the change of the mole number; ν_i is the stoichiometric coefficient of the reaction; $\zeta = n_i/\nu_i$ is the so-called extent of the reaction, which is equal to the number of converted equivalents of the substance involved; $\zeta = 1$, if one equivalent has been converted.

The value

$$A_i = \sum_i \nu_i \mu_i \quad (8.10)$$

was introduced by De Donder.[567] He called it the reaction affinity (not to be confused with the chemical affinity in classic thermodynamics).

It is easy to verify that the reaction affinity is equal to the partial derivative of Gibbs energy with respect to the ζ value at constant T and P, and taken with the opposite sign,

$$A_i = -(\partial G/\partial \zeta)_{t,p} \quad (8.11)$$

It may be regarded as a thermodynamics force X_i. The reaction rate corresponds to the flow J_i in this case.

In a chemical system entropy production can be expressed through the reaction affinity A_i and the reaction rate r_i

$$\sigma = dS/dt = (A_i/T)r_i \quad (8.12)$$

Since $T > 0$, taking into account Inequality 8.3 we find

$$A_i r_i > 0 \quad (8.13)$$

Hence, it follows that the affinity A_i and the rate r_i have the same signs

$$A_i > 0; r_i > 0$$

or (8.14)

$$A_i < 0; r_i < 0$$

i.e., in order for the reaction to proceed in the forward direction ($r_i > 0$), the positive affinity ($A_i > 0$) is necessary, and the negative one ($A_i < 0$) for the reverse reaction ($r_i < 0$). Prigogine et al.[572] have shown that for multistage processes the condition of spontaneous reaction proceeding is defined by the system of inequalities

$$\sum_i A_i r_i > 0 \tag{8.15}$$

where for each step Inequality 8.13 is true, i.e., it is governed by their affinity.

The relationship between the affinity and the reaction equilibrium constant K at arbitrary composition is defined by the expression

$$A_i = RT \ln\{K/\Pi_i(a_i/[a_i])\}^{\nu_i} = A_i^o - RT \ln\{\Pi_i(a_i/[a_i])\}^{\nu_i} \tag{8.16}$$

where a_i is the activity of a corresponding component, $[a_i]$ is the unit of measurement of the activity (for ideal gases these are C_i and $[C_i]$), and A_i^0 is the standard affinity.

The relationship between the rates of the forward and reverse reactions can also be obtained. Using the relation $K_i = k_i/k_{-i}$ (where k_i is the rate for the forward reaction and k_{-i} is the rate for the reverse reaction), which is always true for any elementary step, we derive from Equation 8.16 the equation of De Donder[567]

$$r_i/r_{-i} = \exp(A_i/RT) \tag{8.17}$$

where r_i and r_{-i} are the rates for the forward and reverse reactions, respectively.

Equation 8.17 describes the kinetic irreversibility from the viewpoint of the thermodynamic moving force. In this form it is applicable only to elementary reactions.

8.1.3. APPLICATION OF THE CONCEPT OF AFFINITY TO KINETICS AND CATALYSIS

The concept of affinity was utilized for the explanation of mechanisms of catalytic and noncatalytic processes.[568-570]

The total affinity of the reaction is equal to the sum of affinities of individual steps

$$A_i = \sum_i s_{ik} A_{ik} \tag{8.18}$$

where s_{ik} is a stoichiometric number of kth step in the ith reaction.

Temkin[571] introduced the notion of the mean stoichiometric number

$$\bar{s} = \frac{\sum_k s_{ik} A_{ik}}{\sum_i A_{ik}} = \frac{A_i}{\sum_i A_i} \tag{8.19}$$

Then using Equation 8.17, we obtain the generalized De Donder's relation for the overall reaction

$$r_i / r_{-i} = \exp(A/\bar{s}RT) \tag{8.20}$$

Close to the equilibrium

$$A/\bar{s}RT << 1 \text{ and } r = r_i(A/\bar{s}RT) \tag{8.21}$$

In the limit $A/\bar{s}RT \rightarrow 0$, the rate r_i approaches the equilibrium rate r_{eq}. A linear relation between the rate of ammonia synthesis and the affinity has been established near the equilibrium.[571] It turns out that the reaction stoichiometric number is $\bar{s} = s_{der} = 2$.

Prigogine et al.[572] have proved the linear relation between the rate and the affinity for the reaction of cyclohexane dehydrogenation and benzene hydrogenation

$$C_6H_{12} \rightleftarrows C_6H_6 + 3H_2$$

on either side of the equilibrium. Boudart[570] examined small systematic deviations from the equilibrium in this system and showed that $\bar{s} = 3$. This corresponds to the reaction scheme

$$C_6H_{12} \rightarrow C_6H_{10} + H_2$$

$$3C_6H_{10} \rightleftarrows C_6H_6 + 2C_6H_{12}$$

Far from the equilibrium, the relation (Equation 8.21) is not fulfilled in the general case.

Kreuzer has applied nonequilibrium thermodynamics for the description of kinetics of surface reactions. The kinetics of two-dimensional phase transformations[573] and desorption from the precursor state[574] has been studied.

Boudart[570] applied the concept of affinity to examine the direction of conversion in the system of complex consecutive reactions in catalysis and for chain processes.

In accordance with Inequalities 8.13 to 8.15 and Equation 8.18, $A_i > 0$ and $s_i A_i r_i > 0$ for all steps, i.e., the affinities of all steps in the chain of successive reactions are positive, and their sum is equal to the total affinity $\Sigma s_i A_i = A$. So, the thermodynamic conjugation (proceeding of thermodynamically disadvantageous reactions at the cost of an advantageous one) is impossible in the consecutive set of steps for the stationary state. For example, noncatalytic oxidative dehydrogenation of C_4H_8 according to the scheme

$$C_4H_8 = C_4H_6 + H_2 \ (A_1r_1 < 0)$$

$$H_2 + \frac{1}{2}O_2 = H_2O \ (A_2r_2 > 0)$$

is impossible, though the total affinity is positive

$$C_4H_8 + \frac{1}{2}O_2 = C_4H_6 + H_2O \ (A_1r_1 + A_2r_2 > 0)$$

If only one step with negative affinity is present ($A_i < 0$), the reaction is arrested.

But so-called kinetic conjugation in the chain or catalytic reactions is still possible when the concentration of intermediates is greater than the "equilibrium" one, if they are products of the reaction. The kinetic conjugation is possible for some relationship between the rates of individual steps. Consider a step of the complex reaction[570]

$$X_{\langle 1\rangle} + M_{\langle 1\rangle} \rightleftarrows X_{\langle 2\rangle} + M_{\langle 2\rangle}$$

where $X_{\langle 1\rangle}$ and $X_{\langle 2\rangle}$ are intermediates, and $M_{\langle 1\rangle}$ and $M_{\langle 2\rangle}$ are molecules. If A_i is negative, the concentration $X_{\langle i\rangle}$ should be greater than the equilibrium one or $X_{\langle 2\rangle}$ lower than the equilibrium one to cause the reaction proceeding in the forward direction. For simplicity, suppose that $[M_{\langle 1\rangle}] = [M_{\langle 2\rangle}]$, and the stationary concentration of $X_{\langle 1\rangle}$ is equal to equilibrium $[X_{\langle 1\rangle}]_{st} = [X_{\langle 1\rangle}]_{eq}$. Then, even if the standard affinity of a step $A_i^o < 0$, its affinity can be positive as it seen from the equation

$$\begin{aligned} A_i &= A_i^o + RT\ln\{[X_{\langle 1\rangle}]_{eq}/[X_{\langle 2\rangle}]_{st}\} \\ &= RT\ln\{[X_{\langle 2\rangle}]_{eq}/[X_{\langle 2\rangle}]_{st}\} \end{aligned} \quad (8.22)$$

It is seen that the value of A_i is positive, if $[X_{\langle 2\rangle}]_{st} < [X_{\langle 2\rangle}]_{eq}$. The value of $[X_{\langle 2\rangle}]_{eq}$ in Equation 8.22 is a fictional equilibrium concentration which is reached in the course of the reaction. Similarly, if $[X_{\langle 2\rangle}]_{st} = [X_{\langle 2\rangle}]_{eq}$, then the value of A will be positive when $[X_{\langle 1\rangle}]_{st} > [X_{\langle 1\rangle}]_{eq}$.[560,564]

Temkin and Pyzhov[575] were first to introduce the idea of virtual pressure or fugacity. If certain, but not all, of gas-phase components are equilibrated with the surface, the concept of any gas nonequilibrated but capable of being formed from equilibrated gases is useful in considering conditions on the surface.

Temkin and Pyzhov[575] proposed the following mechanism of ammonia decomposition on metals

$$2NH_3 + 2M \underset{k_{-1}}{\overset{k_1}{\rightleftarrows}} 2M - N + 3H_2$$

$$2M - N \underset{k_{-2}}{\overset{k_2}{\rightleftarrows}} 2M + N_2$$

where M is an atom of a metal and dashed arrows denote virtual equilibrium. The quantity of chemisorbed nitrogen may correspond to the pressure of nitrogen, being in equilibrium according to the scheme: $2NH_3 \rightleftarrows N_2 + 3H_2$. If H_2 and NH_3 are adsorbed weakly, the reaction rate is proportional to the surface coverage by atomic nitrogen Θ_N, and to the virtual nitrogen pressure $[N_2]_{virt}$

$$\Theta_N = a_N[N_2]_{virt} = a_N[NH_3]_{st}/[H_2]_{st}^{3/2}$$

where a_N is the adsorption coefficient of nitrogen atoms. The virtual pressure can be very high: it equals 6000 bar for ammonia decomposition at 673 K. At the iron surface it is sufficient to form the nitride Fe_4N, though the nitride does not form in the atmosphere of N_2.

The concept of the virtual pressure received further development in the works of Kemball[576] and Boudart.[577] De Donder's equation (8.17) enables us to explain the moving forces of catalytic and chain reactions. In accordance with Equation 8.17, a reaction can be shifted to the right if the rate of the forward reaction is far beyond that of the reverse one $r_i >> r_{-i}$. For example, in the chain reaction of bromine with hydrogen

$$Br + H_2 \underset{k_{-1}}{\overset{k_1}{\rightleftarrows}} HBr + H$$

$$H + Br \overset{k_2}{\rightarrow} HBr + Br$$

the first endothermic stage shifts to the right due to the constant removal of hydrogen atoms in the second reaction. It is easy to obtain that in order for the chain reaction to proceed, the conditions $r_1 > r_{-1}$ and $k_1 > k_{-1}$ must be fulfilled. These inequalities provide the kinetic conjugation which permits us to remove the thermodynamic restrictions for the endothermic reaction of Br atom with H_2.

Note that using the energy of exothermic reactions in branched-chain reactions enables us to obtain tremendous concentrations of active intermediates $X_{(i)}$. For instance, in the branched-chain reaction of hydrogen oxidation the concentration of active particles-hydrogen atoms is 10^6 times greater than the equilibrium one. As a result, the endothermic reaction

$$H + O_2 \longrightarrow OH + O - 67 \text{ kJ/mol}$$

becomes possible, which leads to branching. Voyevodsky[555] has pointed out that chain reactions should be considered reactions catalyzed by active intermediates.

The relations of the type in Equation 8.22 are also true for catalysis in stationary conditions. For example, using Auger spectra, Boudart[570] studied ammonia decomposition on W and Mo films at 1000 K and 10^{-4} Pa and found that the stationary surface coverage by nitrogen atoms in the course of the reaction exceeds appreciably its equilibrium value in the absence of the reaction. In the stationary state the rates of the forward and reverse reactions of the second stage are expressed as follows

$$r_2 = ka_{(M-N)st} \qquad (8.23)$$
$$r_{-2} = ka_{(M)st}[N_2]_{st}$$

where $a_{(i)st}$ is the thermodynamic activity of surface intermediates in the stationary state. Were the rates r_2 and r_{-2} in Equation 8.23 equal to each other ("virtual equilibrium"), the virtual nitrogen pressure needed for such an equilibrium would be $[N_2]_{eq} = (k_2/k_{-2})\,(a_{(M-N)st}/a_{(M)st})$. Hence,

$$\frac{r_2}{r_{-2}} = \frac{k_2 a_{(M-N)st}}{k_{-2} a_{(M)st}[N_2]_{st}} = \frac{[N_2]_{eq}}{[N_2]_{st}} \qquad (8.24)$$

For the stationary state the ratio $r_2/r_{-2} = 6400$, i.e., the reaction irreversibility is extremely high. The stationary value of the concentration of nitrogen atoms at the surface is significantly greater than the concentration required for the equilibrium with $[N_2]_{gas}$. Nitrogen desorption, while endothermic, is substantially irreversible. In addition, the catalyst is not poisoned by the nitrogen adsorption, which would be the case for the equilibrated system. Here the kinetic conjugation is achieved by obtaining the overequilibrium nitrogen concentrations at the surface in stages preceding the final thermodynamically disadvantageous one, i.e., due to the interaction with ammonia. N_2 concentration is not involved in the equation for the reaction rate.

Boudart[570] supposed that the dehydrogenation of methylcyclohexane in toluene at the Pt surface[577] passes in a similar way. Here, the toluene concentration does not affect the reaction rate as much, which is attributed to the overequilibrium toluene concentration at the surface in the stationary state. Adding benzene to the mixture almost does not retard the reaction of toluene formation, which also supports the above proposal. Clearly, the equilibrium benzene concentration [B] is much less than the stationary toluene one [T]

$$[B]_{eq} \approx [T]_{eq} << [T]_{st} \tag{8.25}$$

So, the kinetic conjugation shifts the equilibrium of thermodynamically disadvantageous stages both in catalysts and in chain reactions.

8.2. DISSIPATIVE STRUCTURES IN CATALYSIS

8.2.1. STABILITY OF STRUCTURES WHICH ARE FAR FROM THE EQUILIBRIUM

So far we have considered the thermodynamics of irreversible processes for which there are small deviations from the equilibrium when the fluxes and thermodynamic forces are connected linearly in accordance with Equation 8.1. A quite different situation arises for large deviations from equilibrium when the linear relationship fails. For example, Ross et al.[579,580] showed that the principle of minimal entropy production (Equation 8.4) does not hold for such systems. These authors propose another function which is minimized in nonequilibrium stationary conditions. It is defined as the product of the driving force, similar to that of Equation 8.22, and the net of each intermediate and summing over the species.

New stationary states appear in structures which are far from equilibrium. They are stabilized due to the energy exchange with the environment and ordered in space and time. Prigogine et al.[26,572] called them dissipative structures.

The formation of dissipative structures is accomplished by an abrupt conversion from a system with a certain set of parameters to a system with another one. By analogy with thermodynamic phase transitions, such jump transitions are called kinetic phase transitions.

Dissipative structures have been classified into four types: (1) space-inhomogeneous structures, (2) periodic-in-time structures (autooscillations), (3) structures which are periodic both in space and in time (autowaves), and (4) bistable structures (of the type of "trigger"). As we shall see later, all four types of dissipative structures are encountered in heterogeneous catalysis.

Dissipative structures are observed in those cases when the following conditions are satisfied: (1) the system is thermodynamically open, i.e., energy and matter exchange with the environment is possible; (2) dynamic equations of the system are nonlinear; (3) deviations from the equilibrium exceed the critical values; and (4) microscopic processes occur cooperatively. In Section 8.1.1. we presented the general conditions (Equations 8.7 and 8.8) of system stability. Glansdorf and Prigogine[581] have proved that the stability

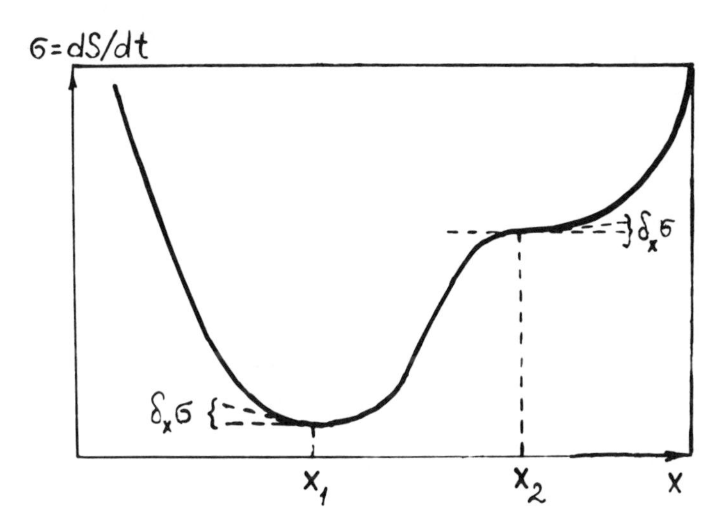

FIGURE 8.1. Entropy production as a function of path coordinate (stationary states of a nonlinear system; x_1 refers to equilibrium; x_2, to unstable stationary state.

of the system in a nonlinear regime is associated not with total entropy production, as in Equation 8.7, but with entropy production induced by the perturbation δX_i. In the expression for entropy production

$$\delta_x\sigma = 1/\rho \sum_i \delta J_i \delta X_i \tag{8.26}$$

where ρ is the density of the media, the right-hand side can be either positive or negative. It is seen from Figure 8.1. that the stationary state is stable only for the case when $\delta_x = \sigma > 0$. Glansdorf and Prigogine have derived the following general condition of stability

$$\delta_x\sigma = (1/\rho) \sum_i \delta J_i \delta X_i \geq 0 \tag{8.27}$$

Nicolis and Prigogine[26] pointed out that for regions far removed from the equilibrium, the situation changes and, generally, the thermodynamic approach along with the relations of the type in Equation 8.27 fail. The character of the chemical kinetics is of the first importance in this region, namely, the retardation of one or other stage due to the potential barrier.

Nevertheless, an application of the condition (Equation 8.27) has allowed us to analyze the stability of chemical systems in a number of cases. A simple chemical reaction of the type of $A + B \rightleftarrows C + D$ has always a stationary solution in the open system (for the closed systems the stationary solution corresponds to the equilibrium). This is not the case for autocatalytic systems.

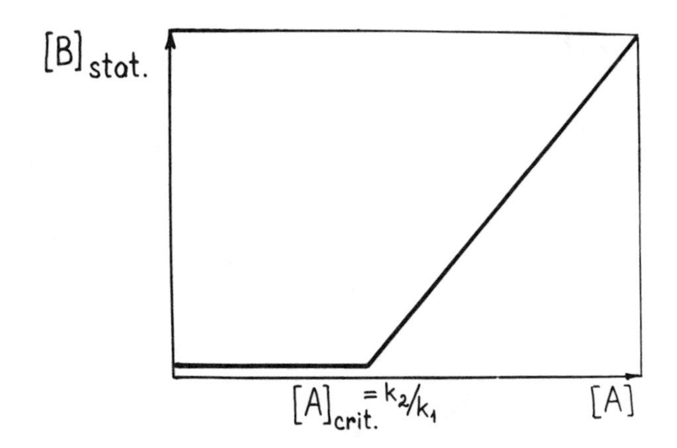

FIGURE 8.2. The stationary concentration of B product as a function of A concentration for the concentrations below and over the critical value.

For a simple autocatalytic reaction $X + Y \rightarrow 2X$ using the relationships in Equations 8.13 and 8.27, we have

$$\delta r \delta A = [V]\,\delta[X]\left(-\frac{\delta[X]}{[X]}\right) = -\frac{[Y]}{[X]}(\delta[X])^2 < 0 \qquad (8.28)$$

where r is the reaction rate and A is its affinity.

According to Equation 8.28, the reaction is not stable. The continued growth of the concentration [X] is observed for the autocatalytic conversion.

When the reaction $A \rightarrow B$ passes through the autocatalytic multiplication of the active sites according to the scheme

$$A + X \underset{k_{-1}}{\overset{k_1}{\rightleftarrows}} 2X,\; X \overset{k_2}{\rightarrow} B \qquad (8.29)$$

the existence and stability of the stationary state depend on the initial concentration of A. If the concentration [A] slightly exceeds the critical value, then the product concentration rapidly reaches its limiting value [B]′. In the theory of differential equations, the value of $[A]_{cr}$, at which the solution changes, is called the branching point, or bifurcation. The stationary concentration of the product [B] vs. the concentration of the input reagent [A] is the point where $[A] = k_2/k_1$ (Figure 8.2).

The variety of phenomena increases in the presence of two intermediates, X and Y. For example, for the autocatalytic system of Lotka-Volterra

$$A + X \overset{k_1}{\rightarrow} 2X;\; X + Y \overset{k_2}{\rightarrow} 2Y;\; Y \overset{k_3}{\rightarrow} B$$

we can obtain the single nonzero stationary solution

$$[X]_o = k_3/k_2 \qquad (8.30)$$
$$[Y]_o = (k_1/k_2)[A]_o$$

and periodic solutions

$$[X]_t = [X]_o + [X]\exp(i\omega t) \qquad (8.31)$$
$$[Y]_t = [Y]_o + [Y]\exp(i\omega t)$$

The concentrations of intermediates X and Y oscillate around the stationary states (Equation 8.30) with the frequency $\omega = \sqrt{[A]}$. The autooscillations, dissipative structures of a new type ordered in time, are observed.

In more complicated systems involving the diffusive exchange of particles with the environment through the interface, spatial dissipative structures are formed which include the spatial periodic structures as well (waves).

Nicolis and Prigogine[26] point out that fluctuations play an important part in the vicinity of bifurcation if the usual deterministic laws are valid between the two points of bifurcation, i.e., the laws of chemical kinetics in this case. It is fluctuations that initiate the kinetic phase transition and select the direction of motion for the branch of the differential equation solution of the system.

In complicated multicomponent systems the exchange of matter and energy between the system and its environment increases.[570] The amplitude of the fluctuation gives rise to bifurcation increase as well. Therefore, sufficiently complex systems are almost always in the metastable state.

8.2.2. POSSIBLE ROLE OF DISSIPATIVE STRUCTURES IN CATALYSIS

Consider now examples of dissipative structures in heterogeneous catalysis. The first type of dissipative structure, namely, spatially periodic structures, has been widely studied with the help of X-ray and electron diffraction methods. The remarkable discovery is the observation of two-dimensional surface phases in catalysis and adsorption. This discovery has led to the gradual departure from the statistical treatment of adsorption. When the adsorption process passes through a series of succeeding two-dimensional phase transitions until the monolayer coverage is reached, it should also reveal itself in the adsorption kinetic characteristics. The phase transitions are the most probable cause of the creation and annihilation of the active sites on the catalyst surface. We must take into account that these processes results in qualitative changes in the kinetic equations.

The literature on two-dimensional phase transitions on the surface is quite voluminous. We will return to these questions again in Chapter 9.

The autooscillations are the second type of dissipative structures. Shchukarev[582] observed the autooscillations in heterogeneous catalysis in the

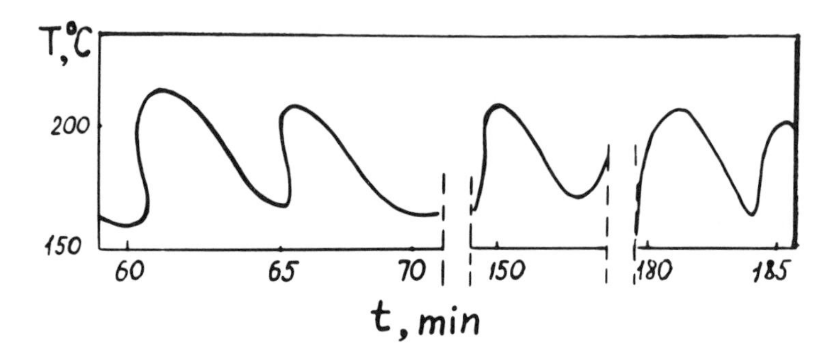

FIGURE 8.3. Autooscillations of cyclohexane oxidation on zeolite NaY; the catalyst temperature as a function of time. Cyclohexane concentration is 3.6 vol%, NaY mass is 300 mg, and the rate of gas mixture 100 cm^3/min.[583]

reaction of H_2 oxidation over a Pt catalyst as long ago as 1915. In the 1970s and 1980s, a host of papers on the autooscillations in heterogeneous catalysis appeared in the literature, mainly concerning catalytic oxidation. Tsitovskajia et al.[583] studied the autooscillations of cyclohexane oxidation over NaY zeolite, see Figure 8.3. The main products of the reaction are CO_2, CO, H_2O, and a small amount of H_2. Ukharsky et al.[584] have shown that peroxide

$$\begin{array}{c} \mathrm{R{-}\underset{\underset{O}{\|}}{C}{-}O{-}O{-}\underset{\underset{O}{\|}}{C}{-}R} \end{array}$$

is an intermediate in this reaction. On further decomposition and reoxidation it converts into a free radical in the gas phase, which initiates the combined heterogeneous-homogeneous process. Fast burning out in the gas phase is followed by slow accumulation of species on the zeolite surface, resulting in cyclic repetition of the process.

To study the causes of autooscillations it is necessary to measure simultaneously both the rate of the catalytic reaction and the rates of creation and decomposition of surface species. IR spectra of surface compounds in the course of the rate of autooscillations of cyclohexane oxidation over KJ zeolite are shown in Figure 8.4.[584] The reaction rate of autooscillations (CO_2 output concentration) at 510 to 530 K is accompanied in the same period by oscillations of IR bands of surface compounds at 1455 and 1720 cm^{-1}. The first band relates to the bending vibrations of CH_2-groups in the molecule of adsorbed cyclohexane, and the second to the carbonyl groups. Hence, it may be deduced that the reaction proceeds through the formation and decomposition of surface carbonyl compound.

Experimental data on autooscillations and corresponding mathematical models are described in a number of review papers.[585,586] Autooscillations and the feedback existing in the reactions are explained by such mechanisms

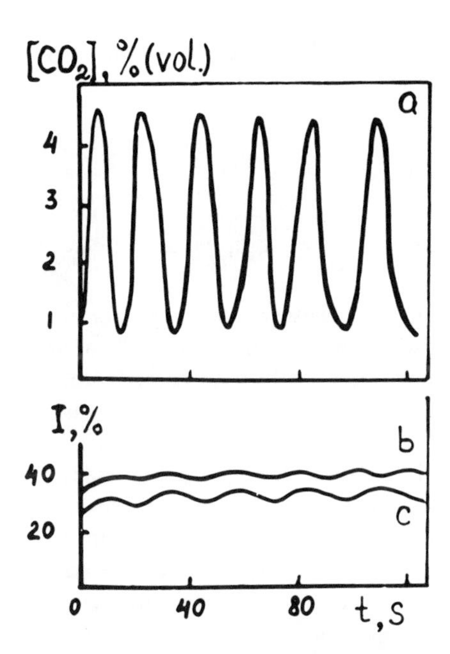

FIGURE 8.4. The variation of CO_2 concentration (a) and the intensities of 1455 cm^{-1} carboxyl-group band (b) and 1720 cm^{-1} carbonyl-group band (c) in autooscillatory cyclohexane oxidation (cyclohexane concentration is 2 vol%, the rate of flow 60 cm^3/min).[584]

as (1) the change of the activation energy of the reaction with surface coverage (for instance, as a result of its charging); (2) the thermokinetic mechanism, i.e., the heating and cooling, if the specific heat of the system is sufficiently large; (3) the homogeneous mechanism; (4) the existence of two or more active sites and adsorbed forms; and (5) the existence of two different phases (for example, oxidized and reduced).

The evidence of two-phases models has been proved recently in a number of works on this subject. Ertl et. al.[587,588] have proved the existence of two types of active sites in the autooscillation reaction of CO oxidation over Pt catalyst using the Video-LEED method which registers simultaneously the patterns from several surface structures. The simple cubic face Pt(100) with a (1×1) structure is unstable and undergoes reconstruction into (5×20), the so-called hex-structure. When CO coverage $\Theta_{CO} < 0.5$, the hex-structure is stable; at larger coverage it reconstructs into (1×1). The probability of impact interaction of O_2 with CO is 10^{-3} to 10^{-4} for the hex-phase and $10^{-1} \div 1$ for the (1×1) phase. The high rate of CO oxidation at (1×1)-phase leads to a Θ_{CO} decrease which results in the reverse reconstruction into a hex-structure, and the total cycle repeats again. The phase transitions proceed through the mechanism of the creation and growth of the nucleating centers.

Using tunnel electron microscopy allows one to study the mechanism of creation of this centers. It consists in the diffusion of CO particles into the overlayer above the islands.[589] Several phases successively change each other in the course of the autooscillations

$$\mathrm{Pt - CO\,(1\times1) + O_2 \rightarrow Pt\,(1\times1) + CO_2 \xrightarrow{CO,O_2} Pt - O\,(1\times1)}$$

$$\mathrm{+\ Pt\,(1\times1)\,(clean) + CO_2 \xrightarrow{CO} Pt - CO\,(hex) \rightarrow Pt - CO\,(1\times1)}$$

The boundary between phases is not abrupt. Forming phases create islands the size of which far exceed the atomic ones,[590] but are less than the size of the monocrystal 90 μm^2. On a Pt(110) surface the autooscillations are observed in a narrower temperature interval and are sensitive to small pressure changes.[591,592] Ertl et al.[593] explained these autooscillations by the phase transitions $(1 \times 1) \rightleftarrows (1 \times 2)$ under the influence of an adsorbate, and Vishnevsky and Elokhin[594] by the oxygen diffusion beneath the Pt surface. The autooscillations in CO oxidation over Pt(210)[595,596] are also attributed to phase transitions. The investigation using Video-LEED and *in situ* work function measurements has revealed that prior to starting the autooscillation regime, faceting of the (210) plane occurs, and it breaks up into pieces of (310) and (110) planes.[597] The autooscillations in CO oxidation over Pd(110) are explained by the periodic displacement of oxygen atoms into and out of the subsurface layer.

Autooscillations on the platinum surface have been observed in other reactions as well, such as NH_3 oxidation[598] and CH_3NH_2 decomposition[599] over polycrystalline Pt, and the reactions NO + CO[600] and NO + H_2[601] on Pt(100). The principle mechanism of autooscillations arising in these reactions has also explained through the formation and decomposition of surface phases.

Autooscillations in the ethylene oxidation rate on Pt[602,603] were attributed to phase changes. Vayenas and Michaels[603] using electrochemical methods have measured directly the oxygen activity in the process of autooscillations and have shown that oscillations arose close to the critical value a^*. Berry[604] has proved that this value corresponds to the formation of the surface oxide PtO. The data on autooscillations in ethene oxidation support the hypothesis of the periodic formation and decomposition of platinum oxide.

Autooscillations on oxide catalysts were observed in propylene oxidation over CuO[605] and CoO · MgO.[606] They were attributed to oxidation-reduction phase transformations of oxides.

At high temperatures the formation and decomposition rate of the two-dimensional surface phase (or cluster) grows, the oscillation period decreases, and in the limit the reaction transforms into the stationary nonperiodic regîme. Hence, raising the oscillation regime at low temperatures may serve as a possible indicator of passing the catalytic reaction through the phase reconstruction on the surface.

Up until now, there has been no convincing explanation of the causes of autooscillations synchronization in heterogeneous catalysis. Strictly periodic autooscillations are often observed not only at the monocrystals, but at powder catalysts as well, containing a large number of grains and microcrystals. Jaeger et al.[607,608] have tried to explain the autooscillation synchronizations in CO oxidation on Pd-containing zeolite by the phase transitions during the oxidation of palladium in combination with the thermal effects. The strongly exothermic reaction O(ads.) + CO(ads.) $\rightarrow$ CO_2 at the surface of Pt particles results in oxygen diffusion into the bulk and in the further phase transition Pd $\rightarrow$ PdO. PdO phase is inactive in catalysis, the catalyst activity drops followed by the slow reduction PdO + CO $\rightarrow$ Pd + CO_2, and then the next reaction cycle proceeds. The oscillations were shown to average at small Pd microcrystal sizes when the reaction passes in the stationary regime. Increasing the size of crystals makes O_2 diffusion between them easier and results in higher thermoconductivity and, as a consequence, in the autooscillations synchronization. For example, for Pd microcrystal of size 10 nm, the change of CO_2 concentration is 95%, whereas for a microcrystal of size 2 nm, it is only 10%.

One of the possible causes of autooscillation synchronization is the interaction through the gas phase.[609,610] Ehsasi et al.[611] used two individual crystals Pd(110) for the study of the synchronization problem. *In situ* measuring of the work function during CO oxidation revealed that autooscillations are synchronized very rapidly in a few seconds through the gas phase. According to Ertl et al.,[592] synchronization in the CO oxidation on Pt(100) proceeds through the surface reactions. At the Pt(110) phase, s_o for O_2 on different surface phases changes less. The reaction is very sensitive to small pressure changes, and synchronization proceeds through the gas phase. The transition from regular oscillations to chaotic ones was observed for CO oxidation on Pd(110).[612] The determined chaos can be described using the so-called strange attractor in which all trajectories are confined to the definite region of three-dimensional space. The transition to chaos passed through the doubling of the oscillation period.

The third type of dissipative structures, i.e., the structures periodic within space and time, or the autowaves, are observed on flat catalyst surfaces in the same cases that autooscillations are. Barelko et al.[12,562,613,614] have found the moving waves in NH_3 oxidations on platinum wire. After short heating of a small part of wire, the reaction transits at this segment of the catalyst from the initial kinetic regime with low catalyst activity to the diffusion one with high activity, and a moving wave can be generated in which high activity segments alternate with low activity ones. Similar results can be obtained when cooling the wire segment and transferring it from the diffusion to the kinetic regimen. The velocity of the wave motion depends on NH_3 concentration in the flow and on the wire temperature. For example, at 50 vol,% NH_3 concentration, the velocity of the wave motion grows from 0.2 to 1.3 cm/s with increasing the wire temperature from 273 to 573 K.

Ertl et al.[615] studied autooscillations in the rate of CO oxidation on Pt(100) using Video-LEED with surface scanning. The method has been improved

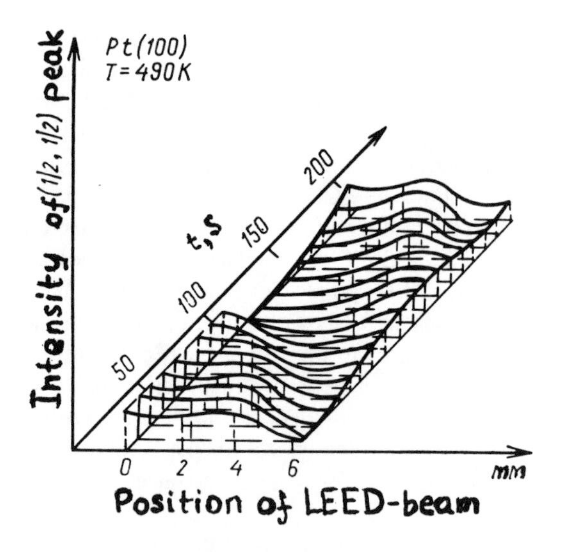

FIGURE 8.5. Variation of the line intensities for LEED signal in autooscillations of the CO oxidation rate on Pt(100) for the structure CO (2 × 2).[615]

to scan the surface structure using the electron beam 0.7 nm in diameter with the time resolution of 0.1 s at constant partial pressures $P_{CO} = 7 \cdot 10^{-3}$ Pa and $P_{O_2} = 7 \cdot 10^{-3}$ Pa.

The diffraction line intensities of the CO (2 × 2) structure on Pt(100) (1 × 1) plane measured in the linear interval of 0.6 cm for different instants of time is shown in Figure 8.5. Along with the autooscillations, the autowaves are clearly seen to propagate over the surface. The autowaves propagating in the phase opposite to the CO (1 × 1) structure were observed on a hex-structure as well. All these data are explained by the above-described phase transitions (1 × 1) ⇄ hex involving CO diffusion between the phase. The autowaves during CO oxidation on Pt(100) were also observed with the aid of scanning photoemission microscopy.[616,617] Vayenas et al.[602] believe that the autowaves in CO oxidation on Pt(100) is the main mechanism for the propagation of autooscillation between Pt particles.

A detached problem is the existence of stationary periodic structures or Turing structures in heterogeneous catalysis. Waves of such a kind have recently been revealed in studies of CO oxidation on Pd(110) using the high-resolution LEED method. Facets with approximately equal width of 50 to 70 lattice periods and similar orientation in a (110) direction are formed even in the absence of rate oscillations in CO oxidation. They look like regular ordered sawtooth-shaped facets (430) separated by flat terraces (110) and strongly represent nonequilibrium frozen structures.

The fourth type of dissipative structures, namely, the existence of two stationary regimes corresponding to the same parameters, has long been known in heterogeneous catalysis. Davis[618] investigated the critical phenomena in

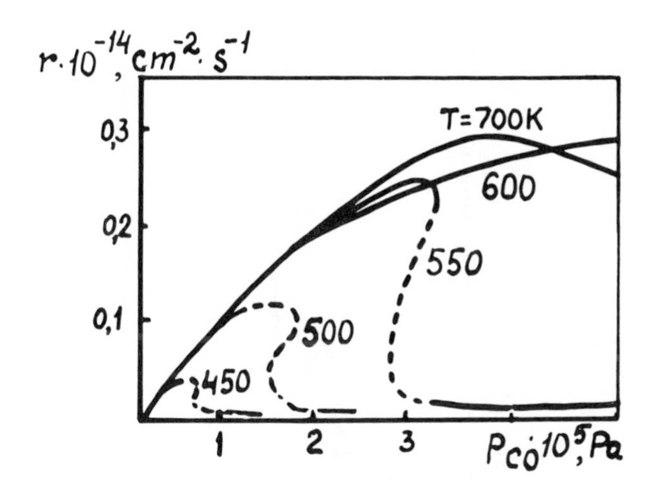

FIGURE 8.6. The stationary rate of CO oxidation on Pt as a function of partial CO pressure at different temperatures (the dashed line corresponds to the unstable state). (From Cox, M. P., Ertl, G., and Imbihl, R., *Phys. Rev. Lett.*, 54, 1725, 1985. With permission.)

the oxidation reactions of H_2, CO, and hydrocarbons on platinum. Frank-Kamenetsky[80] explained these phenomena by the sudden change of the reaction rate on reaching a certain temperature, i.e., by autoheating of the Pt surface and by transferring the reaction into the diffusion region. Since then, the heating mechanism has become dominant in the explanation of critical phenomena over a long period of time.

Nevertheless, further investigations revealed the nonthermal cause of the observed phenomena. Boreskov et al.[619] have found sharp changes in the stationary rate of oxidation of H_2 on Pt, Pd, and Ni for small changes in reagent concentrations. Slin'ko et al.[620] have attributed these changes to the coverage dependence of the activation energy of the reaction. Later, similar effects in CO oxidation on Pt have been explained by models including the competition between the adsorption and impact mechanisms.[621,622] The stationary rate of the process vs. CO pressure at different temperatures is shown in Figure 8.6. in accordance with these models. The two stationary states correspond to predominant oxygen coverage at low P_{CO} and predominant CO coverage at high P_{CO}. The sharp transition from the first, more active regime to the second, less active one proceeds with P_{CO} growth. Similar bistabilities have been observed in CO oxidation on $Cu_2Mo_3O_{10}$[623] and $LaMnO_3$.[624]

Madix et al.[625] have discovered the phenomenon of "catalytic explosion" in HCOOH decomposition on Ni(110). These authors attributed the autoacceleration of this surface reaction to the formation of two-dimensional nucleating centers of clean nickel surface covered earlier by Ni formate. Special attempts have been undertaken to check the possibility of a branched-chain reaction which could be responsible for the "autocatalytic explosion".[626] All possible intermediates arising from HCOOH decomposition such as hydrogen

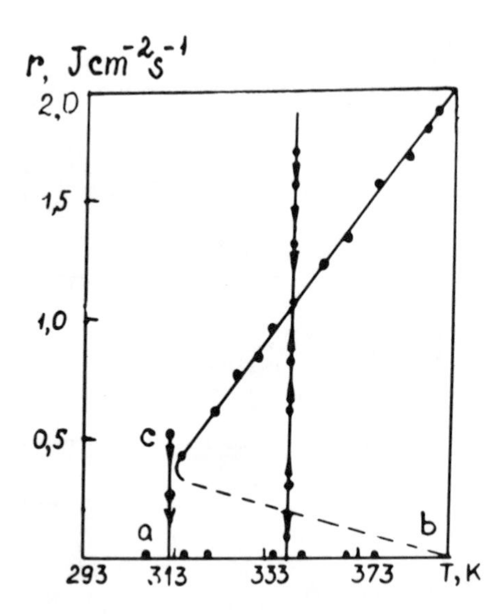

FIGURE 8.7. Temperature dependence of the stationary rate of NH_3 oxidation over Pt wire[619] (arrows show the direction of variation of the reaction rate; the curves *ab* and *cd* correspond to stable states; *bc,* unstable state; $[NH_3]$ = 1 vol%; the rate of flow is 1.5 cm/s).

atoms, CO, etc. have been introduced on Ni surface. All attempts to observe the branched-chain reactions have failed.

Barelko and Volodin[12] and Barelko and Merzhanov[613] observed the critical behavior in ammonia oxidation on thin Pt wires. The thermographic method developed earlier by these authors allowed them to compensate for the heat release in the reaction in 10^{-4} s. Using such a technique revealed that fast transition from the low-temperature to the high-temperature stationary regime (Figure 8.7.) results from the kinetic causes, but not from the "heat explosion". In certain conditions the third, nonstationary regime can be obtained; it is designated by the dashed line in Figure 8.7. NH_3 oxidation on Pt at low temperatures has a long induction period followed by a rapid rate of increase; "the activation energy" relative to the region of rate increase exceeds 400 kJ/mol. "Memory effect" has been observed; if Pt wire was cooled at any point in the induction period to a low temperature and then again heated, the process developed from the same point as if there were no cooling at all. These authors[12,613] attributed the observed phenomena of the branched-chain reaction to the formation of active sites on Pt surface.

Aptekar and Krylov observed critical behavior in calorimetric studies of the propylene oxidation over supported MoO_3/MgO catalyst.

Oxygen adsorption heat vs. oxygen coverage is shown in Figure 8.8 for 3% MoO_3/MgO catalyst at 473 K (Curve 1). The catalyst has been aged in a vacuum and in a hydrogen atmosphere, i.e., partly reduced. With increasing coverage, the adsorption heat decreases from 300 to 85 kJ/mol, last portions

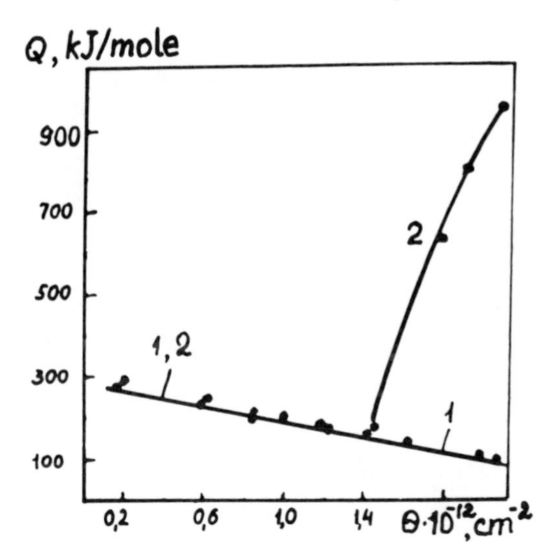

FIGURE 8.8. The differential adsorption heat of O_2 (1) and $O_2 + C_3H_6$ (2) on supported 3% MoO_3/MgO catalyst as a function of oxygen coverage at 473 K.

of O_2 being adsorbed with the noticeable activation energy. Using ESR technique has proved the appearance of the ion-radical O_2^- at adsorption heats of 80 to 100 kJ/mol.

In coadsorption experiments $C_3H_6 + O_2$, the first propylene portion does not adsorb and reacts with oxygen. The adsorption heats are the same as for pure oxygen adsorption, and propylene can be frozen out of the mixture at any point. It continues until the oxygen adsorption heat lowers to ~150 kJ/mol. At this point propylene begins to react with oxygen, the reaction heat increase rapidly (Curve 2), and oxygen consumption grows substantially compared with the individual O_2 adsorption.

So, propylene oxidation begins only when weakly bound O_2^- or possibly O_2^{2-} molecular species arise at the surface. It is not quite understood whether these oxygen species initiate the following propylene oxidation or serve as an indicator of the completion of the surface reconstruction and the formation of a more active surface phase. The fact that the activation energy for oxygen adsorption in the form of O_2^- is about 45 kJ/mol appears to support the second proposal, where the activation energy is required for surface reconstruction.

Further C_3H_6 oxidation proceeds with the participation of strongly bound oxygen. For example, if the intermediate in C_3H_6 oxidation is CO, it begins immediately to react with the more strongly bound (probably atomic) species with the adsorption heat 250 to 300 kJ/mol according to the scheme

$$C_3H_6 + O_2^- \text{ (ads.)} \rightarrow X;\ X + O_2^- \text{ (ads.)} \rightarrow CO + H_2O;$$

$$CO + O \text{ (ads.)} \rightarrow CO_2$$

where X is an intermediate.

The formation of ''dissipative structures'' in catalysis, autooscillations and autowaves, and the multiplicity of stationary states are manifestations of nonequilibrium in catalysis on a macrolevel. The number of works on this subject grows every year and points to the general features of such phenomena.

Chapter 9

PHASE TRANSFORMATIONS IN CATALYSIS

Two- and three-dimensional phases (2-D and 3-D phases, respectively) arising as a result of adsorption, for example, oxide films on metal surfaces, can play an important role in heterogeneous catalysis. As shown in the previous chapter, sharp changes of activity, autooscillations, and autowaves in catalysis can result from phase transformations on the surface or in the catalyst bulk.

The role of phase transitions in heterogeneous catalysis has been a widely debated topic for a long time. Opinions differ as to the equilibrium or non-equilibrium catalyst state in conditions of stationary catalysis. This question is of special importance in connection with the oxidation catalysis passing through the oxidation-reduction mechanism of the catalyst K (so-called Mars-Van Krevelen mechanism)[627]

$$K + \frac{1}{2} O_2 \rightarrow KO$$

$$KO + R \rightarrow K + RO$$

where R is the oxidized molecule.

9.1. THE POSSIBILITY OF PHASE TRANSITIONS IN STATIONARY CONDITIONS

The kinetic and thermodynamic aspects of creation and decomposition of solid phases

$$AB \text{ (solid)} \rightleftarrows A \text{ (solid)} + B \text{ (gas)}$$

were considered by Langmiur using the decomposition of calcium carbonate, e.g.,

$$CaCO_3 \rightleftarrows CaO + CO_2$$

back in 1916.[628]

When the reaction rates of the direct and reverse reactions are equal to each, we have

$$k\Theta_1 = P_B\Theta_2$$

where P_B is the partial pressure of component B; k is a constant; and Θ_1 and Θ_2 are surface coverages of the corresponding solid phases. When thermodynamically equalized, the partial pressure of B molecules does not depend on the relation between AB and A phases and is defined only by the temperature. This result must follow from the equality of the direct and reverse reactions in the equilibrium as well. This condition is fulfilled by the supposition that the reaction passes in the boundary between the phases.

From the physical viewpoint, the formation of a separate A phase is due to the fact that the desorption of B molecules weakens the bonds of adjacent B molecules with the surface, resulting in the nucleating center growth. In the opposite case, when the bonds strengthen or do not change, the next B molecules desorb from sites remote from the first one, thus, a solid solution A + AB occurs instead of the separate phase.

Using CO oxidation as an example Wagner and Hauffe[629] considered kinetic aspects of a possible coexistence of two phases for the metal-oxide catalyst in stationary conditions. If the process proceeds according to the redox scheme, it may be suggested that oxygen reacts with the reduced phase (metal) with a rate of $k_1P_{O_2}$, and CO reduces oxide with a rate of k_2P_{CO}. But when $k_2P_{CO} > k_1P_{O_2}$, the reaction will proceed on the metal surface. Conversely, when $k_2P_{CO} < k_1P_{O_2}$, the reaction proceeds on the oxide surface. Were the rates equal by accident, the changes in concentrations as a consequence of reaction passing would result in violation of the equality $k_2P_{CO} = k_1P_{O_2}$. In the case of the rate inequality, the boundary between phases must displace in one or another direction. This displacement of the boundary does not alter the relation between the rates of intermediate reactions and, thus, continues up to the total disappearance of one of the phases. So in stationary conditions any catalyst consists of the single phase.*

This conclusion has been developed in the work by Boreskov[630] and Bruns.[631] For example, Bruns[631] has shown that for a certain temperature region

$$r_1 = k_o' \exp(-E_1/RT)\, P_{o_2} > k_o'' \exp(-E_2/RT) P_{co} = r_2 \qquad (9.1)$$

and $r_1 \not\equiv r_2$ for another temperature region. The temperature at which $r_1 = r_2$ is defined by the equation

$$T = (E_1 - E_2)/R \ln(k_o' P_{o_2}/k_o'' P_{co})$$

i.e., the transition from the oxide phase to the reduced one occurs at a definite temperature under stationary conditions. The authors'[630,631] opinions are that intermediates cannot exist as individual phases. The more appropriate suggestion is that intermediate reactions lead to the formation of surface compounds or change the composition of the homogeneous solid phase, for example, create the defects in this structure.

* Note that is was Langmuir[628] who pointed out that concrete conditions of reaction passing with the phase-phase formation and decomposition cannot generally be predicted using the phase rule.

These authors[630,631] have also proposed the thermodynamic arguments in addition to their kinetic speculations on the failure of coexistence of two different phases under the stationary conditions of the oxidation-reduction process. Using the phase rule, it is easy show that two crystal phases together with the gas phase are unstable and should be substituted by a single phase in equilibrium. Further investigations[632] with the application of the phase rule to 2-D surface phases resulted in similar conclusions.

Later evolution of catalytic science proved the validity of Langmuir's notions connected with reaction passing in the vicinity of the boundary between phases in the two-phase system. To the contrary, the works[630-632] concluded the existence of a single-phase system under conditions of redox catalysis were invalid. They were based on the mistaken assumption that the catalyst is in an equilibrium state, and the velocity of the oxygen diffusion is infinitely large. In reality, the catalyst is in a nonequilibrium state in the course of the reaction. Several phases usually coexist under the conditions of stationary redox catalysis.

It is interesting that the same conclusion has been drawn by Wagner, one of the authors of the previously mentioned work.[629] Later[633] he considered a redox reaction which proceeds through the Mars-Van Krevelen mechanism on the metal or oxide catalyst, for example, the reaction of hydrogen oxidation

O_2(gas) $\rightarrow$ 2O (adsorbed on the surface or in the catalyst bulk)

O(ads.) + H_2(gas) $\rightarrow$ H_2O(gas)

The first reaction takes place mainly over the reduced catalyst. The phase composition can be characterized by the oxygen activity a, which is proportional to the square root from the oxygen virtual pressure in the gas phase, being in equilibrium with the surface.[633] The second reaction passes over the oxidized catalyst, and its composition is defined by the oxygen activity proportional to the ratio.

Let us denote the metal phase and the oxide phase by I and II, respectively. Then phase I is stable in the stationary state provided that

$$a(\text{stat.T}) < a(\text{I,II}) \tag{9.2}$$

where a(I,II) is the oxygen activity at the transition from state I to II. A similar condition for oxide stability can be written in the form

$$a(\text{stat.II}) > a(\text{I,II}) \tag{9.3}$$

For the reaction of hydrogen oxidation these conditions, i.e., the metal and oxide phase stability, respectively, can be expressed via the partial pressures

$$\left(\frac{P_{H_2O}}{P_{H_2}}\right) K_p < P^*_{O_2} \tag{9.4}$$

$$P_{O_2} > P^*_{O_2} \tag{9.5}$$

where $P^*_{O_2}$ is the equilibrium oxygen pressure in the process of oxide dissociation and K_p is the equilibrium constant of water dissociation. At thermodynamic equilibrium we have

$$P_{O_2} = \left(\frac{P_{H_2O}}{P_{H_2}}\right)^2 K_p \tag{9.6}$$

for the stoichiometric mixture. The relationships in Equations 9.4 and 9.5 are of equal importance and express the stability of one of the phases. In the stationary state, Equation 9.6 and Inequalities 9.4 and 9.5 should be considered yet independently. At 500 K $K_p = 3.3\ 10^{-54}$ Pa, the pressure values of the oxide dissociation are $P^*_{O_2}$ (PdO) $= 10^{-8}$ Pa, and $P^*_{O_2}$ (Pt O) $= 9 \cdot 10^{-2}$ Pa. There is a region of partial pressures at these values of equilibrium parameters where both inequalities are valid.

Figure 9.1. represents the functions of the stationary oxygen activity for phases I and II via the hydrogen partial pressure at constant pressures of oxygen and water. Figure 9.1a corresponds to the case when a(stat.I) $<$ a(stat.II), and Figure 9.1b to the opposite one when a(stat.I) $>$ a(stat.II). In correspondence with the Inequality 9.2, phase I is stable if the oxygen activity is less than a(I,II). Phase II is stable if the oxygen activity in phase II is greater than a(I,II). The regions of phase stability are shown by solid lines, and the regions of instability by dashed ones. The hydrogen pressure at which the stationary oxygen activity in phase I reaches its top value a(I,II) is designated by P_{H_2} (I, min). H_2 pressure at which the stationary oxygen activity in phase II reaches its low critical value is designated by P_{H_2}(II,max). Phase I is stable when $P_{H_2} < P_{H_2}$ (I,min).

The first case (Figure 9.1a) corresponds to the situation when both phases are stable in the intermediate region

$$P_{H_2}(\text{I,min}) < P_{H_2} < P_{H_2}(\text{II,max})$$

and the hysteresis effects are likely to arise. If at first the hydrogen partial pressure is less than P_{H_2} (I,min), phase II is stable, and the catalyst activity corresponds to this phase. Increasing the hydrogen pressure up to P_{H_2} (II,max) does not change the situation, and phase II remains stable. Further P_{H_2} increase results in the phase transition II → I because phase II becomes unstable, and the nucleating centers of phase I appear. If this transition occurs in the course of observation with a further decrease of P_{H_2}, phase I remains stable as long as P_{H_2} is greater than P_{H_2} (I,min). At $P_{H_2} < P_{H_2}$ (I,min), the phase transition I → II occurs.

The second case is shown in Figure 9.1b. Both phases I and II are unstable in the region

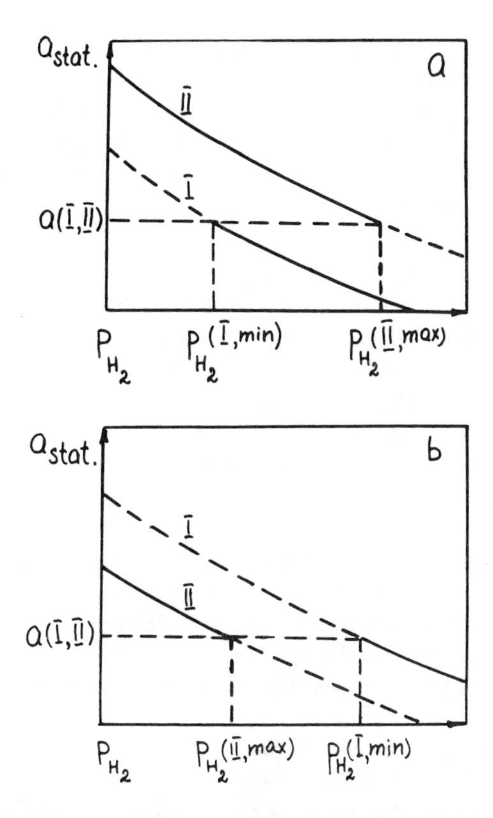

FIGURE 9.1. The stationary activity of oxygen in the I and II phases for reagent mixture $H_2 + O_2 = H_2O$ as a function of hydrogen partial pressure p_{H_2} for p_{O_2} = const and p_{H_2O} = const. (a) a(stat.I) < a(stat.II); (b) a(stat.I) > a(stat.II) (dashed line shows the unstable state).

$$P_{H_2}(II,min) < P_{H_2} < P_{H_2}(I,min)$$

Consider the following possibilities:

1. The nucleating centers of phase II are created when a(stat.,I) > a(I,II), and those of phase I when a(stat.,II) < a(I,II); the phase transitions are possible in the process of preparing and aging of the catalyst.
2. The phases I and II can coexist forming the phase mosaic with the availability of transport processes. The metal ions and electron diffusion from the phase with high oxygen activity to the phases with low activity occurs in the oxidized phases (CoO, FeO, Fe_3O_4). Besides, the oxygen diffusion may proceed through the oxygen vacancies in oxides. Diffusion flux can change along the boundary between the phases in correspondence with local concentration gradients. When the size of phase I and II domains in the mosaic is not high, the value of the parameter a(stat.,I) will be everywhere slightly greater than a(I,II), and a(stat.,II)

slightly less than a(I,II). The rate of creation of the nucleating centers at moderate supersaturations is very small, so further creation of nuclei of another phase is essentially stopped in separate regions of the mosaic despite the local instabilities of phases I and II. The catalyst becomes a two-phase one with $a(\text{stat.,I}) \approx a(\text{stat.,II}) \approx a(\text{I,II})$ in the region between P_{H_2} (II,max) and P_{H_2} (I,min).

In principle, if there are transport processes between the surface spots of different phases, the coexistence of two phases is also possible in the metal-oxygen system in the finite region of partial pressure P_{H_2} (II,max) $< P_{H_2} <$ P_{H_2} (I,min).[633] This consumption is in contradiction with the results of previously mentioned works[630-632] which do not take into account the transport processes between different phases. In the above consideration we assumed that the stationary values of the oxygen activity are completely defined by the nature of the solid phase.

In general, the situation seems to be more complicated. Stationary values of oxygen activity are different for different planes. In addition, oxygen transport between these planes is also observed, occurring by a variety of mechanisms. Therefore, the bulk stationary value of oxygen activity is an average over local ones for different faces.

Rieckert et al.[634,635] proved experimentally that the redox catalysts are two-phase in the stationary state of deep H_2 and C_3H_6 oxidation. The hysteresis loops have been observed on $CuO/Cu_2O/Cu$ and NiO/Ni catalysts when varying O_2 temperature and pressure for the direct and reverse processes. In a broad region of external parameters the oxidized and reduced phases of the catalyst coexist with each other. Oxygen removal from the catalyst and O_2 absorption are independent processes which are affected by particle size, morphology, and impurities. For example, adding Pd to the copper catalyst increases the hysteresis loop and stabilizes the oxygen-deficient Cu_2O phase.

A redox reaction passing through the Langmuir-Hinshelwood mechanism between two adsorbed molecules H_2(ads.) + O_2(ads.) or CO(ads.) + O_2 (ads.) is a limiting case of nonequilibrium where the catalyst bulk does not participate in catalysis. In the next section we consider an intermediate case when the near-surface layer is involved in the process.

9.2. NONEQUILIBRIUM STATE OF NEAR-SURFACE LAYERS IN OXIDIZED CATALYSTS

We have studied CO oxidation over the paraelectric phase of $BaTiO_3$.[636] Donor and acceptor atoms spaced a few tens of nanometers from the surface were proved to participate in the catalytic reaction.

In the temperature interval 400 to 440 K, the active sites of CO oxidation are impurity Mn^{2+} ions contained in $BaTiO_3$ bulk. Along with adsorption and catalytic measurements, it has been proven using *in situ* ESR spectra

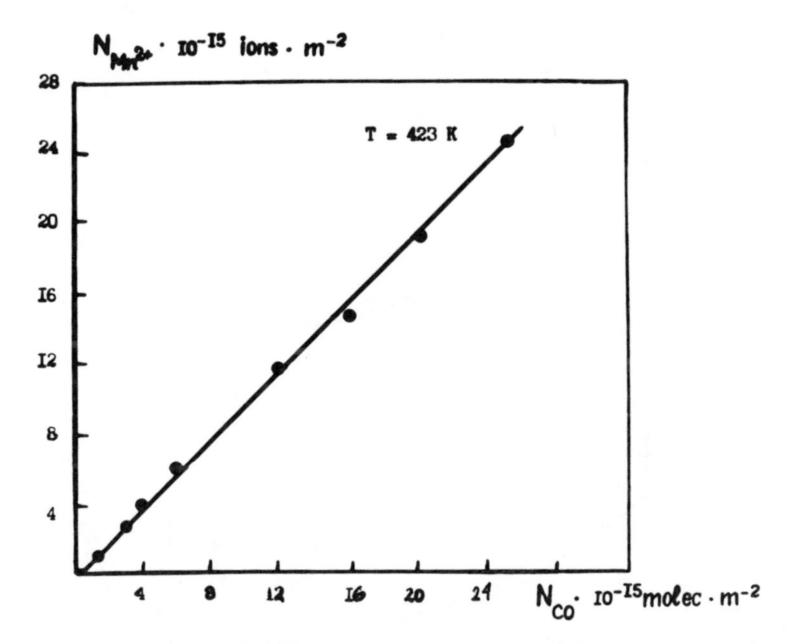

FIGURE 9.2. The concentration of CO molecules adsorbed on $BaTiO_3$ as a function of concentration of impure Mn^{2+} ions appeared after adsorption.[636]

studies of Mn^{2+} ions that they are contained in $BaTiO_3$ as a natural impurity with a concentration about 0.01%. It turned out (Figure 9.2) that in the interval of CO coverage from 0.3 10^{15} to 2.6 10^{16} CO molecules/m^2, CO adsorption proceeds in exact accordance with the stoichiometric equation

$$Mn^{4+} + CO + O_s^{2-} \rightarrow Mn^{2+} + CO_2 \tag{9.7}$$

i.e., adsorption of one CO molecule results in appearance of one Mn^{2+} ion. On the contrary, O_2 adsorption (see Figure 9.3) diminishes the intensity of the ESR spectrum in accordance with the stoichiometry

$$2Mn^{2+} + O_2 \rightarrow 2Mn^{4+} + 2O_s^{2-} \tag{9.8}$$

i.e., two Mn^{2+} ions disappear on adsorption of one oxygen molecule.

At high O_2 coverage, another stoichiometry is observed, two O_2 molecules per one Mn^{2+} ion, which is described by the equation

$$Mn^{2+} + 2O_2 \rightarrow Mn^{4+} + 2O_2^-$$

On exposure to the stoichiometric mixture 2CO + O_2, the catalytic reaction proceeds through the redox mechanism of Equations 9.7 and 9.8

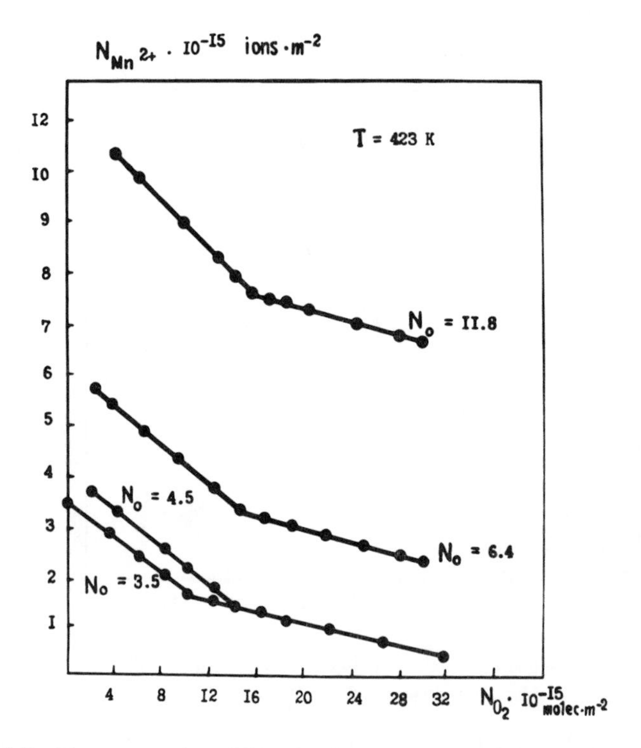

FIGURE 9.3. The concentration of O_2 molecules adsorbed on $BaTiO_3$ as a function of concentration of Mn^{2+} ions disappeared after adsorption for different N_O (N_O is the initial concentration of Mn^{2+} ions, m^{-2}).[636]

The rate of manganese reoxidation (Equation 9.8) is almost 2 orders of magnitude greater than the reduction rate (Equation 9.7). In the stationary regime, the steady and rather low Mn^{2+} concentration is settled. The reaction proceeds predominantly over the oxidized catalyst.

So, it is the bulk atoms and not the surface atoms that are the active sites in this catalytic reaction because the surface Mn^{2+} ions cannot be observed using ESR technique. They may be far removed from the surface (up to 30 nm). At such distances electron and hole exchange of Mn ions with adsorbed (or adsorbing) molecules cannot be explained by the electron tunnel mechanism. The activation energy of CO adsorption on Mn^{4+} is equal to 32 kJ/mol, and that of O_2 on Mn^{2+} is close to 50 to 65 kJ/mol. These values are rather close to the energy levels of local impurity centers (0.5 to 0.6 eV) created by noninteracting with each other's ions and vacancies in $BaTiO_3$ forbidden zone. It is reasonable to suggest that the electron diffusion proceeds via the polaron mechanism. Impure manganese atoms are far apart (10 to 20 nm), and their orbitals do not overlap each other.

At higher temperatures (438 to 473 K), along with manganese impurity Fe and Ti ions also take part in O_2 and CO adsorption and CO oxidation, i.e., the redox process of the base $BaTiO_3$ lattice occurs. In this case electron and hole diffusion is accompanied by oxygen and vacancy.

These investigations were continued in our studies of oxidation of allyl alcohol into acrolein[637]

$$C_3H_5OH + O_2 \rightarrow C_3H_4O + H_2O$$

using the hexagonal phase of MoO_3 doped with approximately 1 wt% of V_2O_5 as a catalyst. This reaction proceeds at 400 to 470 K with 100% selectivity. The reaction rate was measured by reactant concentrations in the gas phase. The concentration measurements of V^{4+} ions in the catalyst were performed simultaneously *in situ* using ESR data. Practically all V^{4+} ions were in the bulk of the catalyst.

Kinetic curves of the catalyst reduction characterized by the change of concentration $[V^{4+}]$ and the curves of allyl alcohol oxidation with the participation of lattice oxygen are opposite. Formation of one V^{4+} ion is accompanied by the disappearance of one molecule of allyl alcohol. On long exposure, $[V^{4+}]$ becomes stationary, and acrolein production practically ceases. The reaction is first order in $[V^{5+}]$ concentration and alcohol pressure with the activation energy of 67 kJ/mol according to the stoichiometric equation

$$V^{5+} \ldots O^- + C_3H_5OH \rightarrow V^{4+} + C_3H_4O + H_2O$$

V^{4+} ions observed in ESR spectra are uniformly distributed over the whole bulk of hexagonal MoO_3. Minimal spacing between V^{4+} ions, in other words, the distance at which V^{4+} spectrum does not broaden, is equal to 1 to 1.5 nm. The crystal size of V/MoO_3 is about 100 nm, and V^{4+} ions lie (approximately) at a depth of 60 nm. Continuous electron exchange between the surface and the catalyst bulk occurs in the course of catalyst reduction. One can envision the following elementary stages in the process

$$CH_2 = CHCH_2OH + O_s^{2-} \rightarrow CH_2 = CHCHO + H_2O + \square_s^{2-} \quad (9.9)$$

$$\square_s^{2-} + V_B^{5+} \ldots O^- \rightarrow \square_s^- + V_B^{4+} \ldots O^- \quad (9.10)$$

$$\square_s^- + V_B^{4+} \ldots O^- \rightarrow V_B^{4+}\square_B + O_s^{2-} \quad (9.11)$$

where $\square$ is the oxygen vacancy and the subscripts S and B are mean "surface" and "bulk", respectively.

The function of the reduction rate vs. V^{4+} concentration enables us to calculate the limiting concentration of vanadium ions participating in the process. Its value is $(5.7 \pm 4) \cdot 10^{19}$ g^{-1} (or $5.7 \cdot 10^{19}$ m^{-2} since the catalyst area is 1 m^2/g) which approximately corresponds to 2 or 3 oxygen monolayers involved in the reaction. The movement of electrons from the surface into

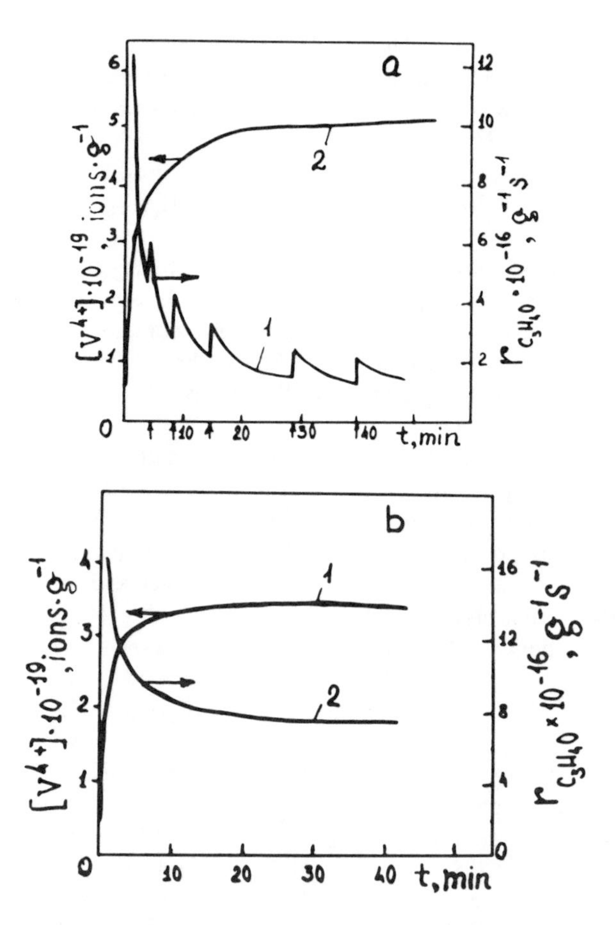

FIGURE 9.4. Kinetic curves (a) of reduction of supported V/MoO_3 catalyst by allyl alcohol and (b) of the establishment of stationary state in conditions of catalytic oxidation of allyl alcohol; (1) the rate of acrolein formation; (2) V^{4+} ion concentration (T = 485 K; $[C_3H_5OH] = 6 \cdot 10^{17}$ cm^{-3}; $[O_2] = 2.6 \cdot 10^{-18}$; peaks on Curve 1 correspond to interruptions of alcohol flow, i.e., a "rest" during 15 min).[637]

the bulk (in Equation 9.10) leads to the movement of vacancies (or reverse oxygen diffusion to the surface, Equation 9.11). Figure 9.4a demonstrates the results obtained at 485 K in the experiment "with a rest". The catalyst was alternately exposed to pulses of alcohol plus helium and pure helium. If the concentration of forming V^{4+} ions is defined only by the amount of reacted alcohol, whether the catalyst was exposed to it continuously or in a pulsed manner, the rate of acrolein formation should increase after "rest" as a result of vacancy diffusion into the catalyst bulk and formation of the active O_2^- center at the surface.

Opposite effects have been observed on oxidation of the catalyst by oxygen. On exposure to oxygen, the concentration of V^{4+} drops in accordance with the second-order law. The activation energy of the oxidation process is equal to 55 kJ/mol at 400 to 450 K. The hexagonal MoO_3 structure seems to facilitate the oxidation process due to the structural channels 0.6 nm in diameter which make possible the diffusion of molecular oxygen.

The catalyst oxidation may be described by the following elementary stages

$$V^{4+}\square_B + O_s^{2-} \rightarrow V_B^{4+} \ldots O^- + \square_s^- \quad (9.12)$$

$$2\square_s^- + O_2 \rightarrow 2O_s^- \quad (9.13)$$

$$O_s^- + V_B^{4+} \ldots O^- \rightarrow O_s^{2-} + V_B^{5+} \ldots O^- \quad (9.14)$$

The transition process to the stationary state is shown in Figure 9.4 for the mixture of alcohol and oxygen under experimental conditions. At 485 K the rate of the catalytic reaction is proportional to the concentration of reduced vanadium from the bulk and to the oxygen pressure in the mixture

$$r = kp_{o_2}\,[V^{5+}]_B \quad (9.15)$$

At low temperature (435 K) the reaction rate does not depend on O_2 concentration. The stationary concentration of the reduced V^{4+} form is inversely proportional to the oxygen pressure in both cases.

The distinction between the functions at 435 and 485 K can be explained by the fact that at high temperature all 6 stages (Equations 9.9 to 9.14) are exhibited in the catalytic reaction; they include the reactions of alcohol (9.9) and oxygen (9.13) interaction with V/MoO_3 surface, electron diffusion (9.10) and (9.14), and oxygen diffusion (9.11) and (9.12). At low temperature, electron exchange between the bulk and the surface is still retained, and the reaction proceeds through the stages in Equations 9.9, 9.13, and 9.14, involving only surface oxygen species.

At the molecular level the results obtained can be represented by the scheme in Figure 9.5.[638] In the process of surface reduction an electron rapidly diffuses to the "active site", i.e., V^{5+} ion, in the catalyst bulk. The resulting vacancy migrates into the subsurface layer at first; then it diffuses slowly to vanadium ions. Unfortunately, the ESR signal of the V^{4+} ion does not enable us to establish its location in the bulk. The planes of crystallographic shear can serve as "collectors" of vacancies (see Figure 9.6).[638] These planes might accumulate a certain number of vacancies in conditions of the stationary process. Another possibility is layer-by-layer reduction of MoO_3 crystal containing dissolved vanadium ions (see Figure 9.7). In the latter case the subsurface layer contains appreciable nonequilibrium vacancy concentration; the boundary between reduced and oxidized layers is located at a great depth. Enhanced diffusion coefficients are often observed in the subsurface layer of

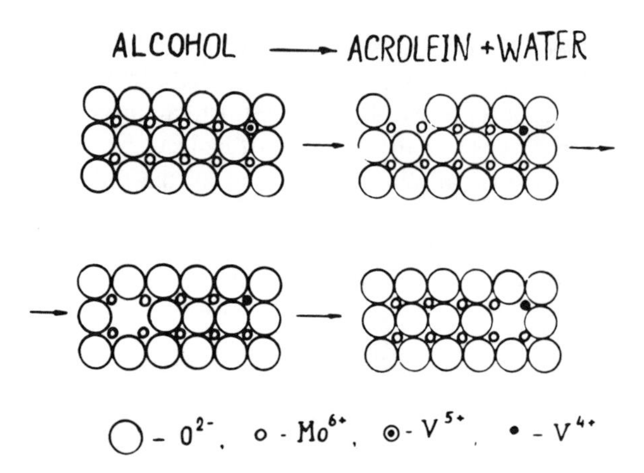

FIGURE 9.5. The scheme of interaction of allyl alcohol with the lattice of V/MoO_3 catalyst.[638]

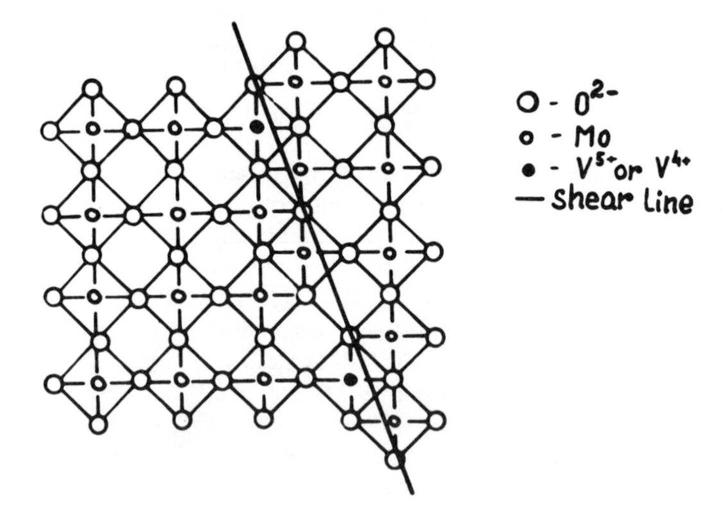

FIGURE 9.6. The scheme of crystallographic shear on reduction of V/MoO_3 catalyst.[638]

a solid. Layer-by-layer reduction of Mo and V oxides has been found by Hodnett and Delmon[639] and Najbar,[640] who studied the concentration of reduced ions via the distance from the surface.

In conditions of stationary catalysis the number of anionic vacancies on the surface is not large, less than 1% of oxygen surface atoms. This conclusion appears to be general in catalysis. The surface of oxide catalysts tends to reconstruct in such a way as to diminish the free energy of the system. Contributory factors for this are orbital hybridization, vibrational instabilities (reduced coordination number, Jahn-Teller effect), and the lowering of electron energy.

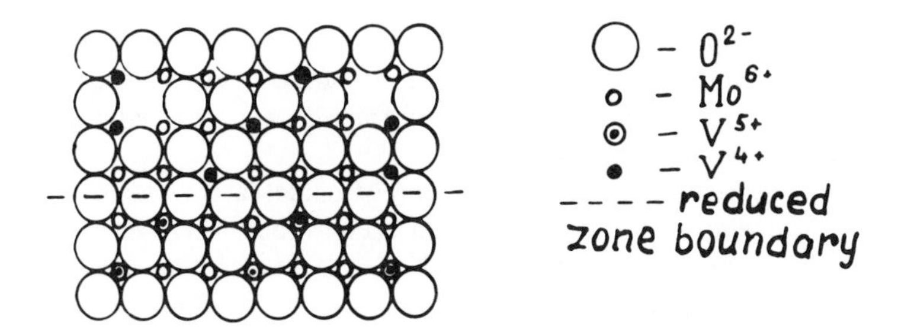

FIGURE 9.7. The scheme of layer-by-layer reduction of V/MoO_3 catalyst.

Cunningham et al.[641] have shown that the energy gain from this reconstruction increases with decrease of the forbidden band width, i.e., it is greater for semiconductors than for dielectrics.

In the light of the data obtained, a well-known discrepancy between the so-called electron theory of catalysis[642] and experiment can be elucidated. On chemisorption and in the course of catalysis, the surface of a semiconductor can be characterized by positive or negative bending of energy bands due to the formation of a layer of charged particles. It follows from the theory that in a large zone bent, charged particles occupy about 1% of sites, and further adsorption in the charged form stops due to high electric fields. It is still known from the experiment that chemisorption often continues up to monolayer and even multilayer coverage. This conflict can be explained by continuous electron exchange between the bulk and the surface and reconstruction of the coordination polyhedrons during adsorption and catalysis.

It has been shown in a number of recent experiments that the quantity of charged adsorbed particles or charged surface sites is not large; using the data on field effect 10^{14} to 10^{17} m^{-2}, i.e., less than 1% of the number of surface atoms.[132] Many researchers explain the temporal change of the number of surface levels of the freshly prepared semiconductor surface by the reconstruction. For example, Jacobi et al.[643] observed the changes in work function and UV angular-resolved photoelectron spectra for the newly made ZnO monocrystal. The changes have been attributed to the migration of surface oxygen ions into the subsurface layer.

Surface reconstruction of transition element oxides is aided by crystal field stabilization. Increasing the coordination number, for example, $\underline{C}_{4v} \rightarrow \underline{O}_h$ gives rise to additional stabilization. As far back as 1965, Stone[644] explained increasing Lewis acidity of supported Cr_2O_3/Al_2O_3 catalysts compared with pure Al_2O_3 by the fact that Cr^{3+} ion attracts additional oxygen ions in its close proximity due to the stabilization energy of the crystal field. Henrich and co-workers[645,646] carried out systematic investigations of the concentration of surface levels on simple oxides using ultraviolet photoelectron spectroscopy (UPS) and X-ray photoemission spectroscopy (XPS). There are sometimes large zone bends (up to 5 eV) and high concentration of charged surface levels (up to 1 electron per 10 surface levels) for freshly prepared monocrystal surface NiO and TiO_2. Then the surface undergoes fast reconstruction, and zone bend in the stationary state does not exceed 1 eV, and the concentration of surface levels 10^{17} m^{-2}.

So, in the stationary state the surface of oxide catalyst is covered by oxygen polyhedrons or a strongly bound chemisorbed layer. In particular, this may be a layer of hydroxyl groups. Nonequilibrium active sites, namely vacancies, can be created in this layer under catalytic conditions. Active sites may also exist as crystal defects such as dislocations, crystal shear planes, and edges and tops of crystallites. The continuous exchange of vacancies and electrons occurs between the surface and the bulk of a catalyst in conditions of redox catalysis. Catalytic stages may also include exchange between weakly adsorbed and chemisorbed layers. In the previously mentioned example of alcohol oxidation over V/MoO_3, the first stage of molecular oxygen interaction with the surface may be the generation of O^- species which rapidly convert into strongly bound O^{2-} ions.

Three types of catalytic reactions passing via the redox mechanism can be distinguished:

1. At low temperatures a charge can be transported for distances of 1 to 2 nm through the electron tunneling mechanism. The reaction has a low activation energy, provided that E_α is not determined by other factors. Then relaxation occurs, resulting in the reconstruction in the close encirclement of the electron donor (or acceptor) ion and, possibly, reconstruction in the vicinity of an adsorbed molecule.
2. At higher temperatures the mechanism of charge transfer is polaronic or zonal. All ions encountered in the path of the polaron undergo relaxation.
3. Further temperature increase gives rise to oxidation and reduction of the base lattice of the catalyst. A deep lattice reconstruction occurs.

A certain stationary state of the catalyst and catalytic process corresponds to each of these mechanisms. Note that even for the third mechanism, i.e., lattice oxidation and reduction, it is not necessary for the catalyst in the stationary state to be totally reduced or oxidized. It has been shown previously that in conditions of oxidative catalysis the exchange of lattice oxygen with that of the gas phase and reaction products may involve atoms for distances of tens of angstroms from the surface, but does not include the whole body of the catalyst.

One more peculiarity connected with the redox catalytic reactions on oxides should be mentioned. These reactions are often run by the Mars-Van Krevelen redox mechanism.[627] In particular, this is the case for the reactions of CO oxidation on $BaMnTiO_3$ and allyl alcohol oxidation on V/MoO_3. In the stationary state the oxidation rate of the catalyst is equal to its reduction

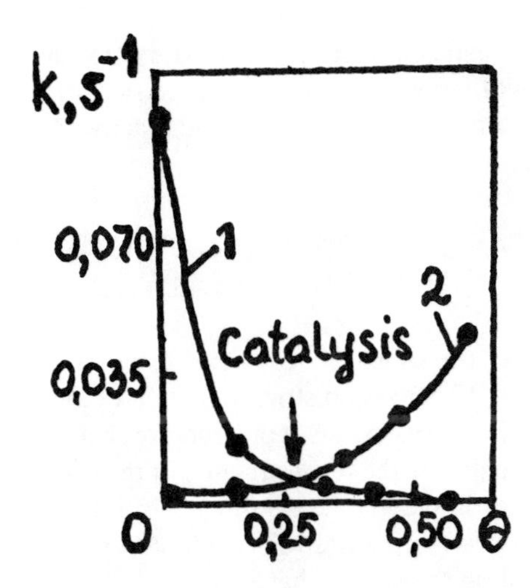

FIGURE 9.8. The rates of interaction stages as a function of surface reduction degree on the surface of mechanically activated SnO_2. (1) CO interaction with the surface; (2) O_2 interaction with the surface.

rate. This rule is often used as a proof of an equilibrium state of the catalyst.[647] However, in the experiments previously described, only the surface layer of the catalyst has been involved in the redox process. A thickness of this layer is defined by the oxygen diffusion rate through the lattice. The catalyst composition is this far from equilibrium. Forzatti et al.[648] demonstrated that in an excess of oxygen, the surface of a $SrTiO_3$ catalyst has been covered by a layer of SrO incorporated in the same crystal phase. Lower layers are depleted in oxygen.

Bobyshev studied the catalytic CO oxidation on SnO_2 powder activated mechanically using a ball mill at the Institute of Chemical Physics, Moscow. The results of his work are demonstrated in Figure 9.8. With an increase in the reduction degree of the oxide surface, the rate of O_2 interaction with the surface (oxidation of the surface) increases, but the rate of CO interaction (reduction of the surfaces), on the contrary, drops. As seen in Figure 9.8, the curves k[CO] and k′[O_2] cross each other at a reduction degree of about 35%. Upon exposure of the catalyst by the stoichiometric mixture 2CO + O_2 with such a reduction degree, the reaction proceeds in the stationary regime with the creation of CO_2.

Without milling, the tin oxide is absolutely inactive even compared on a unit area basis. Heating the catalyst up to 773 K results in its complete deactivation.

So, the equalization of the oxidation and reduction rates of the catalyst in the stationary state does not mean catalyst equilibrium is relative to the gas phase. Only a thin near-surface layer participates in the reaction in this case. The process is stationary, but in nonequilibrium. When the reaction of active sites Z* is sufficiently fast, but restricted to the near-surface layer, similar schemes will probably be applied to the step-wise catalytic processes as well. In the example with milled SnO_2 the active sites Z* were created before the reaction.

9.3. 2D-PHASE TRANSFORMATIONS AND THEIR ROLE IN CATALYSIS

The widespread use of LEED gave rise to discovery of the 2-D phase transformations.[649,650] There is no strict definition of the notion of 2-D phase and corresponding phase transitions at present. We shall use the name "surface phase transition" to denote the reconstruction of the catalyst surface resulting in structural changes in ordered layers of chemisorbed atoms. In some cases the localized form of adsorption is referred to as a phase, no matter whether the clusters formed on the surface have a 3-D crystal lattice or not.

The literature on 2-D phases arising in the course of adsorption and catalysis is rather voluminous. We will concentrate on the formation of oxygen and CO phases during catalytic CO oxidation on metals. This system has already been regarded in Section 8.4.2. Conrad et al.[651] and Engel and Ertl[652] studied this reaction on a Pd(111) face using LEED, UPS, thermodesorption, and isothermal kinetics techniques. Theirs appears to be one of the first works where LEED was directly applied for the study of the catalytic mechanism. The patterns illustrating 2-D phase structure transformations in the process of CO and O_2 adsorption on Pd(111) are shown in Figure 9.9. They help to explain the observed CO oxidation kinetics. These authors intimate that similar structural transformations are inherent for 2-D layers and CO oxidation kinetics on other metals.

At very low surface O(ads.) and CO(ads.) coverages, when Θ is less than 10% of a monolayer, the ordered layers do not form (see Figure 9.9a). Preliminary adsorption has been performed at less than room temperature. On heating, a second-order reaction occurs with $E_\alpha = 105$ kJ/mol

$$CO(ads.) + O(ads.) \rightarrow CO_2$$

When CO coverage $\Theta < 1/3$, and the surface is saturated by oxygen, 2-D islands of CO(ads.) structure $(\sqrt{3} \times \sqrt{3})$ R30° and islands of O(ads.) (2×2) structure (with $\Theta_o = 0.25$) are formed. On CO exposure, more distinct regions of (2×2) O-structure are formed (see Figure 9.9b). The reaction proceeds in this case in the island boundaries and cannot be described by a simple second-order kinetics with $E_\alpha = 105$ kJ/mol.

If saturation by oxygen is achieved, the (2×2) structure with $\Theta_o = 0.25$ is formed (see Figure 9.9c). But adding small amounts of H_2 and CO enables the adsorption of additional amounts of oxygen up to $\Theta = 1/3$, with the formation of a contracted structure $(\sqrt{3} \times \sqrt{3})$ R30°. Further CO ad-

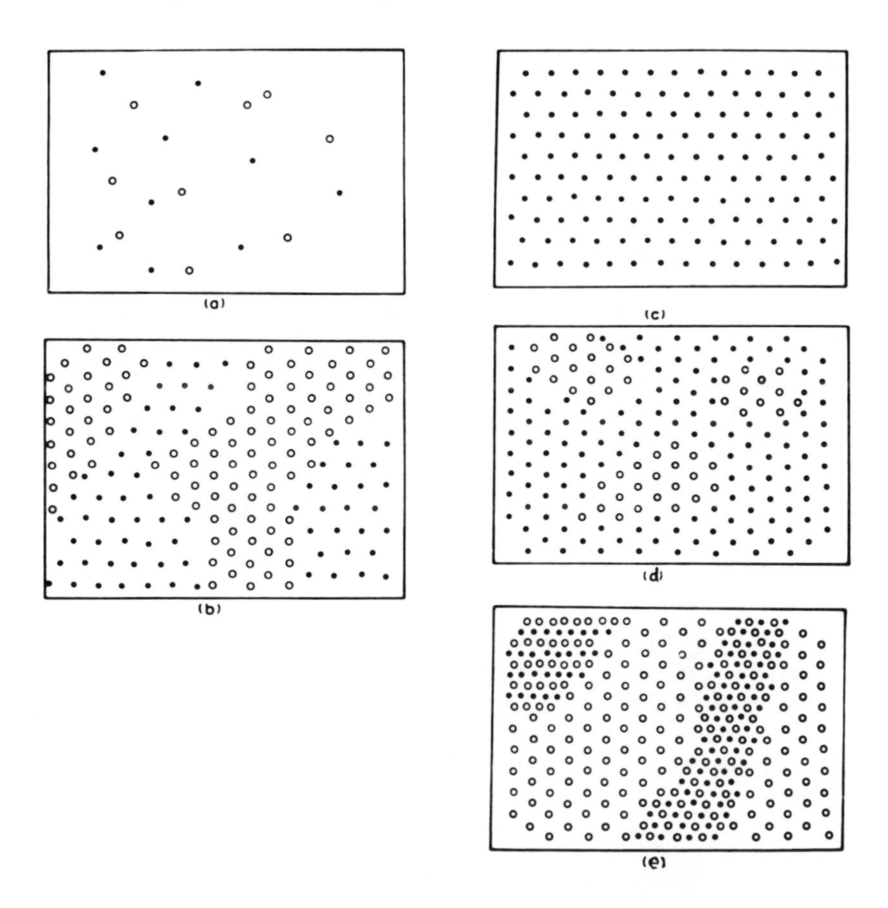

FIGURE 9.9. The adsorption scheme for oxygen (black circles) and CO (white circles) on Pd(111). (a) For very low CO and O coverages; (b) for CO coverage up to $\Theta = 0.33$, and O coverage up to 0.25; (c) for $\Theta_O = 0.25$; (d) formation of islands at $\Theta_O = \Theta_{CO} = 0.33$; (e) formation of mixed structure at $\Theta_O = \Theta_{CO} = 0.5$. (From Enge, T. and Ertl, G., *Adv. Catalysis,* 28, 2, 1979. With permission.)

sorption leads to appearance of the same CO structure (see Figure 9.9d). On reaction of CO(ads.) and O(ads.) with the formation of CO_2, the partly contracted O-structure relaxes to the initial (2 × 2) phase. LEED data confirm the scheme CO → CO(ads.), CO(ads.) + O(ads.) → CO_2. The observed reaction is first order in P_{co} with $E_\alpha = 75$ kJ/mol. Earlier, such kinetics were explained by the Rideal mechanism, though really the reaction does not occur due to the CO molecule collision from the phase with adsorbed oxygen. In this case, CO (or H_2) reconstructs a 2-D phase of adsorbed oxygen in such a way that it becomes reactive toward catalysis. On oxygen saturation followed by saturating CO adsorption a co-operative adsorption with the formation of the mixed (2 × 1) structure with $\Theta_o = \Theta_{co} = 0.5$ is observed (see Figure 9.9e). UPS points to changes in electron structure of O and CO surface species at different coverages. CO_2 formation inside the islands requires significantly lower activation energy.

Using a scanning tunnel microscopy (STM) technique enables us to elucidate the mechanism of phase formation during CO oxidation on Pt(100).[653] It turns out that gas adsorption on the Pt(100) face proceeds via the mechanism of creation and growth of the nucleating centers of a 2-D phase. On exposure of initial hexagonal phase to CO (and NO), the islands of (1 × 1) phase of size 0.2 (at 300 K) to 7 nm (at 460 K) and 0.2 nm in thickness are observed using STM. Then the island growth with the speed of 300 nm/s is observed. The creation of nucleating centers proceeds in accordance with the fluctuation mechanism through the formation of the intermediate [hex Pt(100) + CO(ads.)] which converts successively into the ultimate 2-D phase after reaching the critical concentration $\Theta_{co} = 0.07$

$$\text{hex Pt(100)} + \text{CO(gas)} \rightarrow [\text{hex Pt(100)} + \text{CO(ads.)}] \rightarrow$$
$$(1 \times 1)\text{Pt(100)} + \text{CO(ads.)}$$

Another mechanism has been found for C_2H_4 adsorption on (100)Pt. Under experimental conditions (510 K, $5 \cdot 10^{-7}$ Pa) the nuclei appear on a defect, namely on a step with further growth on the top and lower terraces. The aggregate of parallel belts consisting of successive hills and hollows arises. The growth of a (1 × 1) phase results in increasing the hill width. In this case the process proceeds via two metastable states

$$\text{hexPt(100)} + C_2H_4 \rightarrow [\text{hexPt(100)} + C_2H_4\text{(ads.)}] \rightarrow$$
$$[\text{hexPt(100)} + C_2H_n\text{(ads.)}] + H_2\text{(gas)} \rightarrow$$
$$(1 \times 1)\text{Pt(100)} + C_2H_n\text{(ads.)}$$

where n = 2 or 3.

Chang and Thiel[654] have shown that at a sufficiently low temperature fast adsorption followed by slow diffusion takes place. This gives rise to the formation of ordered but nonequilibrium phases. For example, at 160 to 180 K the adsorption of O_2 on Pd(100) results in the creation of a nonequilibrium c(2 × 2) phase formed by initial ''hot'' O-atoms. On heating, this phase disproportionates with the formation of two surface phases, usually c(2 × 2) and p(2 × 2) ones.

In our work[655,656] the model of the creation and growth of nucleating centers has been utilized for the description of 2-D phase transition kinetics. This model enables us to explain a number of anomalous effects in the kinetics of heterogeneous catalytic reactions.

Consider the adsorption of one of the reagents as a process of the formation of a new phase on the catalyst surface. Following Delmon[657] who described the kinetics of 3-D phase transitions, we shall use his model of the creation and growth of the nucleating centers for the kinetics of a 2-D process. The first stage of the formation of a new phase on the catalyst surface is the creation of nucleating centers, i.e., clusters consisting of a few atoms of a new phase. To describe the kinetics of this process we shall use the idea of homogeneous (fluctuation) nucleation based on the concept of a critical size of the nucleus. The thermodynamic definition of the critical size of a nucleus is given by the equations

$$\begin{aligned} \delta F/\delta N &= O \\ \delta^2 F/\delta N^2 &< O \end{aligned} \qquad (9.16)$$

where F is the Helmholtz energy of the system and N is the number of particles constituting the nucleus.

If the creation of the nucleus is treated as a kinetic process in which the number of particles N varies as a function of time, we may postulate that

$$dN/dt = f(N) \qquad (9.17)$$

where f(N) is a function of the number of particles only. Then the kinetic definition of the critical size can be expressed in the form

$$\begin{aligned} &dN/dt = O \text{ at } N = N_{cr} \\ &dN/dt < O \text{ at } N < N_{cr} \\ &dN/dt > O \text{ at } N > N_{cr} \end{aligned} \qquad (9.18)$$

where N_{cr} is the number of particles in the nucleus having a critical size. Both definitions (Equations 9.16 and 9.18) can be proved to be equivalent.

The kinetics of nucleation may be regarded as a system of consecutive reversible reactions

$$\begin{aligned} &A + A \rightleftarrows A_2 \\ &A_2 + A \rightleftarrows A_3 \\ &A_i + A \rightleftarrows A_{i+1} \\ &A_{cr} + A \rightleftarrows A_{cr+1} \end{aligned}$$

Sometimes the reaction of nuclear coalescence is also included in the system[658]

$$A_n + A_m \rightleftarrows A_{n+m}$$

In the simplest case, if the reaction of nuclear coalescence is neglected, the rate of nucleation in the stationary state can be found to be

$$r = k'c^m \tag{9.19}$$

where k′ and m are constants and c is the concentration of nucleation substance.

The validity of Equation (9.19) has been proven experimentally.[658] The value of a power index depends strongly on a given system; m = 37 for water condensation. In other cases it is substantially less. Data on the formation of critical nuclei in the process of metal oxidation are few in number. The reaction

$$2Pb + M \rightarrow Pb_2 + M$$

has to be a limiting step in the oxidation of lead with a rate constant of $2\ 10^{-3}$ cm^6/s. Here M is a third particle, and m = 2 in this case.[659]

Suppose the rate of nucleation is defined by the equation

$$d\gamma/dt = h\Theta^n \tag{9.20}$$

where $d\gamma/dt$ is the number of nuclei arising in a unit time on a unit area, Θ is the coverage by adsorbed species, n the number of particles required for the nucleation, and h is a constant.

The rate of growth of a 2-D nucleus along either coordinate axis is taken as being proportional to the concentration of adsorbed particles

$$\begin{aligned} r_x &= k_x\Theta \\ r_y &= k_y\Theta \end{aligned} \tag{9.21}$$

Then by time t, the area of the nucleus appearing at the moment τ is equal.

$$S(t,\tau) = k_x k_y \left[\int_\tau^t \Theta\, dt\right]^2 \tag{9.22}$$

The total area of all nuclei is

$$S(t) = h\, k_x k_y \int_o^t \Theta^n(\tau)\left(\int \Theta dt\right)^2 d\tau \tag{9.23}$$

From Equation 9.23 for the reaction rate, we easily obtain the equation

$$dS(t)/dt = 2hk_x k_y\Theta \int_o^t \Theta^n(\tau)\left(\int_\tau^t \Theta dt\right)^2 d\tau \tag{9.24}$$

Equations 9.23 and 9.24 do not take into account the decrease of the reaction rate due to the possible overlap of the nuclei. This effect can be taken into account in the calculation by the introduction of a fictitious conversion degree. Then instead of Equations 9.23 and 9.24, we have

$$\Theta_{np} = 1 - \exp\left[-k\int_{o}^{t}\Theta^{n}(\tau)\left(\int_{\tau}^{t}\Theta dt\right)^{2}d\tau\right] \tag{9.25}$$

$$d\Theta_{np}/dt = k(1 - \Theta_{np})\,\Theta\int_{o}^{t}\Theta^{n}(\tau)\left(\int_{\tau}^{t}\Theta dt\right)^{2}d\tau \tag{9.26}$$

where k is the rate constant and Θ_{np} is the coverage by the new phase. If the coverage of adsorbed particles is constant over the course of the whole process, then Equation 9.26 reduces essentially to the known power-law equation

$$d\Theta_{np}/dt = k(1 - \Theta_{np})\,\Theta^{n+2}t^{2}/2 \tag{9.27}$$

Equation 9.26 or 9.27 describes the kinetics of 2-D phase creation on the catalyst surface. If the assumption is made that in the course of the surface phase transition the active sites are created and exterminated, then an equation of the type in Equation 9.26 or 9.27 describes the nonstationary period of the catalytic process. In this case the time curves of the catalytic activity have inductive pauses and autoacceleration portions.

Now consider the more complicated case of reversible phase transitions. Suppose that adsorption of one of the reagents induces nucleation. A new phase appears at the surface in the process of the increase in nuclei. Adsorption of the second reagent and destruction of the new phase proceeds through the same mechanism, i.e., through the creation and increase of old phase nuclei. For example, the process of catalytic oxidation and reduction may be interpreted in this way. There are several different cases depending on the phase in which the reagent adsorption takes place.

Case 1 — suppose that the reagents A and B adsorb on either phase with equal rate constants for adsorption and desorption. In this case the process may be described by the following system of kinetic equations

$$d\Theta_{A}/dt = k_{ads}^{A}(1 - \Theta_{A} - \Theta_{B})P_{A} - k_{des}^{A}\Theta_{A} - r_{A} \tag{9.28}$$

$$d\Theta_{B}/dt = k_{ads}^{B}(1 - \Theta_{A} - \Theta_{B})P_{B} - k_{des}^{B}\Theta_{B} - r_{B} \tag{9.29}$$

$$d\Theta_{np}/dt = k_{1}(1 - \Theta_{np})\Theta_{A}\int_{o}^{t}\Theta_{A}^{nA}(\tau)\left(\int_{\tau}^{t}\Theta_{A}dt\right)d\tau \tag{9.30}$$

$$- k_{2}\Theta_{np}\Theta_{B}\int_{o}^{t}\Theta_{B}^{nB}(\tau)\left(\int_{\tau}^{t}\Theta_{B}dt\right)d\tau$$

where Θ_A and Θ_B are surface coverages of the components A and B; Θ_{np} is the relative area occupied by the new phase; k_{ads}^{A} and k_{ads}^{B} are the rate constants for adsorption and desorption; k_1 and k_2 are effective rate constants for the creation and destruction of the new phase; r_A and r_B are consumption rates of A and B components; n_A and n_B are the critical number of type A and B molecules required for nucleation; and P_A and P_B are partial pressures of A and B.

Equations 9.28 and 9.29 describe the ordinary kinetics on the homogeneous surface, and Equation 9.30 is a formal one describing phase transitions on the surface.

Consider a first-order reaction in A for the case when the new phase is a catalyst, i.e.,

$$r_A = k\Theta_{np}\Theta_A \qquad r_B = 0 \tag{9.31}$$

We have for this case in the stationary state

$$\Theta_A = \frac{a_A P_A}{[1 + (k/k^A_{des}\Theta_{np}]\{1 + a_B P_B + [a_A P_A/(1 + k\Theta_{np}/k^A_{des})]\}} \tag{9.32}$$

$$\Theta_A = \frac{a_B P_B}{1 + a_B P_B + a_A P_A/(1 + k\Theta_{np}/k^A_{des})} \tag{9.33}$$

where a is the adsorption coefficient.

If $k\Theta_{np} < k^A_{des}$, Equations 9.32 and 9.33 do not differ from that of the homogeneous surface. For simplicity, suppose that this condition is fulfilled. Now we can find the stationary solutions of Equation 9.30 by setting $d\Theta_{np}/dt$ equal to zero. Then the following condition is obeyed in the stationary state

$$k\,(1 - \Theta_{np})\Theta_A^{nA+2} = k_2\Theta_{np}\Theta_B^{nB+2} \tag{9.34}$$

As seen from Equation 9.34, in this case the system has a single stationary state. The stationary surface portion occupied be the new phase is equal to

$$\Theta_{np} = k_1\Theta_A^{nA+2}/(k_1\Theta_A^{nA+2} + k_2\Theta_b^{nB+2}) \tag{9.35}$$

The kinetic equation has the simplest form when $n_A = n_B = n$ and $k\Theta_{np} < k^A_{des}$. Then we have

$$r = \frac{k_1 a_A P_A}{1 + a_A P_A + a_B P_B}\;\frac{\alpha\,(P_A/P_B)^{n+2}}{1 + \alpha\,(P_A/P_B)^{n+2}} \tag{9.36}$$

where

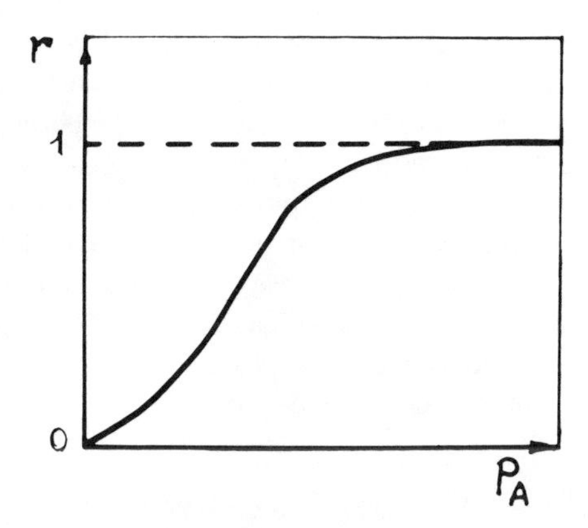

FIGURE 9.10. Schematical dependence of the surface coverage by 2-D phase on the pressure of input product that results in Equation 9.27.

$$\alpha = \frac{k_1}{k_2} (a_A/a_B)^{n+2} \tag{9.37}$$

At P_B = const, the function r vs. P_A can be conceived of as a curve with the initial first-order interval followed by the regions of sharp increase and saturation (see Figure 9.10). For the first-order reaction in B, we have, similarly,

$$r = \frac{k\, a_B P_B}{1 + a_A P_B + a_B P_B} \frac{1}{1 + \alpha\, (P_B/P_A)^{n+2}} \tag{9.38}$$

It is seen from Equation 9.38 that the reaction rate is extremely dependent on P_B. At low P_B the rate grows proportionally to P_B, but then sharply drops, practically to zero. The reason is that the reaction rate increases with the growth of P_B at a constant number of active sites, but then adsorption of the B component results in the destruction of the active catalyst surface. It is seen from Equations 9.36 and 9.38 that the reaction rate for the simplest cases of kinetics of heterogeneous reactions taking into account phase transitions on the catalyst surface, can be written in the stationary state as follows

$$r = r_1 r_2 \tag{9.39}$$

where r_1 is the rate of change of the phase composition and r_2 is the reaction rate on the unaltered catalyst surface. The characteristic feature of such kinetics is the presence of a reaction with an anomalous high order $n \geqslant 3$. An equation of the type in Equation 9.39 can explain experimental curves with inductive periods and intervals of autoacceleration and abrupt retardation of the reaction.

Such kinetic characteristics make possible the existence of regions with multiple stationary states and limiting cycles for open systems.

Case 2 — Now consider the case when A and B components adsorb predominantly on the catalytically active phase, which has been formed in the course of the reaction. In this case increasing the portion of the surface occupied by the new phase results in the growth of catalytic activity. At the same time the process of origination and extermination of new sites is also accelerated. The complicated behavior of the system is determined by the competition of these processes. In the case under consideration the adsorption kinetics can also be described by Equations 9.28 and 9.29, implying that Θ_A and Θ_B are now relative areas of the new phase covered by A and B molecules. The following equation is valid for the kinetics of phase transitions:

$$\frac{d\Theta_{np}}{dt} = k_1(1 - \Theta_{np})\Theta_{np}\Theta_A \int_o^t (\Theta_{np}\Theta_A)^{n_A} \left(\int_\tau^t \Theta_{np}\Theta_A dt\right) d\tau$$

$$- k_2\Theta_{np}^2\Theta_B \int_o^t (\Theta_B\Theta_{np}) \left(\int_\tau^t + \Theta_{np}\Theta_B\, dt\right) d\tau \qquad (9.40)$$

The following condition which is similar to that of the previous case is satisfied in the stationary state:

$$k_1(1 - \Theta_{np})\Theta_{np}^{n_A+2}\Theta_A^{n_A+2} = k_2\Theta_{np}^{n_B+2}\Theta_B^{n_B+2} \qquad (9.41)$$

Let $n_1 = n_A - n_B$, then we have

$$\psi_1 = (1 - \Theta_{np})\Theta_{np}^{n_1}\Theta_A^{n_1} = k_2/k_1(\Theta_B/\Theta_A)^{n_B+2}\Theta_{np} = \psi_2 \qquad (9.42)$$

The stationary states correspond to the solution of Equation 9.42 for Θ_{np}. It follows from Equation 9.42 that the system may have up to three stationary states depending on the value of n_1. The scheme of the graphic solution of Equation 9.42 is shown in Figure 9.11a. The lower (1) and upper (3) points correspond to stable states, but the middle point to an unstable one.

Case 3 — Suppose that component A adsorbs on the new phase, but B on the old one. The new phase is assumed to be catalytically active. In the stationary state Θ_{np} is defined from the condition

$$\psi_3 = k_1(1 - \Theta_{np})\Theta_{np}^{n_A+2} = k_2\Theta_{np}(1 - \Theta_{np})^{n_B+2}\Theta_B^{n_B+2} = \psi_4 \qquad (9.43)$$

The scheme of the graphic solution of Equation 9.43 is shown in Figure 9.11b. It is seen that four stationary states are possible in this case.

The above consideration points to the possible connection of 2-D phase transitions with nonstationary phenomena in heterogeneous catalysis. The high kinetic orders which arise in equations for the reaction rate can be connected with the creation of nuclei of the surface or bulk phase.

The actual mechanism of phase transition cannot be reduced to the simple motion of the boundary between phases, but is much more complicated.[660]

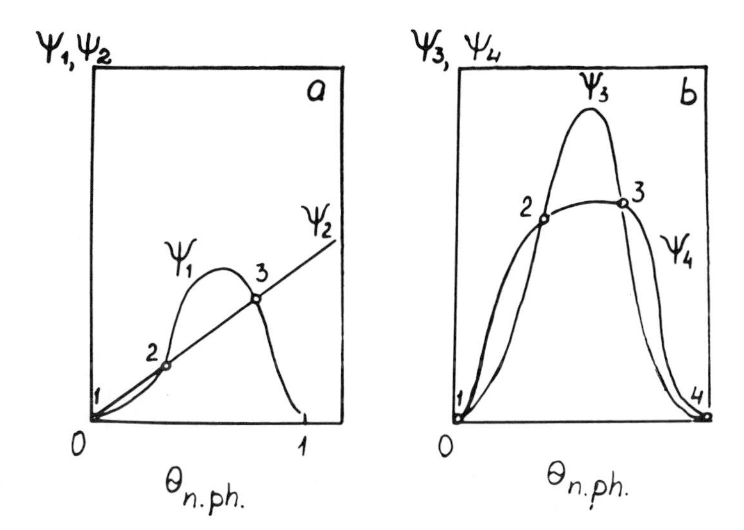

FIGURE 9.11. Graphic solution of Equations 9.33 and 9.34 (figures denote the stationary solutions).[655]

The processes occurring at the metal-oxide, metal-gas, or oxide-gas boundaries play a leading role. Besides, it should be taken into account that the boundary structure of nuclei A in matrix B may differ from that of nuclei B in matrix A due to different interphase energies.[661,662] Figure 9.12 illustrates the growth of A phase surrounded by B phase and vice versa.[662] Different energy required for the nucleus origination results in different patterns of the processes A → B and B → A. Due to this fact hysteresis is observed in the curves for the degree of the phase transition vs. temperature. The system is two-phase in the hysteresis loop.

So, our model[655,656] is free from self-contradictions. It should only be noted that the kinetic equation (9.30) obtained above describes the whole process approximately, and it is inadequate for mathematical treatment of the motion of an individual interphase boundary. Besides, the competition scheme of nuclear growth suggested by us may be applied not only to the formation of surface oxides, but to the formation of 2-D structures of ordered atoms on adsorption of a mixture, for example, carbon oxide and hydrogen or CO and oxygen on Pt.[651] It is evident that in these cases the boundaries of "spots" covered by different components cannot be conceived as a single homogeneous boundary between phases. In general the two terms in Equation 9.30 may be thought of as relating to the different motion segments of the interphase boundary.

The concentration of components at the phase boundary may vary due to the difference of the diffusion coefficients or strong changes in the catalytic activity in the course of the phase transition. The boundary between phases by itself can be not only inhomogeneous, but even unstable. Martin and

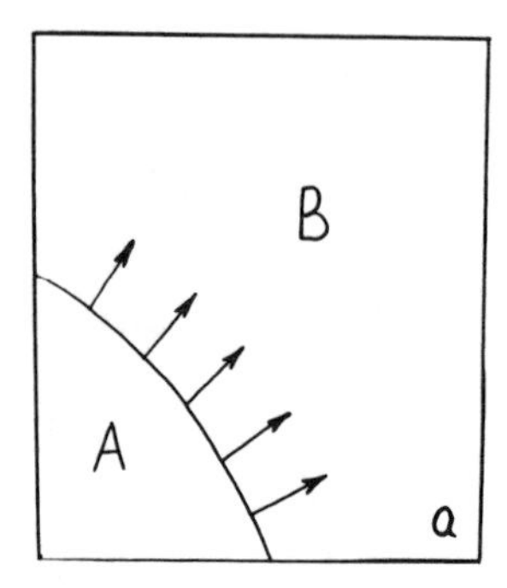

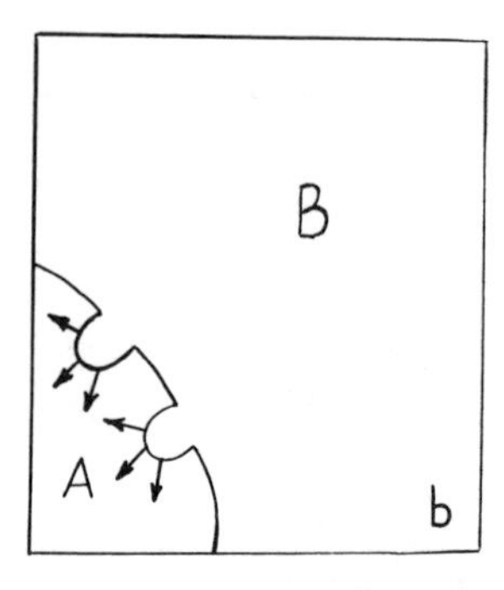

FIGURE 9.12. The scheme of nucleating growth of the A phase in B (a) and vice versa (b). (From Porter, S. K. et al., *J. Chem. Soc. Farraday Trans.*, 79(9), 2043, 1983. With permission.)

Schmalzried[663] have found that if new additional molecules appear at the boundary between phases, for example, in the process of the interaction of the reduced phase with oxygen,

$$M^{2+} + \frac{1}{2} O_2(\text{gas}) + 2e^- \rightarrow MO + \square_{M^{2+}}$$

(where $\square_{M^{2+}}$ is the cation vacancy), the interphase boundary is morphologically stable. In the reverse process,

$$MO + \square_{M^{2+}} = M^{2+} + \frac{1}{2} O_2(\text{gas}) + 2e^-$$

resulting in the formation of "subtraction structures", the boundary is morphologically unstable. In the 3-D case the morphological instabilities give rise to deep pores and loose surface.

Naumovets et al.[664] have proved experimentally that the existence of nonequilibrium frozen defects leads to the conversion of the two-phase system into individual islands of separated phases, but the total energy of the system decreases due to the energy gain in atomic interactions of the inhomogeneous system. Using LEED, these authors observed the evolution of an initially abrupt, straight boundary between two coexisting 2-D phases: c(2 × 2) at $\Theta = 0.25$ and c(1 × 3) at $\Theta = 0.5$ for the system Li/W(011). As time elapses, the islands of the first phase inside the second and vice versa appear.

The data testify that 2-D phase transitions may be the cause of abrupt kinks in the coverage, adsorption and reaction rate curves vs. pressure temperature, or the gas-phase composition. We have considered the mechanism of phase transition through the formation and growth of nuclei. But other mechanisms can exist as well, namely, the so-called kinetic or nonequilibrium phase transitions. Their possible role in heterogeneous catalysis has been the subject of considerable discussion in the last few years.

Many researchers take advantage of the model of Ziff-Gulari-Barshad (ZGB). These authors[665] considered similar kinetic transitions for the square homogeneous lattice. The adsorption probability CO $\rightarrow$ CO–Z and $O_2 \rightarrow 2OZ$ for each collision of a molecule with active site z has been taken to be unity. Similarly, when a CO molecule or O atom strikes the adjacent site: CO-Z + O-Z $\rightarrow CO_2$ + 2Z, the reaction probability is also taken to be unity. The mere fact that a CO molecule occupies one site on the surface, but an O_2 molecule two, results in sharp kinks in the coverage vs. gas composition curves, and, as a consequence, in the reaction rate variations. For example, if the CO molar fraction in the mixture y_1 is less than 0.389, the surface is fully covered by oxygen; at $y_2 > 0.525$ the surface is covered by CO, and only in the interval $y_1 < y < y_2$ is the surface covered by the mixture CO + O and can the catalytic reaction proceed. The kinetic phase transitions occur at the points y_1 and y_2. Using the ZGB model, these authors have found conditions at which periodic interaction of the metal alternately with CO and O_2 results in greater production output than interaction with the mixture CO + O_2.

Further development of the ZGB model has been used in the investigation of other lattice types, for instance, the hexagonal one, the account of the desorption rate diffusion, and the finite reaction probability (<1).[666-670] The stationary process of CO oxidation is possible only in the middle region of Θ_{CO} values. At the extreme points of this interval, kinetic phase transitions shows an abrupt change of catalytic activity. Recent experimental work[671] agrees in general with the ZGB theory.

Experimental data on the kinetics of 2-D phase transitions have been few in number for a long time, but recently more extended studies have been undertaken[672] due to the progress of two methods: Video-LEED (dynamic LEED studies) and tunnel electron microscopy. The first of these methods has an evident disadvantage since it can be used only in a high vacuum. Besides, the vast majority of investigations have been performed at the abrupt conversion of the equilibrium into the nonequilibrium phase by way of rapid temperature increase or decrease. Such a procedure does not enable us to study initial stages of phase formation, i.e., the stage of nucleation.

The kinetics of 2-D phase transformations has been studied in detail for adsorbed O_2 layers at W(110), S/Mo(110), O_2 and CO/Ir(100) and Pt(100), as well as for the kinetics of the reconstruction of the clean Si(111) surface. In the last case the silicon surface transforms from a (2 $\times$ 1) structure at low temperature to a (7 $\times$ 7) structure at a high one. In all cases the kinetics corresponds to the growth of the already formed islands of the new phase according to the law $L \sim t^n$, where L is the linear size of the island and $n = 0.33$ to 0.5. There are also examples of the cooperative effects in 2-D phase transitions. For example, the conversion of the structure (1 $\times$ 1) into (1 $\times$ 5) has been observed on clean Pt(100) followed by the transition into the hexagonal phase as a result of the shift of linear atomic layers.

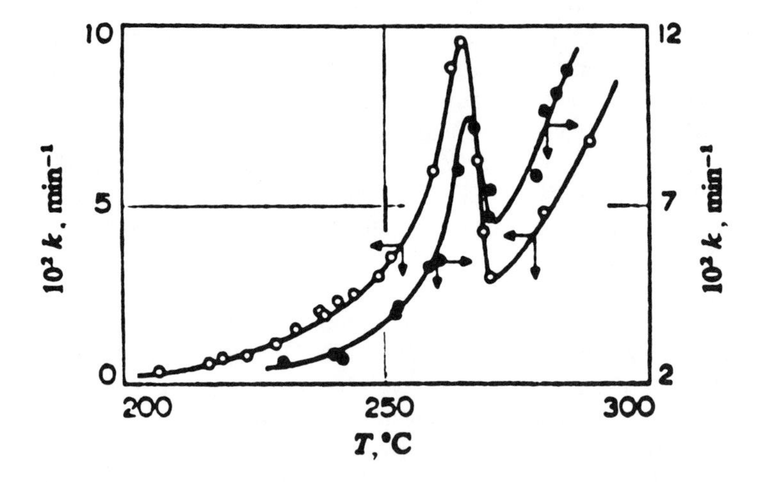

FIGURE 9.13. Temperature dependence of the rate constant of NiO reduction by hydrogen in the vicinity of the Neel point. (From Delmon, B. and Roman, A., *Trans. Farraday Soc.*, 67, 971, 1971. With permission.)

9.4. CATALYSIS IN THE VICINITY OF THE PHASE TRANSITION POINT

The problem of catalytic transformations in the vicinity of the phase transition point is worthy of separate consideration. There is some information in the literature that the formation of the new phase in the process of catalysis and adsorption can promote or inhibit the catalytic reaction or radically change its rate. Singularities in the temperature dependence of a catalytic reaction in the vicinity of the phase transition point should be expected for rapid non-diffuse phase transitions like those of martensite or for transitions of the type antiferromagnetic ⇄ paramagnetic and ferroelectric ⇄ paraelectric.

The theory of such processes has been considered by Suhl.[673] The degrees of freedom of particles in thermally activated motion are coupled to the parameter order associated with a phase transition of the bath, fluctuation in this parameter causing a modulation of the activation barrier near T_o. The data by Delmon and Roman[674] on the rate variation of NiO reduction by heterogen are presented in Figure 9.13. It is seen that the abrupt change of the NiO reduction rate takes place in the vicinity of the Neel point when NiO transits from the ferromagnetic to the paramagnetic state.

Systematic studies of the catalytic and adsorption properties of ferro-electrics in the vicinity of the Curie point T_o (the transitions ferroelectric ⇄ paraelectric) have been undertaken at the Institute of Chemical Physics using ESR.[636] The Curie point of barium titanate (403 K) is offset by several degrees on adsorption of different gases, donors and acceptors shifting it to opposite directions, namely, NO and oxygen increase T_o, but CO and propylene de-

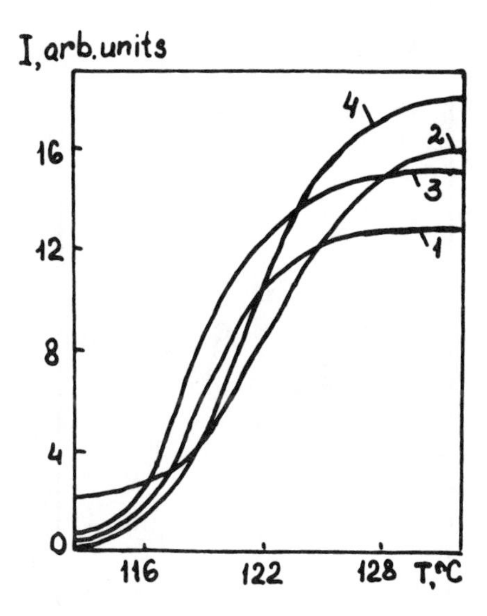

FIGURE 9.14. The transition: ferroelectric ⇄ paraelectric in $BaTiO_3$. Variation of Fe^{3+} ESR signal intensity in $BaTiO_3$ in vacuum and after adsorption of different gases.[636] (1) Vacuum; (2) O_2; (3) C_3H_6; (4) CO.

crease it (see Figure 9.14). ESR lines of Mn^{2+} and Fe^{3+} ions contained in $BaTiO_3$ as natural impurities serve as indicators of the ferroelectric transitions in this case. These lines are observed in the cubic paraelectric phase and are not present in the tetragonal ferroelectric one. As a result of sample polycrystallinity, the transition is spread by several degrees.

Ferroelectric phase transitions, in turn, affects catalysis and adsorption; for instance, adsorption of reagents and CO oxidation occur at the paraelectric $BaTiO_3$ phase but do not occur at the ferroelectric one. The temperature dependence of the $BaTiO_3$ reduction rate by carbon monoxide is shown in Figure 9.15. In the course of catalytic reduction the Mn^{4+} ion transforms into Mn^{2+}. This process was traced using ESR spectra. Arrhenius dependence inflects at point T_o. The catalytic process of CO oxidation proceeds through the mechanism of alternative oxidation and reduction of the catalyst. Kinetic behavior alters in the region of the ferroelectric phase transitions; the rate of catalyst oxidation can be proportional not only to the pressure of O_2, i.e., the gas-oxidizer, but to CO, i.e., the gas-reducer, as well. The role of CO in this case resides in the fact that it shifts Tc in the direction favorable for adsorption and catalysis.[636] Kiselev et al.[675] observed anomalies on adsorption of H_2O and NH_3 on VO_2 in the vicinity of the point of semiconductor-metal transitions.

Sharp decrease of the sticking coefficient that was observed by many researchers at the point of phase transition is one such anomaly. The oxygen sticking coefficient vs. coverage at Ni(100) is shown in Figure 9.16.[676] It is

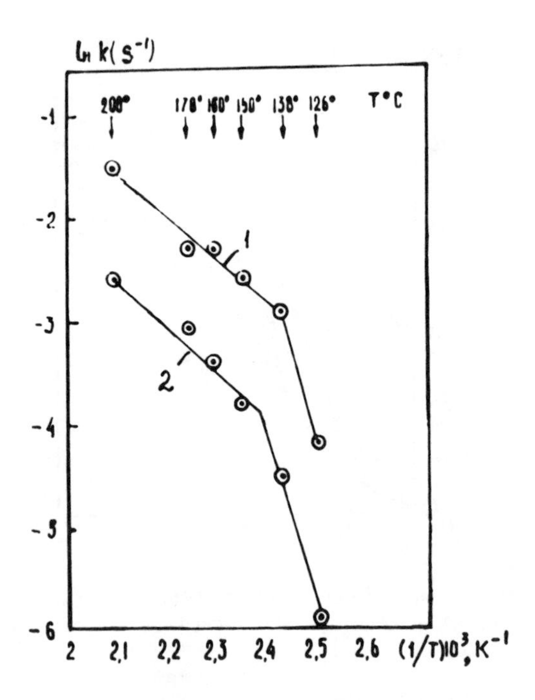

FIGURE 9.15. Temperature dependence of $Ba(Mn)TiO_3$ reduction rate by carbon monoxide using kinetic measurements of Mn^{2+} ESR signal (1) and CO absorption (2) (the kink point corresponds to the Curie temperature).[636]

seen that the transition from one structure to another is accompanied by the minimum, which is likely due to passing through an unordered state.

Oxidation reactions on an oxide surface in the presence of a small doping amount of metal may serve as examples of nonequilibrium phase formation in catalytic conditions. Ilchenko et al.[677] revealed that a small amount of platinum activates oxide catalysts of oxidation, for example, V_2O_5.

Using the method of scanning calorimetry, Bychkov et al.[678] studied *in situ* methane oxidation into CO_2 and H_2O over supported Pt/V_2O_5 catalyst with a very small Pt contamination of 0.15%. Heating Pt/V_2O_5 catalyst up to 823 K in the reaction mixture resulted in the formation of well-known V_2O_4 stable phase. The presence of this phase is confirmed by the appearance of a heat effect of the reverse phase transition at 341 K: V_2O_4 (monoclin.) $\rightleftarrows$ V_2O_4 (tetragon.). The stable phase was completely inactive in the catalytic process at 548 K.

At the same time the reduction by methane at 548 K results in the formation of the metastable phase of reduced vanadium oxide. Under these conditions 20 to 25% of oxygen is removed from the catalyst, and its effective stoichiometric formula is close to V_2O_4 as well. However, the heat effect that is characteristic for the transition V_2O_4 (monoclin.) $\rightleftarrows$ V_2O_4 (tetragon.) is not observed. The stable catalytic activity of the metastable phase remains

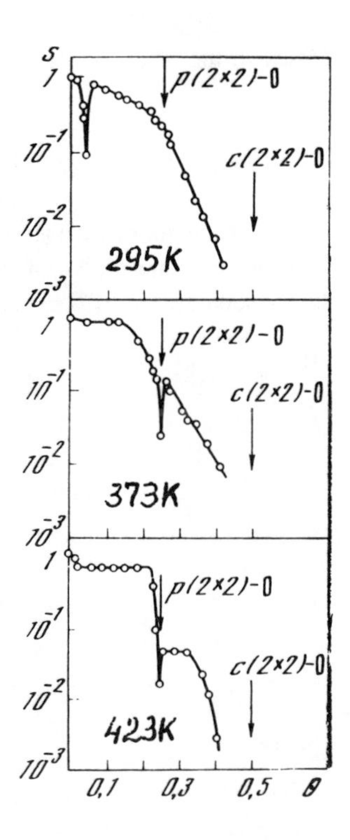

FIGURE 9.16. The sticking coefficient s as a function of coverage Θ on Ni(100) at different temperatures (arrows indicate the appearance of 2-D phase).[676]

unchanged on cooling the catalyst to 458 K. The temperature of this reoxidation by oxygen is almost 200°C lower than that of the stable phase. The low-temperature (323 to 623 K) oxidation stage of the metastable phase has anomalous high heats ΔH (340 to 500 kJ/mol of O_2). On heating, the metastable phase V_2O_4 transforms into the stable one with high heat liberation.

Study of the surface of the Pt/V_2O_5 catalyst with the aid of XPS shows that the surface of the oxide phase after the catalytic process contains vanadium atoms in the oxidation degrees 5+, 4+, and even lower than 4+ for both stable low-active and metastable high-active phases of V_2O_4. Surface compositions of active and inactive phases are little different from each other. As to the platinum state, XPS data indicate that on heating in a methane atmosphere the catalytic activity of the supported Pt/V_2O_5 catalyst aries simultaneously with the disappearance of the oxidized Pt forms (with the electron binding energy of $4D_{5/2}$ state 317.1 eV) and with the appearance of metallic Pt (with the binding energy of 314.1 eV).

Hence, CH_4 dissociation at Pt clusters occurs in the reaction of CH_4 oxidation over Pt/V_2O_5 catalyst

$$Pt + CH_4 \rightarrow Pt - H + Pt - CH_2$$

Atomic hydrogen diffuses in the V_2O_5 phase and reduces it to the non-equilibrium V_2O_4 phase

$$Pt - H + V_2O_5 \rightarrow Pt + V_2O_4 + OH \rightarrow (H_2O)$$

The nonequilibrium V_2O_4 phase makes easier oxygen transport to the boundary of Pt/V_2O_5 where the methane oxidation takes place. Adsorption sites of O_2 and CH_4 are likely separated in this case.

Chapter 10

NONEQUILIBRIUM STATES OF CATALYST

In Section 9.2. we gave a set of examples that demonstrated the formation of a nonequilibrium near-surface layer in the course of catalytic reaction. The reaction of the active site deactivation compared with the rate of the catalytic transformation keeps it in the active state.

In the simplest case, interaction between the catalyst C and the reagent R can be represented by the scheme

$$C + R \xrightarrow{1} CR^* \xrightarrow{2} C^* + P \xrightarrow{3} C + P \qquad (10.1)$$

In this scheme Step 1 is the formation of an intermediate complex CR*, Step 2 is the formation of the reaction product P and the catalyst C* in which atomic spacing and valence angles are different from the initial catalyst state C, and Step 3 is the return (relaxation) of the catalyst to the initial state. One can conceive a more complicated scheme, for example, the scheme in Equation 10.2 with the reverse catalyst transition from the unrelaxed state C* to the relaxed one C

$$\begin{array}{lllllll} C + R & \searrow\!\!\!\nwarrow & & \nearrow\!\!\!\swarrow & CR^* & & \\ \downarrow\uparrow & & CR^* & & & \nearrow\!\!\!\swarrow & CP \\ C^* + R & \nearrow\!\!\!\swarrow & & \searrow\!\!\!\nwarrow & CP^* & \rightleftarrows & C^* + P \rightleftarrows C + P \end{array} \qquad (10.2)$$

The relation between the rate r_2 of the catalytic Step 2 and the rate r_3 of the relaxation Step 3 in Equation 10.1 defines the behavior of the whole process. When $r_3 > r_2$, the reaction step can be omitted in the kinetic scheme. If $r_2 > r_3$, the specific role of unrelaxed and, to some extent, nonequilibrium active structures of the catalyst should be taken into account.

If the activity of structure C* is greater than that of C, the catalytic process proceeds dominantly via an "excited" C* structure. We have already pointed out that characteristic catalytic times are of the order of 10^{-2} to 10^{2} s. Hence, if the relaxation time is close or exceeds this value, the relaxation step can significantly affect the catalytic one. As we demonstrated in Section 7.1, this results in nonequilibrium catalytic effects. Electron transport processes between the catalyst surface and its bulk have the shortest times in a solid.

10.1. NONEQUILIBRIUM ELECTRON TRANSITIONS IN ADSORPTION AND CATALYSIS

The simplest adsorption stage is the trapping of an electron or hole by the surface. It can have different characteristic times. Studies of the electric field effect in silicon, germanium, and other covalent semiconductors revealed the existence of "fast" surface states with the cross-section of $\sigma \approx 10^{-13}$ to 10^{-15} cm^3, which corresponds to the time 10^{-8} to 10^{-10} s, and "slow" surface states with $\sigma \approx 10^{-18}$ to 10^{-21} cm^3 and corresponding time 10^{-4} to 10^{-1} s. "Superslow" traps with the cross-section $\sigma \approx 10^{-20}$ to 10^{-30} cm^3 have been observed. They are likely connected with the adsorption states. It is precisely these superslow states that define surface charging and potential. In this case the trapping times vary in the interval 10^{-1} to 10^3 s at low temperatures, which equal or exceed the characteristic times of the catalytic reaction.[132] For such times the suggestion of the electron equilibrium in the system, as was assumed in the so-called "electron theory of catalysis" on semiconductor surfaces, which was in general use in 1950 to 1980, seems to be incorrect.[642]

The kinetic of slow conduction relaxation in the process of charge carrier trapping has been a subject of a number of works (see, for example, the review in our book[132]). An expansion of the function $\Delta\sigma_s(t)$ in a series of exponents is the most common. The first exponent accounts for the reconstruction in the closest surrounding of the center which has captured an electron; the second one is due to more distant fragments, etc. One of the most successful approximations is the so-called Kocs law

$$\Delta\sigma_s(t) \sim \Delta\sigma_o \exp(-t/\tau_s)^\alpha \tag{10.3}$$

where τ and a are parameters.

Usually for dry media or in a vacuum $a \approx 0.3$. Kiselev and Krylov[132] have proved the validity of Kocs law with a large number of experimental data concerning the relaxation at a Ge or Si surface. The characteristic relaxation time τ_s varies with increase in temperature according to the law

$$\tau_s = \tau_o \, \mathrm{esp}(\Delta E_\tau/kT) \tag{10.4}$$

where ΔE_τ depends strongly on environment.

The ΔE_τ value cannot be considered as the true activation energy. Strictly speaking, the function $\Delta\sigma_s(t)$ in Equation 10.3 is nonexponential, and, therefore, ΔE_s can be regarded as some effective parameter. Weak dependence of τ_s on T for Ge and several other semiconductors points out that the tunnel mechanism can be a possible mechanism of charge exchange between slow states and the conduction zone of a semiconductor.

Frenkel was first to show that an electron capture increases the probability of defect origination exp (ϵ_t/kT) times, where ϵ_t is the well depth of the trapping center in the forbidden band of semiconductors. The energy consumed in the process of reaction of the defect is compensated by the energy released in the process of electron capture. The greater ϵ_t is, the less effective the multiphonon energy dissipation is, and, in contrast, the more effective, the process to create a defect transfer into vibrational energy of the defect is.

Vinetsky and Kholodar[679] have shown that the temperature threshold (T_{thr}) of the defect formation in an ionic crystal obeys the relation

$$kT_{thr} > (\epsilon_o - \epsilon_p - \epsilon_\alpha)/\ln (N^2/RV) \qquad (10.5)$$

where ϵ_o is the energy of the defect formation, ϵ_p is the depth of a polaron level, ϵ_α is the energy of the trapped electron measured from the bottom of the conduction band, R is the density of states in the zone ($\sim 10^{19}$ cm^{-3}), N is the number of lattice nodes ($\sim 10^{22}$ cm^{-3}), and V is the crystal volume.

It has turned out that in many crystals the defects arise at rather low temperatures, $T_{thr} \leq 150$ K. The energy of the defect formation is even lower at the surface; therefore, $T^{s}_{thr.} < T^{B}_{thr.}$ (S and B superscripts mean ''surface'' and ''bulk'', respectively).

Trapping by the adsorbed molecule is a special case of electron capture by a defect. If electron levels at the trapping center are deep enough, the value of released energy is greater than half the forbidden band width, and is usually more than the energy of lattice phonons by a factor of 10. The probability of energy transfer in the process of simultaneous creation of a large number of phonons is low. Therefore, when the adsorbed molecule traps an electron, its vibrational modes usually become excited. As a rule, the lifetime of local vibrations of the adsorbed complex is much longer than vibrational relaxation times in condensed media. In the course of relaxation, chemical transformation of the adsorbed molecule or its desorption is possible.[132]

Kiselev et al.[680] studied experimentally the dissipation of energy released in the process of trapping the charge carriers in slow adsorption states of Ge arising on adsorption of polar molecules: H_2O, D_2O, CH_3OH, NH_3, and benzoquinone. The relaxation has been found to be accelerated on adsorption of polar molecules at the Ge surface and accompanied by the simultaneous increase of ΔE relaxation value and the frequency factor τ_0^{-1}. In this case the relaxation occurs due to electron-vibrational interaction in the whole complex: adsorbed molecule plus active surface site plus adjacent bulk atoms of a solid.

The total potential energy of the system as a function of the configuration is shown in Figure 10.1.[680] The charge transfer from the free (Curve I) to the localized state (Curve II) in the trapping process results in significant atomic rearrangement at the center of localization. The bond strengthening is accompanied by a change of the coordinate $Q_1 \rightarrow Q_2$. The larger is the shift in ΔQ, the less is the overlap integral for electron wavefunction (dashed curves) which defines the transition probability from I to II. In isolated complexes the energy of electron transition gives rise to excitation of high vibrational levels. In this case the activation energy ΔE_τ is determined by electron excitation into the conduction band (E_{CS}), i.e, $\Delta E_\tau = E_{CS} - F$, where the subscript S refers to ''surface'' and F is the Fermi level. If an adsorbed molecule is surrounded by another adsorbed molecule, the energy exchange between them requires substantially fewer lattice phonons participating in the

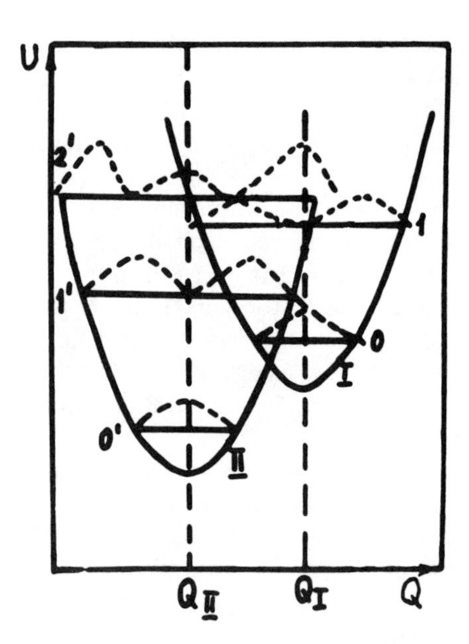

FIGURE 10.1. The total energy in the system "electron-lattice" as a function of the configuration coordinate. (From Kislev, V. F., Kozlov, S. N., and Levschin, N. L., *Phys. Status Solids*, 6, 93, 1981. With permission.)

process. The overlap integral increases with the increase of the vibrational level number (see Figure 10.1), and transitions between higher vibrational levels, for example, $I_1 \rightarrow II_1$ or $I_1 \rightarrow II_2$, become more probable than $I_o \rightarrow II_o$ as in the first case (in the absence of other adjacent molecules). The activation energy increases: $\Delta E_\tau = E_{CS} - F + h\nu$, where $h\nu$ is the energy transferred to the ensemble of adsorbed molecules. Due to the larger overlap integral and more effective energy dissipation, the frequency factor τ_o^{-1} increases more effectively than exp $(-\Delta E_\tau/kT)$. This is also true for all experimentally studied molecules at the Ge surface. A linear constraint between $\lg\tau_o^{-1}$ and ΔE_τ is attributed to characteristic properties of energy dissipation. The probability of energy dissipation increases with the increase in surface coverage.

Later, the same group of researchers[681] proved the direct participation of trapped electrons and holes in catalytic reaction at the surface. Upon light exposure of modified silicon (Si + F or Cl) by light with $h\nu > 2E_g$, the linear dependence of the yield of HCOOH dehydrogenation

$$\begin{array}{l} \diagdown \;\; F \qquad\quad O \\ -Si \;\; : O\!-\!C \!\!\diagup\!\!\!\diagup \quad + e^- + h^+ + h\nu \rightarrow H_2 + CO \\ \diagup \qquad\quad | \quad \diagdown \\ \qquad\qquad H \quad\; H \end{array}$$

where hν is the quantum of vibrational excitation of the adsorption complex, was observed on a number of recombined charge carries. An average quantum yield increases with the growth of $h\nu - 2E_g$. This result cannot be understood in the context of "the electron theory of catalysis at semiconductors",[642] but it can be explained by the electron-vibrational capture model, i.e., excitation of vibrational modes of adsorbed complexes.[132]

In a number of studies carried out at the Institute of Chemical Physics, it was shown that electron or hole transport to adsorbed molecules can be performed at a considerable distance from the solid bulk. Adsorption does not require the immediate overlap between orbitals of adsorbed molecules and a surface atom-active site. It has been shown in these studies that a granule of oxidation catalyst can consist of a rather thick envelope of a certain composition and the kernel of another one, i.e., the nonequilibrium structure is formed. It can be suggested that a thickness of such an envelope should correspond to the thickness of a layer of "net charge transfer" via the tunnel or polaron mechanism from an electron donor or acceptor in the catalyst bulk to the surface molecules. Net charge transfer leads to the following motion of atomic particles, for example, oxygen diffusion in the oxide that often limits the rate of catalytic oxidation.

The well-known phenomenon of adsorboluminescence was discovered by Roginsky, Kadushin, and Rufov[131] at the Institute of Chemical Physics. Il'ichev and Rufov[682] studied in detail its characteristic properties on O_2 adsorption at the oxide surface. Oxygen adsorption on MgO is the most extensively studied case. On the basis of these experiments, the diffusion model of oxygen adsorboluminescence on magnesium oxide has been proposed.[683] On oxygen adsorption, a substantial proportion of O_2 is in a weakly bound mobile state. Mobile oxygen interacts with F-centers when they come within short distances of each other. An electron tunnels into the O_2 molecule, resulting in O_2^--ion-radical formation. Then O_2^- migration occurs, followed by the secondary electron tunneling into the V_1-center of MgO, with the formation of the excited $(O_2Z)^*$ complex accompanied by emission of a luminescent quantum. The process can be represented by the following scheme

$$O_2 + Z \xrightarrow{k_1} O_2Z$$

$$O_2 + Z_o \xrightarrow{k_2} (O_2 \ldots Z_o)_{dif}$$

$$(O_2 \ldots Z_o)_{dif} + Z_2 \xrightarrow{k_3} O_2Z_{2'}$$

$$(O_2 \ldots Z_o)_{dif} + F_o \xrightarrow{\omega} (O_z^- \ldots Z_o)_{dif} + F^*$$

$$(O_2^- \ldots Z_o)_{dif} + V_1\text{-center} \xrightarrow{k_4} (O_2Z)^* \rightarrow O_2Z + h\nu$$

where k_1 is an adsorption rate constant, k_2 is that of adsorption of mobile oxygen, and k_3 is the rate constant of mobile oxygen trapping by defects.

The activation energy of the oxygen diffusion over the MgO surface is equal to 15 kJ/mol, the probability of elementary diffusion jump per unit time exceeds $5 \cdot 10^{-1}$, and the diffusion coefficient $D > 10^{-10}$ cm^2/s.

Electron transfer probability depends on the distance between the F-center and O_2 molecule, according to the well-known law which is typical for electron tunneling at a large distance

$$w(r) = w_o \exp(-r/r_o) \qquad (10.6)$$

where r_o is a radius of electron ψ-function of the acceptor.

It has been calculated that on oxygen adsorption the luminescent emission depletes the donors up to a depth of 2 nm from the MgO surface. Cross-section is of the order of 10^{-12} cm^2, and it is likely defined by the characteristic properties of electron tunneling to mobile oxygen and by the following relaxation.

The characteristic time of solid excitation induced by oxygen adsorption can be estimated from these works. The calculated value is ~100 s. Actually, these times are significantly less. Thus, the earlier reported value is $\sim 10_5^-$ s which is still a very long time. Roginsky and Rufov observed a luminescence of a solid in the course of hydrogen oxidation and N_2O decomposition over MgO.[684] Eremenko[685] observed a similar effect during N_2O decomposition on thorium.

10.2. MULTIPHASE CATALYSTS

10.2.1. NONEQUILIBRIUM PROCESSES IN MULTIPHASE CATALYSTS

In Chapter 9 we already considered the nonequilibrium phase transitions occurring in the process of catalytic reaction. It was shown that in several cases adsorption and a catalytic reaction can proceed according to the mechanism of the phase transformation, i.e., nucleation and growth even in conditions of stationary catalysis. Despite the requirements of equilibrium thermodynamics (the phase rule), simple catalytic systems can convert into two-phase systems in the course of catalysis.

Multicomponent catalysts are, as a rule, multiphase systems. Spenser[686] has shown that in the multicomponent system the previously mentioned thermodynamic restrictions concerning the multiphase state, in accordance with the phase rule, are absent. Such restrictions are true only for the two-component system. In many cases the multiphase state is of fundamental importance in devising active and selective catalysts.

One of the benefits of the multiphase system is the possibility of intermediate species transport from one phase to another, thus, shifting the unfavorable equilibrium conditions.

In Section 8.1.3 we presented Equation 8.22. It follows from this equation that the endothermicity of intermediate stages can be overcome on account of substantial concentration increase of the intermediate X_1 in the left-hand side of the stoichiometric equation, or due to substantial concentration decrease of the intermediate X_2 in the right-hand side of the equation. The second case seems to be of widespread occurrence in heterogeneous catalysis.

One of the pioneer investigators of this problem was Weisz,[687] who studied in detail hydrocarbon transformations over so-called bifunctional catalysts. In particular, it was shown that in the reforming reaction over supported $Pt/Al_2O_3 \cdot SiO_2$ catalyst, *n*-paraffin dehydrogenation into *n*-olefins occurs at platinum, but further skeletal isomerization of *n*-olefins into *iso*-olefins proceeds at Al_2O_3 or aluminosilicate. High rates in this reaction can be reached as a result of spatial separation of different steps. Suppose that the reaction runs as follows

$$\text{A(n-paraffin)} \underset{k_{-1}}{\overset{k_1}{\rightleftarrows}} \text{B}(\textit{n}\text{-olefin}) \overset{k_2}{\rightarrow} \text{C}(\textit{iso}\text{-olefin}) \tag{10.7}$$

where the equilibrium constant for the first step is $K = k_1/k_{-1} << 1$. If both steps are of the first order, the summary reaction rate is

$$-\frac{d[A]}{dt} - \frac{k_2[A]}{1/K + k_2/k_{-1}} = k'[A] \tag{10.8}$$

A proper choice of rate constants k_2 and k_1 may substantially increase output C according to the scheme (Equation 10.7) in comparison with a simple consecutive scheme $A + B \rightarrow C$. It is possible to arrange by choosing one phase for stage $A \rightarrow B$, and the other for stage $B \rightarrow C$, provided that a high removal rate of B intermediate from the reaction zone $A \rightleftarrows B$ is realized. In this case the maximal rate of B production is limited by the rate of diffusion. Taking into account the rate of diffusion, Weisz found the following expression for the reaction rate

$$-\frac{d[A]}{dt} = \frac{d[C]}{dt} = k'\left(1 + \frac{k'}{K}\frac{L}{D}\right)^{-1}[A] \tag{10.9}$$

where D is the diffusion coefficient and L is the diffusion distance for B molecules which is of the order of the size of catalyst particles or interparticle spacing.

Increasing the diffusion coefficient as well as decreasing the distance between dissimilar catalyst particles facilitates the attainment of complete conversion. Even higher selectivity can be obtained for more complicated cases with a set of consecutive and parallel stages proceeding on different phases.

Weisz[687] pointed out that the simplest method to check the joint action of different components is the method of mechanical mixing. It consists of the comparison of a separate action of catalysts X and Y with the action of catalyst mixture X + Y, which is composed of equal amounts of X and Y. If the reaction rate in the last case is higher than the sum of rates for the individual catalyst, it implies that the process is multistep, with the participation of different phases. The order of conversions can be established in experiments with different layer-by-layer catalyst mixtures. Using this technique, Weisz has proved that the reactions of isomerization, hydrocracking, and paraffin hydrogenolysis are multistep. He found the optimal selectivity conditions for different products and defined the realization criterion of the multistep reaction using an expression of the type in Equation 10.9. An estimation with a typical gas-phase diffusion coefficient $D = 2\ 10^{-3}\ cm^2/s$ and a reaction rate value $dN/dt = 10^{-6}\ mol/(s \cdot cm^3)$ gives

$$P_B > 2.3\ 10^9 \left(\frac{T}{273}\right) \frac{dN}{dt} \frac{L^2}{D} \tag{10.10}$$

where P_B is the partial pressure of the intermediate B in bars and L is the characteristic size of catalyst particles. For mechanically mixed particles with $L = 10^{-4}$ cm, the reaction can pass even if the intermediate is created in the gas phase with a partial pressure $P_B = 10^{-7}$ bar. Using similar criteria, different active and selective catalysts with particles of optimal size have been developed for such complicated catalytic processes as hydrocarbon reforming and cracking, the transformation of xylene into ethylbenzene, and others.

Thus, the endothermicity in catalysis may be overcome both due to the generation of a superequilibrium concentration of active sites and due to continuous removing intermediates to other reaction zones.

Weisz[687] considered nontrivial multistage reactions of the type $A \rightleftarrows B \rightarrow C$ from the viewpoint of thermodynamics as well. The summary variation of the free energy $\Delta G_{A \rightarrow C}$ must, of course, be negative, but an intermediate stage can, in his opinion, have a positive value. In the case of simple monomolecular transformation we have

$$P_{B.eq.} = P_A \exp\left(-\Delta G_{A \rightarrow B}/RT\right) \tag{10.11}$$

and Equation 10.10 transforms into the condition

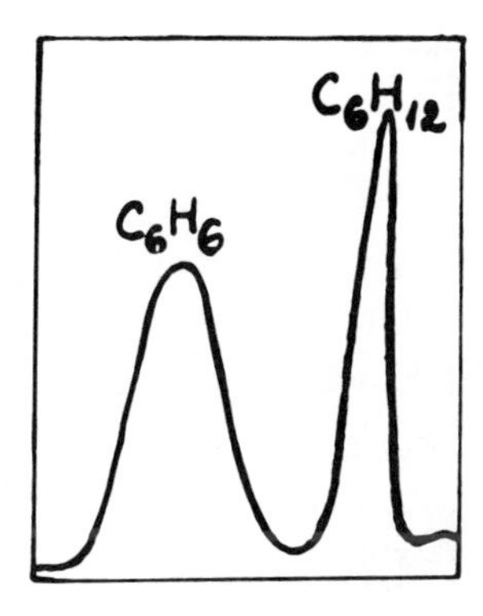

FIGURE 10.2. Catalytic chromatogram obtained for cyclohexane dehydrogenation into benzene over Pt/Al_2O_3 catalyst.[688]

$$\Delta G_{A \to B} < RT \ln \left[4.4 \; 10^{-9} \left(\frac{273}{T} \right) D \; P_A/(dN/dt)L^2 \right] \qquad (10.12)$$

One can conceive of even a more trivial case, i.e., a reversible thermodynamically unfavorable reaction without intermediates: $A \rightleftarrows B + C$. In this case the shift of unfavorable equilibrium may occur as a result of continuous migration of X component to other surface sites with higher adsorption binding energies. Roginsky et al.[688] were the first to investigate the so-called chromatographic regime in the catalytic reaction of cyclohexane dehydrogenation into benzene over supported Pt/Al_2O_3 catalyst

$$C_6H_{12} \rightleftarrows C_6H_6 + 3H_2$$

At 400 to 500 K this thermodynamically unfavorable reaction can be almost absolutely shifted to the right in the pulse regime when adding cyclohexane portions to the flow of inert gas. In the reaction zone the concentration of intermediates is significantly lower due to continuous removal of reaction products. A typical "catalytic chromatogram" shown in Figure 10.2 represents the concentrations of reaction (products) just behind the reaction without special separation. It is seen that benzene and cyclohexane outputs are separated. The conversion degree of cyclohexane into benzene vs. lg1/c in the chromatographic regime (Curve 1) is shown in Figure 10.3 in composition with the equilibrium Curve 2 and Curve 3 for the stationary dynamic regime.

The theory of catalytic reactions in the chromatographic regime was developed by Roginsky and co-workers.[689] In particular, it was shown that especially high enhancement of endothermic reactions should be expected at small equilibrium constants and, on the contrary, high rate constants. Doping the catalyst with adsorbent which adsorbs dominantly only one of the reaction

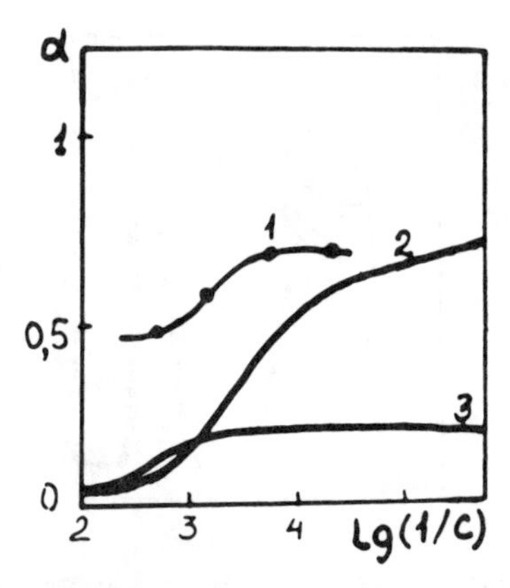

FIGURE 10.3. The conversion degree α as a function of lg (1/c) in the reaction of benzene dehydrogenation. (1) Chromatographic regime; (2) equilibrium curve; (3) in a flow setup.[688]

products (for example, C_6H_6 in the reaction of cyclohexane dehydrogenation) can result in still further shifting of the reaction to the right. So, in this case the benefits of the multiphase system are also evident.

Ponec[690] has pointed out that the activity and selectivity of some reactions can be enhanced using zeolites as adsorbents of the reaction products. For example, zeolites have been used for removal of water in some reactions of condensation and for removal of lower alcohols in exchange reactions of alcohols with esters.

In conditions of stationary heterogeneous catalysis the gradients of catalyst composition inside each of the phases can arise along with the concentration gradients of intermediates migrating in pores between different phases or via the mechanism of surface diffusion. Schmalzried[691] studied in detail the mechanism of gradient origination for redox reactions of oxides. He considered a crystal of MO oxide having two boundaries with different partial oxygen pressures P'_{O_2} and P''_{O_2} (Figure 10.4). In such oxides the O^{2-} ions remain fixed, but there are fluxes of cations, cation vacancies, and holes due to the gradient of oxygen concentration. The motion is aided by the existence of some initial concentration defects.

In solid solutions $M_{\langle 1\rangle}M_{\langle 2\rangle}\Theta$ the ions $M^{2+}_{\langle 1\rangle}$ and $M^{2+}_{\langle 2\rangle}$ migrate with different velocities. This leads to segregation in the initially homogeneous crystal. Using the solution $Co_{0.1}Mg_{0.9}O$ as an example, it has been experimentally shown that at different O_2 pressure at two crystal boundaries it becomes enriched with Co on one side, and with Mg on another.[691] If we have a combined crystal, for example, spinel AB_2O_4 with a narrow interval of homogeneity, then phase decomposition instead of segregation occurs.

$$AB_2O_4 + A^{2+} \rightleftarrows 2AO + 2BO + (\square + 2h)$$

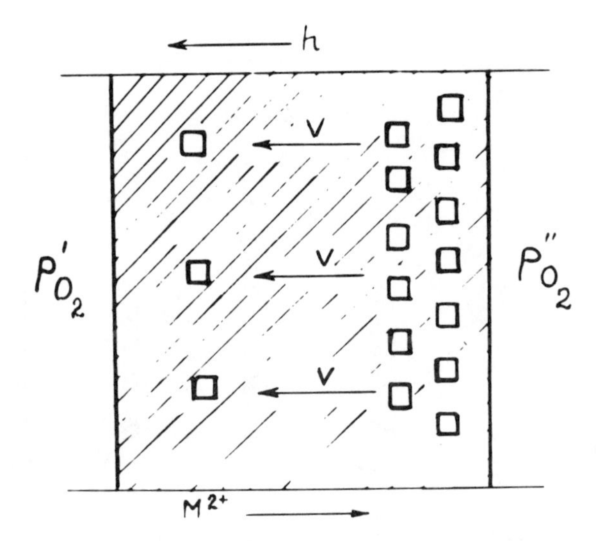

FIGURE 10.4. The scheme of MO oxide crystal restricted from both sides by gas phases with different oxygen pressure (p'_{O_2} and p''_{O_2}) (V is a cation vacancy; h, hole). (From Schmalzried, M., *React. Solids,* 1(2), 117, 1986. With permission.)

as a result of accumulation of the more mobile AO component. The kinetic decomposition of spinel was actually observed experimentally, including the formation of periodic layers with different phases of a type of Liesegang rings.[692] It should be noted that such a kinetic decomposition has been observed in the presence of the gradient of oxygen potential as well as other gradients (temperature, mechanical compression, etc.). When the AB_2O_4 crystal is maintained at surrounding, stationary oxygen, the crystal is absolutely stable, neither oxidation nor reduction occurs.

In heterogeneous redox catalysis the stationary gradients of oxygen and vacancy concentrations can originate in multiphase catalysts only in the presence of crystallographically coherent oxide lattices. This condition seems to be of fundamental importance in the development of active and selective oxidation catalysts.

10.2.2. CATALYSTS OF PARTIAL OXIDATION

The growth of catalytic selectivity for propylene oxidation into acrolein and acrylic acid is shown schematically in Figure 10.5 as a function of time. In the 1950s the best catalyst was Cu_2O with a maximum selectivity of 50%, in the 1960s bismuth molybdate (with a selectivity of 85% and conversion of 82%). At the present time four-, five-, and six-component oxide catalysts are employed. The necessity of several components in active and selective catalysts of partial oxidation is not accidental.

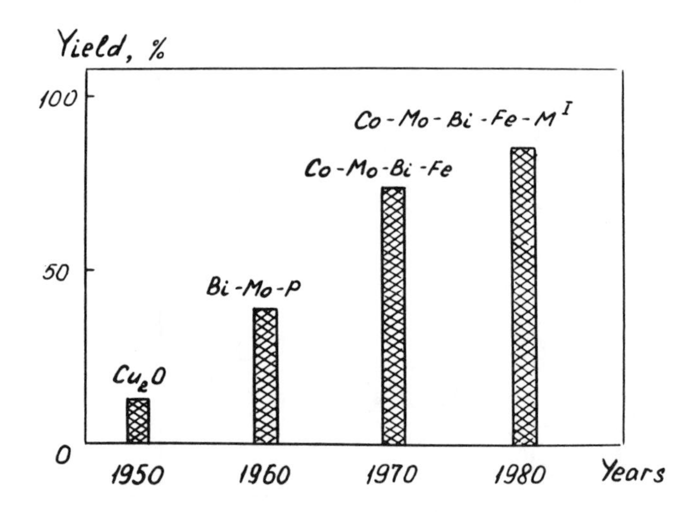

FIGURE 10.5. The growth of catalyst selectivity in the reaction of propylene oxidation into acrolein and acrylic acid with an increase of catalyst components number for the past 40 years.

We have studied the activity and selectivity of oxide catalysts of partial oxidation as well as the role of multiphase systems in detail.[693-697] In the complex Bi-Fe-Co-Mo-O catalyst of partial propylene oxidation into acrolein, the role of individual phases has been understood with the aid of X-ray structure investigation and Mössbauer spectroscopy *in situ* (in the course of the catalyst reaction).[696-697] From the very beginning after preparation the catalyst consists mainly of four phases: β-$CoMoO_4$, $Fe_2(MoO_4)_3$, β-$Bi_2(MoO_4)_3$, and an excess MoO_3 phase. $Fe_2(MoO_4)_3$ and $Bi_2(MoO_4)_3$ lattices are structurally close to each other.

Experiments have been carried out for mixtures of C_3H_6 and O_2 at 3:1, 2:1, and 1:1, respectively. In the course of the catalytic reaction the content of the $Bi_2(MoO_4)_3$ phase does not vary, the concentration of $CoMoO_4$ increases to some extent, and the concentration of $Fe_2(MoO_4)$ substantially decreases due to its reduction into β-$FeMoO_4$. The last form originates only in the presence of the β-$CoMoO_4$ phase. The study of the Bi-Fe-Mo-O catalyst without $CoMoO_4$ revealed that the $Fe_2(MoO_4)_3$ phase is not reduced. The pure $Fe_2(MoO_4)_3$ phase also is not reduced in the course of the reaction. Thus, the iron-molybdate phase is the most labile under catalytic conditions; its reduction occurs due to the crystallographic coherency between β-$CoMoO_4$ and β-$FeMoO_4$ lattices, and the β-$FeMoO_4$ phase has been observed in the catalyst using Mössbauer spectroscopy. X-ray analysis fails to identify it because of the very close coincidence between parameters of the β-$FeMoO_4$ and β-$CoMoO_4$ lattices. Some enhancement of β-$CoMoO_4$ reflexes is due to the formation of β-$FeMoO_4$.

The output of the reaction product, i.e., acrolein, becomes stationary much more rapidly than the stationary phase composition. It can be explained by the fact that the reaction occurs in the vicinity of the boundary between the oxidized and reduced phases of iron molybdates; the reaction rate depends on the boundary length rather than on the surface area of each phase. This length only slightly changes at different reduction degrees.

The process of catalytic oxidation and reduction can be schematically written in the following form

$$C_3H_6 + 2Fe^{3+} + MoO_3 + O^2 \rightarrow C_3H_4O + 2Fe^{2+} + MoO_2 + H_2O,$$

$$O_2 + 2Fe^{2+} + M^{2+}MoO_3 \rightarrow O^{2-} + 2Fe^{3+} + M^{2+}\ MoO_4$$

It is followed by the electron and oxygen exchange at the interphase boundary

$$(2Fe^{2+} + MoO_2) + (2Fe^{3+} + M^{2+}MoO_4) \rightarrow$$
$$(2Fe^{3+} + MoO_3) + (2Fe^{2+} + M^{2+}MoO_3)$$

Co^{2+} ion from $CoMoO_4$ seems to take no part in the redox process of the acrolein synthesis because the change of $CoMoO_4$ by $MgMoO_4$ with a similar crystal lattice only slightly affects the catalytic process.

Temperature dependencies of the isomer shift (IS) value in the Mössbauer spectrum of iron molybdate as well as the summary rate of the propylene oxidation r are demonstrated in Figure 10.6.[697] The theoretical IS temperature dependence in Debye approximation is depicted by solid lines. Arrows mark the temperature boundary between selective and deep propylene oxidation; it corresponds to 680 ± 20 K. At lower temperatures, partial oxidation of C_3H_6 into acrolein dominates; at higher ones total oxidation occurs.

As seen from Figure 10.6, in air the IS temperature dependence (Curve 1) is in a agreement with the theoretical curve. The same is valid for experiments *in situ* at 300 to 520 K. However, at T ⩾ 520 K, the isomer shift undergoes strong anomaly, deviates from the theoretical value, and reaches a constant value of 0.52 ± 0.02 mm/s in the region of selective catalysis. In the interval 520 to 680 K the IS described by Curve 2 varies in a reversible way. At 680 K (the beginning of deep oxidation) substantial changes occur in the catalyst. A portion of β-$FeMoO_4$ increases, and the value of $L = [Fe^{2+}]/([Fe^{2+}] + [Fe^{3+}])$ changes from 10 to 35%. Thereafter the temperature IS dependence is described by a straight line 3 with a slope differing radically from the theoretical value. The last dependency remains unchanged in the temperature interval 300 to 750 K in experiments *in situ* as well as in air. Irreversible changes of iron molybdate at 680 K are confirmed by X-ray data analysis which reveals its structural changes.

The results obtained are explained in the following way. The anionic vacancy formation occurs at the boundary of the iron molybdate in the course of selective oxidation due to lattice oxygen elimination and to the formation of anionic vacancies □

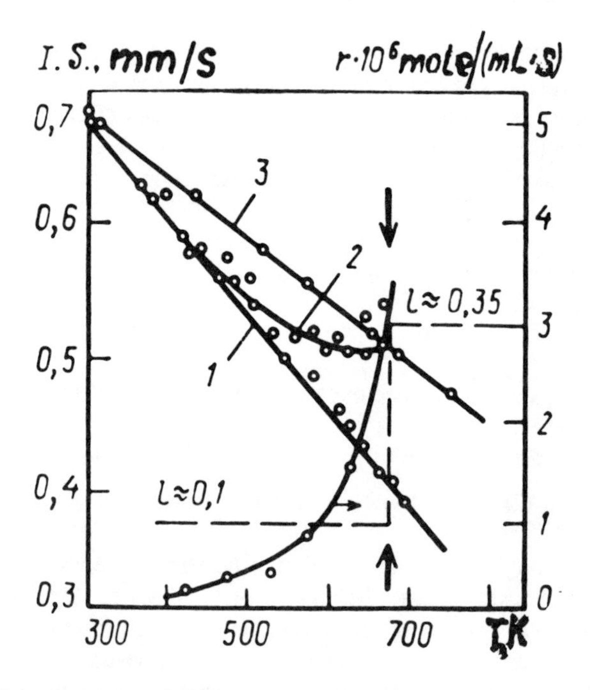

FIGURE 10.6. Temperature dependencies of the isotopic shift in the Mössbauer spectrum and of the summary oxidation rate of propylene r and the ratio $L = [Fe^{3+}]/[Fe^{2+} + Fe^{3+}]$ for Co-Bi-Mo-Fe-O catalyst. (1) In air; (2) *in situ* in the region of selective propylene oxidation into acrolein; (3) *in situ* after experiments at 680 K (arrows denote the boundary between the partial and total oxidation). (From Krylov, O. V. et al., *J. Catal.*, 25, 948, 1984. With permission.)

$$Fe_2Mo_2O_{12} \rightleftarrows [Fe_2Mo_3O_{12-x}\square_x]^* + x/2O_2$$

Thermally activated vacancy ionization results in the appearance of an excess negative charge that affects the electron state of iron ions in the molybdate. Spectral changes of the whole iron molybdate crystal are observed experimentally, which is indicative of a rapid migration of electron density over the excited $[Fe_2Mo_3O_{12-x}\square_x]^*$ crystal in a time of 10^{-7} to 10^{-8} s. The iron ion transits from an initial $3d^5$ state into a $3d^{5+\Delta}$ state. The last is intermediate between the states of Fe^{3+} in oxidized molybdate and Fe^{2+} in the reduced one.

At the interphase boundary between iron and cobalt molybdates, $Fe_2(MoO_4)_3$ reduces to $FeMoO_4$. The close coincidence between β-$FeMoO_4$ and β-$CoMoO_4$ structures aids the formation of β-$FeMoO_4$ nuclei. The superequilibrium concentration of vacancies $\square_x$ arises at the boundary between Fe^{2+} and Fe^{3+} molybdates in the course of the reaction. These vacancies can diffuse to the surface and serve as sites for oxygen dissociation adsorption from the gas phase with the formation of lattice oxygen O^{2-} (or O_2). Oxygen adsorption likely occurs at the β-$FeMoO_4$ phase. The nonequilibrium polarized

$[FeMo_3O_{12-x}\square_x]^*$ structure is a ''transport channel'' which supplies the necessary O_L amount to active sites responsible for propylene oxidation into acrolein. According to the above model,[696] the latter sites are concentrated at the interphase boundaries. Irreversible changes of the $Fe_2(MoO_4)_3$ phase occur in the region of deep oxidation. The amount of β-$FeMoO_4$ phase which cannot be reoxidized grows abruptly. The total C_3H_6 oxidation mainly involves surface molecular oxygen with the formation of CO and CO_2, but not O_L.

Based on the data obtained, the following conclusion about the role of different phases in a Bi-Fe-Mo-Co-O catalyst of partial oxidation can be drawn.

The β-$FeMoO_4$ phase is necessary for oxygen adsorption

$$O_2 + 4Fe^{2+} \rightarrow 2O^{2-} + 4Fe^{3+}$$

The defective structure of partially reduced $Fe_2(MoO_4)_3$ is a good conductor for oxygen ions. An O^{2-} oxygen ion migrates to the boundary between the $Fe_2(MoO_4)_3$ and $Bi_2(MoO_4)_3$ phases where adsorption and activation of propylene, followed by the reaction of formed allyl with oxygen, occur

$$C_3H_6 \xrightarrow{Bi_2(MoO_4)_3} C_3H_5 + H(OH_{surf.})$$

$$C_3H_5 + O^{2-} \longrightarrow C_3H_4O + H(OH^-) + e$$

Oxygen and electron exchange take place at the interphase boundary. The $CoMoO_4$ phase stabilizes the nonequilibrium phase β-$FeMoO_4$ in a process of reaction which is structurally coherent with it.

The active sites in the reaction of oxygen and propylene with the catalyst are concentrated in the vicinity of $FeMoO_4/Fe_2(MoO_4)_3$ and $Fe_2(MoO_4)_3/Bi_2(MoO_4)_3$ interphase boundaries, respectively. The concentration of active intermediates at these sites can be substantially increased in comparison with the monophase catalyst that, in turn, increases the activity. The activity of the multicomponent catalyst is two orders of magnitude higher than the activity of bismuth molybdate catalyst, while the selectivity is practically the same. The relative arrangement of different phases in the operating catalyst is shown in Figure 10.7.

So, the enhanced activity and selectivity of the complex oxidation catalyst are explained by separation of the following steps: the hydrocarbon activation, oxygen adsorption, and its diffusion between different phases. Nonequilibrium vacancy concentrations together with concentration gradients of lattice oxygen appear in the catalyst. To prevent the undesirable reaction of deep oxidation, the absence of active O_2^- and O^- forms is required in the vicinity of an activated molecule.

In order for the deep oxidation process at molybdate catalysts to be understood, the Weisz method of comparison of the activity of the complex

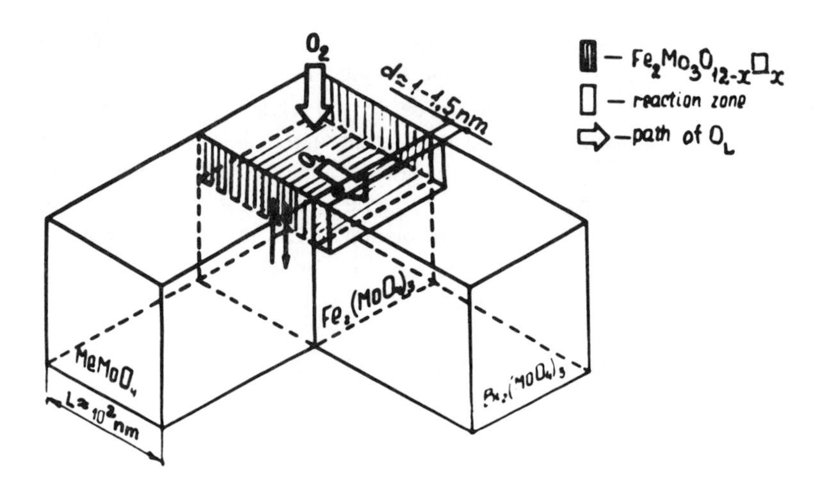

FIGURE 10.7. Relative position of phases in the operating Co-Bi-Mo-Fe-O catalyst in the reaction of propylene oxidation into acrolein ($Fe_2MoO_9O_{12-x}\square_x$ phase-containing vacancies is shown by hatching; here □ is an oxygen vacancy in the anion lattice; hollow arrows show the path of oxygen through the catalyst; black arrows denote electron exchange). (From Krylov, O. V. et al., *J. Catal.*, 25, 948, 1984. With permission.)

catalyst with the activity of separate components has been used.[698] In the process of propylene oxidation over a $CoMoO_4$ + MoO_3 or over a $CoMoO_4$ + $Bi_2(MoO_4)_3$ mixture the overadditive propylene conversion compared with the sum of conversions over each component has been observed. The outputs of acrolein and CO_2 are practically additive, but CO output increases substantially (tenfold and more). This is explained by the fact that there are intermediates at one or both phases which can desorb into the gas phase and convert into CO on adsorption at another phase.

Experiments with a two-layer arrangement in the reactor have been performed on individual components to investigate their part. The overadditive increase of propylene conversion was not observed when $CoMoO_4$ was first, and $Bi_2(MoO_4)_3$ behind it along the way in the reagent mixture (5% of C_3H_6 in air). In the opposite arrangement CO output increases more than tenfold. This is indicative of the formation of active intermediates X at bismuth molybdate followed by their conversion at $CoMoO_4$

$$C_3H_6 + O_2 \xrightarrow{Bi_2(Mo_4)_3} X \xrightarrow{CoMoO_4} CO + H_2O$$

Further experiments for freezing out this X intermediate in the ESR cavity have shown that X is a radical.

These data are supplemented by the data obtained with the aid of Mössbauer spectroscopy[696,697] that elucidate the part of different phases of the Bi-Co-Mo-Fe-O catalyst. Coverage of the $CoMoO_4$ phase by another one is necessary to prevent the undesirable reaction of total oxidation. Nevertheless, in a number of works on this subject a statement is made that in stationary conditions the catalyst is always monophase. Thus, Sleight[669] relates the selectivity of a modified oxide catalyst to the vacancy formation in the catalyst lattice. Forzatti et al.[700] have pointed out that highly selective Te-containing catalysts are monophase and form $CdTeMoO_4$, $CoTeMoO_4$, $MnTeMoO_4$, and $ZnTeMoO_4$ phases.

In the majority of papers that apply X-ray analysis of oxide catalysts, these catalysts have been shown to be multiphase. Even though a catalyst is initially monophase, it becomes multiphase in the course of the reaction because of segregation inducted by concentration gradients. For example, in contrast to the results obtained earlier,[700] the same authors[701] found that a Te-Mo-O catalyst of propylene ammoxidation converts in the course of the reaction into a mixture of phases (TeO_2, $TeMo_5O_{16}$, MoO_2, and Mo_6O_{11}).

An Fe-Mo-O catalyst of partial methanol oxidation into formaldehyde includes an excess of a MoO_3 phase together with an $Fe_2(MoO_4)_3$ phase. Stabilization of these phases on an α-Al_2O_3 support has a beneficial effect on the establishment of constant catalyst activity.[702] Application of Mössbauer spectroscopy[703] revealed that in an Fe-Mo-O catalyst with a ratio Mo/Fe = 2:4, there are α-$FeMoO_4$ and a metastable phase, with the Fe^{3+} ion being in a high-spin state in a distorted tetrahedron.

Many researchers report that selective catalysts containing V_2O_5 are multiphase. V_2O_5 doped with TiO_2 is a selective catalyst for oxidation of aromatic hydrocarbons. In this system the reduction of V_2O_5 is promoted by the structural correspondence between VO_2 and TiO_2 phases.[704] However, in V_2O_5 there are also crystallographic directions similar to those of the anatase lattice. For example, the (010) plane of V_2O_5 is coherent with (001), (100), and (011) anatase planes. On V_2O_5 reduction, a V_6O_{11} phase coherent with V_2O_5 is formed as well.[705] Yabrov et al.[706] have found the formation of a solid solution of V^{4+} ions in TiO_2 in the course of *o*-xylene oxidation. Yet catalysts with a higher (90%) content of this solid turned out to be less active. Cole et al.[707] have drawn a conclusion that atoms at the boundary between V_2O_5 and V_6O_{13} are the active sites.

The vanadium-phosphorus catalyst of selective butylene oxidation into maleic anhydride is usually considered to be monophase. However, Courtine[705] and Bordes and Courtine[708] have found in this catalyst the microdomains of a $VOPO_4$ phase along with $(VO)_2P_2O_7$. In this case the active sites also appear to be located at the interphase boundary. The multiphase state of the operating vanadium-phosphorus catalyst has been confirmed with the help of Raman spectroscopy *in situ*.[709]

A two-phase system arises in the reaction of stationary oxidation of benzene and xylene over V_2O_5-MoO_3; these are $V_9Mo_6O_{40}$ and V_2O_5 phases enriched with Mo. Ioffe and co-workers[710] note that in contrast to the results of Yabrov et al.[706], the thermal equilibrium state of the catalyst cannot be reached under these conditions. Berry and Brett[711] have found the segregation of a Sb_2O_4 phase in the reaction of hydrocarbon oxidation over $Sn_{1-x}Sb_xO_2$ and $V_{1-y}Sb_yO_4$ catalysts.

Courtine[705] reviewed the whole aggregate of data on selective oxidation catalysts and arrived at the conclusion that multiphase systems are of fundamental importance in redox catalysts. At nearby planes of the oxidized and reduced phases the atoms are in excited states compared with normal ones in the lattice bulk. Whereas in the oxide reduction process a stress at such boundaries is compensated for by the mechanical relaxation (for instance, by a crystallographic shift), in the stationary state part of the energy is transferred to the reagents. The ground energy state is turned back on reoxidation. Supported oxide systems also operate due to the formation of coherent boundaries, at least in small areas. One can easily understand that these views agree very closely with the results of our work.

Thus, in many cases (but not always) two- and multiphase systems arise in the process of oxidation catalysis over complex oxide catalysts.[712] Cullis and Hucknail[703] have made an even more radical statement that solid solutions do not play an important part in catalysis. Multicomponent catalysts are always multiphase.

One of the possible explanations of higher activity of a multicomponent system is also an increase in electron and ion conductivity of each phase in the multiphase system. As shown by Wagner,[713] an increase in conductivity is observed, even though one of the phases is dielectric. For example, AgI conductivity increases in the presence of a small amount of Al_2O_3 that introduces additional defects into the system due to the interphase boundaries.

As already demonstrated, in the stationary process the origin of concentration gradients of intermediates is of principal importance for multiphase systems. Such gradients allow an unfavorable shift in equilibrium conditions of intermediate stages. As an example of such an intermediate, take an oxygen ion migrating through the lattice in oxides. The phenomenon of interphase diffusion of active particles received the name spillover.

Delmon[714,715] suggested the new term "remote control" to designate the enhancement of the catalytic effect in multiphase systems. Other names are "synergy effect" and "phase cooperation". He gives numerous examples of such synergy effects, for example, in thiophene hydrogenolysis over supported $(MoS_z + CoS)/\gamma\text{-}Al_2O_3$ catalyst, in isobutylene oxidation into metacrolein over $SnO_2 + Sb_2O_4$, in dehydration of *N*-methylformamide over $MoO_3 + Sb_2O_4$. He also intimated that our studies[694-696] were among the first to suggest an important part of multiphase systems in catalysis.

The effect of "remote control" is of practical use when preparing a catalyst in the form of several layers from different substances. Another practical application is to carry out a redox reaction with spatial separation of reagents using membranes. For example, DiCosimo and co-workers[716] performed oxidative dehydrodimerization of propylene into diallyl $C_3H_5\text{–}C_3H_5$ over $Bi_2O_3 \cdot La_2O_3$ catalyst in the form of a disk separating C_5H_6 and oxygen.

At 873 K the selectivity for diallyl reaches 76% compared with 38% without separation.

10.2.3. CATALYSTS OF REACTIONS WITH HYDROGEN

Nonequilibrium effects are also extensively distributed in catalysis on metals. Gryaznov et al.[717,718] have shown that the application of catalytic membranes fabricated of palladium alloys increases selectivity in the reaction of butane dehydrogenation into butadiene because of hydrogen removal on the other side of the membrane. Oxidation of diffusion hydrogen increases the butadiene yield.

The part of surface diffusion in activity and selectivity of heterogeneous catalysts seems to be much greater than has been thought up to now. The hydrogen spillover discovered by Neikam and Vannice[719] in the Pt/WO_3 system provides H-atom transport from the metal to the oxide. In the absence of metal the hydrogen adsorption fails to occur on oxide. This phenomenon must be of fundamental importance in a reaction with hydrogen participation since it shifts the equilibrium of H-atoms.

Spillover was also observed for CO and NCO species. Matyshak et al.[720] studied the diffusion of NCO species from the surface of Pt or Rh to γ-Al_2O_3 support and vice versa. These species are intermediates in the selective conversion of NO + CO into $N_2 + CO_2$.

The Fischer-Tropsch synthesis, which likely proceeds via the redox mechanism, can also be explained by the formation of spatially separated active sites and by passing different reaction steps at different phases or at the interphase boundary.

Wang and Eckerdt[721] showed that after the reaction, CO + H_2 over supported Fe/SiO_2 catalyst it consists of the phases Fe_3O_4, Fe_3C, and metallic Fe. Bianchi et al.[722] applied the method of Mössbauer spectroscopy to check the hypothesis according to which the activity of Fe/Al_2O_3 catalyst is connected with the formation of carbides. It was shown that the reaction rate peaks on the reduced catalyst within the first hour of operation, and then drops. A mixture of χ-$Fe_{2.5}$ and ϵ-$Fe_{2.2}$ carbides is formed simultaneously. Using the response method, it has been yet shown that other carbon-containing surface compounds are the intermediates of this reaction. At the end of reduction the catalyst consists of 74% Fe^{o} and 26% Fe^{2+}.

According to the views of Rozovsky,[723] the synthesis of hydrocarbons and alcohols from CO + H_2 and the formation of surface carbides are conjugated reactions. If the catalyst phase transformation and the catalytic reaction have a common stage, the catalytic reaction changes the rate of phase transformation. Point surface defects of iron carbide are the active sites in this reaction. The scheme of the reaction of synthesis can be represented as follows

$$\begin{array}{l} Fe_{a.c.} + H_2 \rightleftarrows Fe(H_2) \xrightarrow{CO} Fe\ C\ OH_2 \longrightarrow FeC + H_2O \\ \qquad\qquad\qquad\qquad\qquad\qquad \downarrow\uparrow \\ \qquad\qquad\qquad\qquad\qquad Fe\ \ C\ OH \longrightarrow \text{Products of synthesis} \end{array}$$

It would seem reasonable that in reactions at iron catalyst the interphase boundaries between carbide and metal iron should be the phases of enhanced activity where the concentration of point defects seems to be especially high.

As a result of a number of studies of Fischer-Tropsch synthesis at iron oxides with the aid of Mössbauer spectroscopy and X-ray diffraction (XRD) *in situ,*[724-728] a set of successive phase transformations has been established: $Fe_2O_3 \rightarrow Fe_3O_4 \rightarrow \epsilon'\text{-}Fe_{2.2}C \rightarrow \varkappa\text{-}Fe_{2.5}C$. It is interesting that the stationary operating catalyst consists of Fe_3O_4 oxide and iron carbides, but does not contain the metal phase. Discussion about the possible role of carbides in the Fischer-Tropsch synthesis is being continued. Only the carbide phase has been found at the active iron catalyst promoted by potassium. Intermediate carbide $\epsilon'\text{-}Fe_{2.2}C$ is probably the most active phase in the reaction.[726] The correlation between the initial rate of Fischer-Tropsch synthesis and the rate of ϵ'-carbide formation has been observed.[725]

In our work[727] it was shown that in the course of the Fischer-Tropsch synthesis $\alpha\text{-}Fe_2O_3$ transforms into layer-by-layer structure consisting of Fe_3O_4 particles with an intercalated carbide phase. Such structures consisting of sequential oxide and carbide layers with a thickness of 2 to 4 nm are much more active than catalysts, which were preliminary subjected to shear deformation, and consist of separate oxide and carbide regions. The interphase carbide-oxide boundaries appear to be possible active sites for CO dissociation. The length of these boundaries is much greater in intercalated particles than that of separated carbide and oxide ones. Using XRD *in situ,* it was found that the formation of the carbide phase coincides with the beginning of olefin production in the reaction outputs.

The study of catalytic properties of specially synthesized Co_3C and Co_4C carbides revealed that they are at a considerable disadvantage in relation to supported Co-catalysts. The latter are 1 or 2 orders of magnitude more active, and their selectivity is shifted to the formation of heavier hydrocarbons. Catalytic activity of Co_4C carbide is rather stable. At 468 K, the conversion parameters remain practically unchanged within 10 h of operation. However, the phase composition changes significantly. Many carbide lines disappear as a result of carbide transformations into other phases. It seems that in the case of oxide catalysts, the reaction likewise proceeds at interphase boundaries which may displace, though the total number of active sites does not substantially vary.

Colley et al.[728] studied the reaction of the Fischer-Tropsch synthesis using the XRD method *in situ* at supported Co/MgO, Co/Al_2O_3, and $Co\text{-}Mn/Al_2O_3$ catalysts. The formation of a new, earlier unknown Co(bcc) metal phase has been observed.

Sachtler[729] describes a number of cases when MnO and other oxides (ZrO_2, TiO_2, and La_2O_3) alter the selectivity of CO hydrogenation, so that predominant formation of methanol, ethanol, and other oxygen-containing compounds is observed. One of the possible explanations of the influence of the oxidized phase reduces to CO adsorption at the boundary between Mn and Fe-containing phases

$$\begin{array}{c} \cdot\cdot\cdot O = C \\ \qquad\qquad | \\ Mn - O - Fe \end{array}$$

The tilted CO molecule has more opportunities to react with a growing chain.

Another explanation is based on the two-center model according to which not only the formation of CH_4, but the growth of a carbon chain as well occurs at the metal phase. The chains are terminated at the oxide phase. When the content of oxide is large, the distribution of products is shifted in the direction of low-molecular hydrocarbons.

REFERENCES

1. **Langmuir, I.,** *Chem. Rev.*, 6, 451, 1929.
2. **Taylor, J. B. and Langmuir, I.,** *Phys. Rev.*, 40, 463, 1932; 44, 423, 1937.
3. **Zeldovitsch, Ya. B. and Roginsky, S. Z.,** *Acta Physicochim. URSS*, 1, 559, 1934.
4. **Scheve, J. and Schulz, I. W.,** *Proc. 4th Int. Congr. Catalysis*, Akademia Kiado, Budapest, 1971, 483.
5. **Gardner, R. A.,** *Proc. 4th Int. Congr. Catalysis*, Akademia Kiado, Budapest, 1971, 466.
6. **Gagarin, S. G. and Kolbanovsky, Ju. A.,** *Dokl. Akad. Nauk SSSR*, 278, 669, 1984.
7. **Johnson, B., Larsson, R., and Rabensdorf, B.,** *J. Catal.*, 102, 26, 1966.
8. **Larsson, R.,** *J. Catal.*, 88, 509, 1984; *Chem. Scr.*, 12, 78, 1977.
9. **Kobozev, N.,** *Zh. Fiz. Khim.*, 21, 413, 1947.
10. **Semenov, N. N. and Voyevodsky, V. V.,** *Heterogeneous Catalysis in Chemical Industry*, Goskhimizdat, Moscow, 1955, 233 (in Russian).
11. **Kon', M. Ja., Shvets, V. A., and Kazansky, V. B.,** *Dokl. Akad. Nauk SSSR*, 203, 624, 1972.
12. **Barelko, V. V. and Volodin, Ju. E.,** *Kinet. Katal.*, 17, 683, 1976.
13. **Krylov, O. V.,** *Kinet. Katal.*, 26, 263, 1985.
14. **Roginsky, S. Z.,** Gostekhteoretizdat, *Probl. Kinet. Katal.*, 4, 123, 1940 (in Russian).
15. **Lennard-Jones, J. E. and Strachan, C.,** *Proc. R. Soc. London*, A150, 442, 1935; A156, 6, 1936.
16. **Bonch-Bruevich, V. L.,** *Usp. Fiz. Nauk*, 40, 369, 1950.
17. **Zwanzig, R. W.,** *J. Chem. Phys.*, 32, 1173, 1960.
18. **Cabrera, M. B.,** *Discuss. Faraday Soc.*, 28, 16, 1959.
19. **McCarrol, B. and Ehrlich, G.,** *J. Chem. Phys.*, 38, 523, 1962.
20. **McCarrol, B.,** *J. Chem. Phys.*, 39, 1317, 1963.
21. **Mazhuga, V. V. and Sokolov, N. D.,** *Dokl. Akad. Nauk SSSR*, 168, 625, 1966; *Probl. Kinet. Katal.*, Nauka, Moscow, 14, 254, 1970 (in Russian).
22. **Roginsky, S. Z., Tretjakov, I. I., and Schechter, A. B.,** *Zh. Fiz. Khim.*, 23, 50, 1949; Dokl. Akad. Nauk SSSR, 51, 516, 1954.
23. **Carton, G. and Turkevich, J.,** *J. Chem. Phys.*, 51, 516, 1954.
24. **May, J. W.,** *Adv. Catal.*, 21, 152, 1970.
25. **Rozovsky, A. Ja.,** *Dokl. Akad. Nauk SSSR*, 151, 1390, 1963; *Kinet. Katal.*, 5, 609, 1964.
26. **Nicolis, G. and Prigogine, I.,** *Self-Organization in Non-Equilibrium Systems*, John Wiley & Sons, New York, 1977.
27. **Hacken, H.,** *Synergetics*, Springer-Verlag, Berlin, 1978.
28. **Boudart, M.,** *Adv. Catal.*, 20, 153, 1969.
29. **Krylov. O. V.,** *Kinet. Katal.*, 21, 79, 1980.
30. **Kondrat'ev, V. N. and Nikitin, E. S.,** *Chemical Processes in Gases*, Nauka, Moscow, 1981 (in Russian).
31. **Somorjai, G. A.,** *Chemistry in Two Dimensions*, Cornell University Press, Ithaca, 1981, 556.
32. **Smith, J. N.,** *Experimental Methods in Catalytic Research*, Vol. 3, Anderson, R. B., Ed., Academic Press, New York, 1976, 149.
33. **Palmer, R. L. and Smith, J. N.,** *Catal. Rev.*, 12, 279, 1975.
34. **Cavanagh, R. R. and King, P. S.,** *Chemistry and Physics of Solid Surfaces*, Vanselow, R. and Howe, R., Eds., Springer-Verlag, Berlin, 1984, 141.
35. **Gerber, R. B.,** *Chem. Rev.*, 87, 29, 1987.
36. **Savkin, V. V., Kisljuk, M. U., and Skljarov, A. V.,** *Kinet. Katal.*, 28, 1409, 1987.
37. **Gibson, W. M.,** *Chemistry and Physics of Solid Surfaces*, Vol. 5, Vanselow, R. and Howe, R., Eds., Springer-Verlag, Berlin, 1984, 427.

38. **Vjatkin, A. F.,** *Vestn. Akad. Nauk SSSR,* NY, 90, 1986.
39. **Winters, H. F.,** *J. Appl. Phys.,* 43, 4809, 1972.
40. **Bodrov, A. E., Dalidchik, F. I., Kovalevski, S., and Shub, B. R.,** *Ion Beam Method in Studies of Collision Processes of Molecules With the Solid,* Report, Institute of Chemical Physics, Moscow, 1988.
41. **Bykov, V., Dalidchik, F., Kovalevsky, S., and Shub, B.,** *Proc. 6th Symp. Heterog. Catalysis,* Sofia, Bulgaria, 1987, 348.
42. **Bykov, V. and Kovalevsky, S.,** *Poverkhnost,* N9, 140, 1986.
43. **Darko, T., Baldwin, D. A., and Shamir, N.,** *J. Chem. Phys.,* 76, 6408, 1982.
44. **Rebentrost, F., Kompa, K. L., and Ben-Shaul, A.,** *Chem. Phys. Lett.,* 77, 394, 1981.
45. **Demtröder, W.,** *Laser Spectroscopy,* Springer-Verlag, Berlin, 1982.
46. **Kliger, D. S.,** *Ultrasensitive Laser Spectroscopy,* Academic Press, New York, 1985.
47. **Sarkisov, O. M. and Cheskis, S. G.,** *Usp. Khim.,* 54, 396, 1985.
48. **Letokhov, V. S.,** *Laser Photoionization Spectroscopy,* Nauka, Moscow, 1981 (in Russian).
49. **Vedeeva, G. V., Zasavitsky, I. I., Koloshnikov, V. G. et al.,** *Report at the 2nd Soviet-French Symposium on Optical Apparatus,* Physical Institute of the Academy of Science USSR, Moscow, 1981 (in Russian).
50. **Evenson, K. M., Broida, H. P., Wells, S., and Mahler, R. J.,** *Phys. Rev. Lett.,* 21, 1038, 1968.
51. **Gershenzon, Ju. M. and Livshits, B. L.,** *Kvantovaya Elektron.,* 6, 933, 1979.
52. **Braun, V. R., Krasnoperov, L. N., and Panfilov, V. N.,** *Kvantovaya Elektron.,* 7, 1895, 1980.
53. **Gershenzon, Ju. M., Il'in, S. D., Kishkovich, O. P., Kurbatov, V. A., Kucherjavy, S. I., Lebedev, Ya. S., and Rosenshtein, V. B.,** *Khim. Fiz.,* 2, 488, 1983.
54. **Broude, S. V., Gershenzon, Ju. M., and Il'in, S. D.,** *Dokl. Akad. Nauk SSSR,* 233, 366, 1975.
55. **Pakhomycheva, L. A., Sviridenkov, E. A., Suchkov, A. F., Titova, L. V., and Churilov, S. S.,** *Pis'ma Zh. Teor. Eksp. Fiz.,* 32, 60, 1970.
56. **Sarkisov, O. M., Sviridenkov, E. A., and Suchkov, A. F.,** *Khim. Fiz.,* 1, 1155, 1982.
57. **Zharov, V. P. and Letokhov, V. S.,** *Laser Optic-Acoustic Spectroscopy,* Nauka, Moscow, 1984 (in Russian).
58. **Fairbahks, W. M., Hansch, T. W., and Schawlow, A. L.,** *J. Opt. Soc. Am.,* 65, 199, 1975.
59. **Becker, K. H., Haaks, D., and Tartarczyk, T.,** *Z. Naturforsch,* 29a, 829, 1974.
60. **Johnson, P. M.,** *Acc. Chem. Res.,* 13, 20, 1980.
61. **Kubiak, G. D., Hurst, J. E., Rennagel, H. G., Zare, R. N., and McCleland, G. M.,** *J. Chem. Phys.,* 79, 5163, 1983.
62. **Kubiak, G. D., Sitz, G. O., and Zare, R. N.,** *J. Chem. Phys.,* 83, 2538, 1985.
63. **Akhmanov, S. A. and Koroteev, N. I.,** *The Methods of Nonlinear Optics in the Spectroscopy of Light Dispersion,* Nauka, Moscow, 1981 (in Russian).
64. **Bray, R. G., Hochstraser, R. M., and Sung, H. M.,** *Chem. Phys. Lett.,* 33, 1, 1975.
65. **Letokhov, V. S.,** *Laser Application in Spectroscopy and Photochemistry,* Nauka, Moscow, 1983 (in Russian).
66. **Thorman, R. P., Anderson, D., and Bernasek, S. L.,** *Phys. Rev. Lett.,* 44, 743, 1980.
67. **Bernasek, S. L.,** *Chem. Rev.,* 87, 91, 1987.
68. **Skljarov, A. V., Rozanov, V. V., and Kisljuk, U.,** *Kinet. Katal.,* 19, 416, 1976.
69. **Chen, Y. R.,** *Annu. Rev. Phys. Chem.,* 40, 327, 1989.
70. **Tom, H. W. K., Aumiller, G. D., and Brito-Cruz, C. H.,** *Phys. Rev. Lett.,* 60, 1438, 1989.
71. **Superfine, R., Guyot-Sionnest, P., Hunt, J. H., Kao, C. T., and Shen, Y. R.,** *Surf. Sci.,* 200, L455, 1988.
72. **Gadzuk, J. W. and Luntz, A. C.,** *Surf. Sci.,* 144, 429, 1984.

73. **Zhizhin, G. N., Moskaleva, M. A., Shomina, E. V., and Yakovlev, V. A.,** *Surface Polaritons,* Agranovich, V. M. and Mills, D. A., Eds., North-Holland, Amsterdam, 1982, 93.
74. **Zhizhin, G. N., Moskaleva, M. A., Shafranovskii, P. A., and Shub, B. R.,** *Acta Phys. Hung.,* 61, 19, 1987.
75. **Shafranovsky, P. A., Sinev, M. Ju., and Shub, B. R.,** *Kinet. Katal.,* 29, 1420, 1988.
76. **Sutsu, L. T., White, H. W., and Wragg, J. L.,** *Surf. Sci.,* 249, L343, 1991.
77. **Gershenzon, Ju. M., Rozenshtein, V. B., Spassky, A. I., and Kogan, A. M.,** *Dokl. Akad. Nauk SSSR,* 205, 624, 1972.
78. **Gershenzon, Ju. M., Kovalevsky, S. A., Rozenshtein, V. B., and Shub, B. R.,** *Dokl. Akad. Nauk SSSR,* 219, 1400, 1974.
79. **Vasiljev, N. M., Kovalevsky, S. A., Ryskin, M. E., and Shub, B. R.,** *Kinet. Katal.,* 23, 1199, 1982.
80. **Frank-Kamenetsky, D. A.,** *Diffusion and Heat Conduction in Chemical Kinetics,* Nauka, Moscow, 1967.
81. **Shafer, K. and Klingenberg, M.,** *Z. Elektrochem. Ber. Bunsenges. Phys. Chem.,* 58, 829, 1954.
82. **Millikan, R. C.,** *J. Chem. Phys.,* 38, 2855, 1963.
83. **Dayennette, L., Margotin-Maclou, M., Juejuen, H., Carion, A., and Henry, L.,** *J. Chem. Phys.,* 60, 697, 1974.
84. **Kovacz, M., Rao, D. R., and Javan, A.,** *J. Chem. Phys.,* 48, 3339, 1968.
85. **Rozenshtein, V. B.,** The Study of Heterogeneous and Chemical Relaxation of Some Molecules, Ph.D. thesis, Institute of Chemical Physics, Moscow, 1975 (in Russian).
86. **Villermaux, J.,** *J. Chem. Phys.,* 61, 1023, 1964.
87. **Shäfer, K., Rating, W., and Euhan, A.,** *Ann. Phys.,* 42, 176, 1942.
88. **Shäfer, K.,** *Z. Elektrochem.,* 58, 455, 1954.
89. **Shäfer, K. and Gerstacker, H.,** *Z. Elektrochem.,* 58, 455, 1954.
90. **Hunter, T. F.,** *J. Chem. Soc. A,* N11 1804, 1967.
91. **Fedotov, N. G., Sarkisov, O. M., and Vedeneev, V. I.,** *Materials of the 3rd All-Union Symp. on Combustion and Explosion,* Nauka, Moskow, 1972, 102.
92. **Gershenzon, Ju. M., Egorov, V. I., Rozenshtein, V. B., and Umansky, S. Ja.,** *Chem. Phys. Lett.,* 20, 77, 1973.
93. **Morgan, J. E. and Schiff, H. J.,** *Can. J. Chem.,* 41, 903, 1963.
94. **Abouaf, R. and Legay, F.,** *J. Chem. Phys.,* 63, 1393, 1966.
95. **Vasiljev, G. E., Makarov, E. F., Panin, I. G., and Tal'roze, V. L.,** *Dokl. Akad. Nauk SSSR,* 191, 1077, 1970.
96. **Heidher, R. F. and Kasper, J. V. V.,** *Chem. Phys. Lett.,* 15, 179, 1972.
97. **Gershenzon, Ju. M., Rozenshtein, V. B., and Egorov, V. I.,** *Khim. Vys. Energ.,* 7, 542, 1973.
98. **Black, A., Wise, H., Schechter, S., and Sharpless, R. L.,** *Chem. Phys. Lett.,* 6, 3526, 1974.
99. **Nalbandjan, A. B. and Voyevodsky, V. V.,** *The Mechanism of Oxidation and Combustion of Hydrogen,* Academy of Science URSS, Moscow, 1949 (in Russian).
100. **Gershenzon, Ju. M. and Rozenshtein, V. B.,** *Dokl. Akad. Nauk SSSR,* 221, 644, 1975.
101. **Kovalevski, S. A.,** The Reaction of Vibrational Energy of Nitrogen and Deuterium on Some Surface, Ph.D. thesis, Institute of Chemical Physics, Moscow, 1978 (in Russian).
102. **Armbruster, N. H.,** *J. Am. Chem. Soc.,* 64, 3545, 1942.
103. **Kozhushner, M. A., Kustarev, V. G., and Shub, B. R.,** *Dokl. Akad. Nauk SSSR,* 237, 871, 1979.
104. **Koshushner, M. A., Kustarev, V. G., and Shub, B. R.,** *Surf. Sci.,* 81, 261, 1979.
105. **Zhdanov, V. P.,** *Teor. Eksp. Khim.,* 16, 229, 1980.
106. **Reuter, G. E. H. and Sondheimer, E. H.,** *Proc. R. Soc.,* 195, 336, 1948.
107. **Vasiljev, N. M., Kustarev, V. G., and Shub, B. R.,** *Kinet. Katal.,* 21, 781, 1980.

108. **Price, D. L., Sinka, S. K., and Gupta, R. P.,** *Phys. Rev.*, 74, 2573, 1974.
109. **Kustarev, V. G.,** The Relaxation of Exited Atoms and Molecules Near the Boundary Between Two Media, Ph.D. thesis, Institute of Chemical Physics, Moscow, 1981 (in Russian).
110. **Vavilov, V. S., Kiselev, V. F., and Mukashev, B. N.,** *Defects in Silicon and at its Surface,* Nauka, Moscow, 1990 (in Russian).
111. **Lavrenko, V. A.,** *Atomic Hydrogen Recombination on Solid Surfaces,* Naukova Dumka, Kiev, 1973, 204.
112. **Moore, C. V.,** *Florescenka,* Duilbult, G. G., Ed., Academic Press, New York, 1967, 133.
113. **Heidner, R. F. and Kasper, J. V. V.,** *J. Chem. Phys.*, 51, 4163, 1969.
114. **Gershenzon, Ju. M., Rozenshtein, V. B., and Umansky, S. Ja.,** *Dokl. Akad. Nauk SSSR,* 223, 629, 1975.
115. **Leonas, V. L.,** *Prikl. Mat. Teor. Fiz.*, 124, 1963.
116. **Shaitan, K. V.,** *Zh. Fiz. Khim.*, 51, 336, 1977.
117. **Karlov, N. V. and Shaitan, K. V.,** *Zh. Eksp. Teor. Fiz.*, 71, 464, 1976.
118. **Dibatt, H., Aboaf-Marguin, L., and Legay, F.,** *Phys. Rev. Lett.*, 29, 145, 1972.
119. **Tinti, D. S. and Robinson, G. W.,** *J. Chem. Phys.*, 49, 3229, 1968.
120. **Mott, N. F. and Jackson, M.,** *Proc. Roy. Soc.*, A137, 703, 1932.
121. **Winter, T. G.,** *J. Chem. Phys.*, 38 (N 8), 2761, 1963.
122. **Sharma, R. D. and Braun, C. A.,** *J. Chem. Phys.*, 50, 924, 1969.
123. **Cross, R. J. and Gordon, R. G.,** *J. Chem. Phys.*, 45, 3571, 1966.
124. **Gray, C. G.,** *Can. J. Phys.*, 46, 135, 1968.
125. **Vasiljev, N. M., Kovalevsky, S. A., Kozhushner, M. A., and Shub, B. R.,** *Kinet. Katal.*, 22, 1199, 1981.
126. **Shljapintokh, V. Ja., Karpuchin, O. N., Postnicov, L. M., Zakharov, N. V., Vichutinsky, A. A., and Tsepalov, V. F.,** *Chemiluminescent Methods in Studies of Slow Chemical Processes,* Nauka, Moscow, 1975 (in Russian).
127. **Rufov, Ju. N.,** *Probl. Kinet. Katal.*, 16, 212, 1975 (in Russian).
128. **Reeves, R. R., Mannella, G., and Harteck, P.,** *J. Chem. Phys.*, 32, 946, 1960.
129. **Harteck, P., Reeves, R., and Mannella, G.,** *Can. J. Chem.*, 38, 1648, 1961.
130. **Weinraub, M. P. and Mannella, G.,** *J. Chem. Phys.*, 51, 4973, 1969.
131. **Rufov, Ju. N., Kadushin, A. A., and Roginsky, S. Z.,** *Dokl. Akad. Nauk SSSR,* 171, 905, 1966.
132. **Kiselev, V. F. and Krylov, O. V.,** *Electronic Phenomena in Adsorption and Catalysis,* Springer-Verlag, Berlin, 1987.
133. **Krylova, I. V.,** *Physical and Chemical Nature of the Exoelectronic Emission,* D.Sci. thesis, Moscow University, Moscow, 1973 (in Russian).
134. **Tabachnik, A. A., Umansky, S. Ja., and Shub, B. R.,** *Khim. Fiz.*, 2, 938, 1983.
135. **Tabachnik, A. A. and Shub, B. R.,** *Khim. Fiz.*, 2, 1242, 1983.
136. **Ryskin, M. E. and Shub, B. R.,** *React. Kinet. Catal. Lett.*, 17, 41, 1981.
137. **Ryskin, M. E. and Shub, B. R.,** *Khim. Fiz.*, 1, 212, 1982.
138. **Ryskin, M. E., Shub, B. R., Pavlichek, S., and Knor, Z.,** *Chem. Phys. Lett.*, 99, 140, 1983.
139. **Futcho, A. H. and Craut, F. A.,** *Phys. Rev.*, 104, 356, 1956.
140. **Wieme, W. and Wieme-Lenaerts, J.,** *Phys. Lett.*, 47, 37, 1974.
141. **Slovetsky, D. I. and Todesaite, R. D.,** *Khim. Vys. Energ.*, 7, 297, 1973.
142. **Kolts, J. H. and Setser, D. W.,** *J. Chem. Phys.*, 68, 4848, 1978.
143. **Kolts, J. H., Brashears, H. C., and Setser, D. W.,** *J. Chem. Phys.*, 67, 2931, 1977.
144. **Allison, W., Dunnig, F. B., and Smith, A. C. H.,** *J. Phys.*, B5, 1175, 1972.
145. **Palkina, L. A., Smirnov, B. M., and Shibisov, M. I.,** *Zh. Eksp. Teor. Fiz.*, 56, 340, 1969.
146. **Nikitin, E. E. and Umansky, S. Ja.,** *Advances in Science and Technology,* Institute of Scientific and Technological Information, Moscow, 1980 (in Russian).

147. **Gordeev, E. S., Tabachnik, A. A., Umansky, S. Ja., and Shub, B. R.,** *Khim. Fiz.*, 7, 975, 1989.
148. **Trukhin, A. N.,** *Fiz. Tverd. Tela,* 23, 993, 1982.
149. **Beckerle, J. D., Yang, Q. Y., Johnson, A. D., and Ceyer, S. T.,** *J. Chem. Phys.*, 86, 7236, 1987.
150. **Lin, C. L. and Kaufman, F.,** *J. Chem. Phys.*, 55, 3760, 1971.
151. **Wright, R. A. and Winkler, A. K.,** *Active Nitrogen,* Academic Press, New York, 1968, 523.
152. **Campbell, I. V. and Trush, B. A.,** *Proc. R. Soc. London,* A296, 201, 1967.
153. **Carleton, N. P. and Oldenberg, O.,** *J. Chem. Phys.*, 36, 3460, 1962.
154. **Chemansky, D. E.,** *J. Chem. Phys.*, 51, 689, 1969.
155. **Tabachnik, A. A.,** Heterogeneous Deactivation of $Ar(^3P_{0,2-})$ Atoms and $N_2(A^3;\Sigma_\nu^+$, v = 0.1) Molecules, Ph.D. thesis, Institute of Chemical Physics, Moscow, 1984 (in Russian).
156. **Vidaud, P. H. and von Engel, A.,** *Proc. R. Soc.*, 1515, 531, 1969.
157. **Vidaud, H., Wayne, R. P., Yaron, M., and von Engel, A.,** *J. Chem. Soc. Faraday Trans.*, 72, 1185, 1976.
158. **Clark, W. G. and Setser, D. W.,** *J. Phys. Chem.*, 84, 2225, 1980.
159. **Meyer, J. A. and Setser, D. W.,** *J. Phys. Chem.*, 74, 2238, 1970.
160. **Hays, G. N., Oskam, H. I., and Tracy, C. T.,** *J. Chem. Phys.*, 59, 2027, 1974.
161. **Badger, R. V., Wright, A. C., and Whitlock, R. F.,** *J. Chem. Phys.*, 42, 4345, 1965.
162. **Wallace, L. and Hunten, D. M.,** *J. Geophys. Res.*, 13, 4813, 1968.
163. **Wayne, R. P.,** *Adv. Photochem.*, 7, 311, 1969.
164. **Kearns, D. R.,** *Chem. Rev.*, 71, 396, 1971.
165. **Khan, A. U.,** *J. Chem. Phys.*, 65, 2219, 1976.
166. **Kautsky, H., de Brujin, H., Neuwirth, R., and Baumaister, W.,** *Chem. Ber.*, 66, 1588, 1933.
167. **Wilsson, R. and Kearns, D. R.,** *Photochem. Photobiol.*, N 19, 181, 1974.
168. **Scheffer, I. R. and Ouchi, M. D.,** *Tetrahedron Lett.*, N 3, 233, 1970.
169. **Vladimirova, V. I., Rufov, Ju. N., and Krylov, O. V.,** *Kinet. Katal.*, 18, 809, 1977.
170. **Krylov, O. V.,** in *Kinetics Problems of Simple Chemical Reactions,* Nauka, Moscow, 1973, 115 (in Russian).
171. **Clark, I. D. and Wayne, R. P.,** *Chem. Phys. Lett.*, 3, 93, 1969.
172. **Arnold, S. I., Finlayson, N., and Ogryzlo, E. A.,** *J. Chem. Phys.*, 44, 2529, 1969.
173. **Giachardy, D. J., Harris, G. W., and Wayne, R. P.,** *J. Chem. Soc. Faraday Trans.*, 72, 619, 1976.
174. **Becker, K. H., Groth, W., and Schurath, V.,** *Chem. Phys. Lett.*, 8, 259, 1971.
175. **Thomas, R. G. O. and Thrush, B. A.,** *Proc. R. Soc. London,* A356, 287, 1977.
176. **O'Brien, R. J. and Myers, G. H.,** *J. Chem. Phys.*, 53, 3832, 1970.
177. **Grigor'ev, E. I.,** Photosensibilized Creation and Quenching of Singlet Oxygen in Gas-Solid Heterogeneous Systems, Ph.D. thesis, Karpov Physico-Chemical Institute, Moscow, 1982 (in Russian).
178. **Ryskin, M. E.,** *Heterogeneous Deactivation of Singlet Oxygen,* Ph.D. thesis, Institute of Chemical Physics, Moscow, 1983 (in Russian).
179. **Elias, L., Ogryzlo, E. A., and Schiff, H. I.,** *Can. J. Chem.*, 37, 1680, 1959.
180. **Vidaud, P. H., Wayne, R. P., and Yaron, M.,** *Chem. Phys. Lett.*, 38, 306, 1976.
181. **Kaufman, F. and Kelso, J. R.,** *J. Chem. Phys.*, 28, 510, 1958.
182. **Sazonova, N. V. and Levchenko, L. P.,** *Kinet. Katal.*, 6, 765, 1965.
183. **Chester, M. A. and Somorjai, G. A.,** *Surf. Sci.*, 52, 21, 1975.
184. **Schrader, M. E.,** *Surf. Sci.*, 78, 1227, 1978.
185. **Legare, P., Hilaire, L., Sotto, M., and Maire, G.,** *Surf. Sci.*, 91, 175, 1980.
186. **Eley, D. D. and Moore, P. B.,** *Surf. Sci.*, 76, L599, 1978.
187. **Kureneva, T. Ja., Ryskin, M. E., and Shub, B. R.,** in *Materials of the 7th Soviet-Japan Seminar on Catalysis,* Irkutsk, 1983, 78. (in Russian).

188. **Tretjakov, I. I., Skljarov, A. V., Shub, B. R., and Roginsky, S. Z.,** *Dokl. Akad. Nauk SSSR*, 189, 1302, 1969.
189. **Tretjakov, I. I., Skljarov, A. V., and Shub, B. R.,** *Kinet. Katal.*, 11, 166, 1970.
190. **Conrad, H., Ertl, G., and Kuppers, G.,** *Surf. Sci.*, 76, 323, 1978.
191. **Conrad, H., Doyen, G., Ertl, G., Kuppers, J., Sisselmann, W., and Haberland, H.,** *Chem. Phys. Lett.*, 88, 281, 1982.
192. **Gordon, E. B. and Ponomarev, A. N.,** *Kinet. Katal.*, 8, 663, 1967.
193. **Schäfer, K. and Grau, G. G.,** *Z. Elektrochem.*, 53, 203, 1949.
194. **Shub, B. R.,** Heterogeneous Relaxation of Inner Molecular Energy and Nonequilibrium Processes Solid Surfaces, D.Sci. thesis, Institute of Chemical Physics, Moscow, 1983 (in Russian).
195. **Feuer, P.,** *J. Chem. Phys.*, 39, 1311, 1963.
196. **Goldansky, V. I., Namiot, V. A., and Khokhlov, R. V.,** *Zh. Eksp. Teor. Fiz.*, 70, 2349, 1976.
197. **Hunter, T. F.,** *J. Chem. Phys.*, 51, 2641, 1968.
198. **Zhdanov, V. P. and Zamaraev, K. I.,** *Catal. Rev. Sci. Eng.*, 23, 373, 1982.
199. **Krylov, O. V., Kisljuk, M. U., Gezalov, A. V., Shub, B. R., Maksimova, N. D., and Rufov, Ju. N.,** *Kinet. Katal.*, 13, 898, 1972.
200. **Lin, S. N., Lin, H. D., and Knittel, D.,** *J. Chem. Phys.*, 64, 441, 1976.
201. **Vasiljev, N. M., Kovalevsky, S. A., Kozhushner, M. A., and Shub, B. R.,** *Kinet. Katal.*, 19, 782, 1978.
202. **Zhdanov, V. P.,** *Teor. Eksp. Khim.*, 16, 229, 1980.
203. **Vasiljev, N. M., Kovalevsky, S. A., Kozhushner, M. A., and Shub, B. R.,** *Kinet. Katal.*, 22, 5, 1981.
204. **Stepanov, B. E. and Tapilin, V. M.,** *React. Kinet. Catal. Lett.*, 4, 1, 1976.
205. **Seikhaus, W., Szwarz, J., and Olander, J.,** *Surf. Sci.*, 33, 445, 1972.
206. **Spruit, M. E. M., Kuipers, E. W., Geuzebrock, F. H., and Kleyn, A. W.,** *Surf. Sci.*, 215, 421, 1989.
207. **Weinberg, W. H.,** *J. Colloid Interface Sci.*, 47, 372, 1974.
208. **Smith, D. L. and Merrill, R. P.,** *J. Chem. Phys.*, 53, 3588, 1970.
209. **Steinruck, H. P. and Madix, R. P.,** *Surf. Sci.*, 185, 36, 1987.
210. **Hurst, J. E.,** *Chem. Phys. Lett.*, 43, 1175, 1979.
211. **Lilienkamp, G. and Toennies, J. P.,** *J. Chem. Phys.*, 78, 5210, 1983.
212. **Palmer, R. L. and O'Keefe, D.,** *Chem. Phys. Lett.*, 16, 529, 1970.
213. **Lapujalade, J. and Legay, Y.,** *J. Chem. Phys.*, 63, 1389, 1975.
214. **Gibson, K. D., Sibener, S. J., Hall, B. M., Mills, D. L., and Black, J. E.,** *J. Chem. Phys.*, 83, 4256, 1985.
215. **Bartolini, V., Franchini, A., Santoro, G., Toennies, J. P., Wohl, G., and Zhang, G.,** *Phys. Rev.*, B40, 3524, 1989.
216. **Dabiri, A. E., Lee, T. J., and Stickney, R. E.,** *Surf. Sci.*, 26, 522, 1971.
217. **Bradley, T. L., Dabiri, A. E., and Stickney, R. E.,** *Surf. Sci.*, 29, 590, 1972.
218. **Goodman, F. O.,** *Surf. Sci.*, 30, 525, 1972.
219. **Comsa, G., David, R., and Schumacher, B. J.,** *Surf. Sci.*, 85, 45, 1979.
220. **Van Willigen, M.,** *Chem. Phys. Lett.*, A28, 80, 1968.
221. **Balooch, M., Cardillo, M. J., Miller, D. R., and Stickney, R. E.,** *Surf. Sci.*, 46, 358, 1974.
222. **Comsa, G. and David, R.,** *J. Chem. Phys.*, 78, 1582, 1983.
223. **Harris, J., Holloway, S., Rahman, I. S., and Yang, T. S.,** *J. Chem. Phys.*, 89, 4427, 1988.
224. **Ohno, Y., Toya, T., Ishi, S., and Nagni, K.,** *Appl. Surf. Sci.*, 33/34, 238, 1988.
225. **Rettner, C. T., Michelsen, H. A., Auerbach, D. J., and Mallins, C. B.,** *J. Chem. Phys.*, 94, 7499, 1991.
226. **West, L. F. and Somorjai, G. A.,** *J. Chem. Phys.*, 57, 5143, 1972.

227. **Campbell, C. T., Ertl, G., Kuipers, H., and Segner, J.,** *Surf. Sci.*, 107, 107, 1981.
228. **Tenner, M. G., Kuipers, E. W., Langhout, W. Y., Kleyn, A. W., Nicolson, G., and Stolte, S.,** *Surf. Sci.*, 236, 151, 1990; *J. Vac. Sci. Technol.*, A8, 2692, 1990; *J. Chem. Phys.*, 90, 7406, 1990.
229. **Ionov, S. I., Lavilla, M. E., Mackey, R. S., and Bernstein, R. B.,** *J. Chem. Phys.*, 90, 7406, 1990.
230. **Palmer, R. L., Smith, J. N., Saltsbury, H., and O'Keefe, D. R.,** *J. Chem. Phys.*, 53, 1666, 1970.
231. **Robota, H. J., Vielhaber, N., Lin, M. C., Segner, J., and Ertl, G.,** *Surf. Sci.*, 155, 101, 1985.
232. **Brusdoglins, G. and Toennies, J. P.,** *Surf. Sci.*, 126, 647, 1983.
233. **Cowin, J. P., Yu, C. F., Siebener, S. J., and Wharton, L.,** *J. Chem. Phys.*, 79, 3537, 1983.
234. **Yu, C. F., Whaller, K. B., Hogg, C. S., and Siebener, S. J.,** *Phys. Rev. Lett.*, 51, 2210, 1983.
235. **Berndt, R., Toennies, J. P., and Wolf, C. J.,** *J. Chem. Phys.*, 92, 1468, 1990.
236. **King, D. S., Mantell, D. A., and Cavanagh, R. R.,** *J. Chem. Phys.*, 82, 1046, 1985; 84, 5131, 1986.
237. **King, D. S. and Cavanagh, R. R.,** *J. Chem. Phys.*, 76, 5634, 1982.
238. **Segner, J., Robota, H., Vielhaber, W., Ertl, G., Frenkel, F., Hager, J., Kliger, W., and Walther, H.,** *Surf. Sci.*, 131, 273, 1983; *Phys. Rev. Lett.*, 46, 152, 1982; *Chem. Phys. Lett.*, 90, 225, 1982.
239. **Mantell, P. A., Man, Y. F., Ryalli, S. B., Haller, J. L., and Fenn, J. B.,** *J. Chem. Phys.*, 78, 6338, 1983.
240. **Assher, W., Guthrie, W. L., Lin, T. H., and Somorjai, G. A.,** *J. Chem. Phys.*, 78, 6992, 1983.
241. **Prybyla, J. A., Heinz, T. F., Misewich, J. A., and Loy, M. M. T.,** *Surf. Sci. Lett.*, 250, L173, 1990.
242. **Assher, M. and Somorjai, G. A.,** *Dynamics of Surface,* Proc. 17th Jerusalem Symp., North-Holland, Dordrecht, 1984, 117.
243. **Mödl, A., Robota, H., Segner, J., Vielhaber, W., Lin, M. C., and Ertl, G.,** *J. Chem. Phys.*, 76, 5634, 1982.
244. **Laudersdale, J. C., McNutt, J. F., and McCardy, C. W.,** *Chem. Phys. Lett.*, 107, 43, 1984.
245. **Tanaka, S. and Sugano, S.,** *Surf. Sci.*, 143, L371, 1984.
246. **Muhlhausen, C. W., Williams, L. R., and Tully, J. C.,** *J. Chem. Phys.*, 83, 2594, 1985.
247. **Hamza, A. V., Ferm, P. M., Budde, F., and Ertl, G.,** *Surf. Sci.*, 199, 13, 1988.
248. **Cavanagh, R. R. and King, D. S.,** *J. Vac. Sci. Technol.*, A1., 1267, 1983; A2, 1036, 1984.
249. **Hayden, B. E., Krietzschmar, K., and Bradschaw, A. M.,** *Surf. Sci.*, 125, 366, 1983.
250. **Mödl, A., Robota, H., Segner, J., Vielhaber, W., Lin, M. C., and Ertl, G.,** *J. Chem. Phys.*, 83, 4800, 1986.
251. **Bialkowski, S. E.,** *J. Chem. Phys.*, 78, 600, 1983.
252. **Tang, S. L., Lee, M. B., Beckerle, J. D., Hines, M. A., and Ceyer, S. T.,** *J. Chem. Phys.*, 82, 2826, 1985; 84, 6488, 1986; *J. Vac. Sci. Technol.*, A3, 1665, 1985.
253. **Tully, J. C.,** *Kinetics of Interface Reactions, Proc. Works. Campobello,* Grunze, M. and Kreuzer, H. J., Eds., Springer-Verlag, Berlin, 1987, 37.
254. **Gadzuk, J. W., Landmaan, U., Kuster, E. J., Cleveland, C. L., and Barett, R. N.,** *Phys. Rev. Lett.*, 49, 426, 1982.
255. **Jacobs, D. C., Kolasinski, K. W., Shane, S. F., and Zare, R. N.,** *J. Chem. Phys.*, 87, 5038, 1987; 91, 3182, 1989.
256. **Barker, J. A. and Auerbach, D. J.,** *Surf. Sci. Rep.*, 4, 1, 1984.

257. **Kubiak, G. D., Sitz, G. O., and Zare, R. N.,** *J. Chem. Phys.*, 81, 6397, 1984.
258. **Zacharias, H. and David, R.,** *Chem. Phys. Lett.*, 115, 205, 1985.
259. **Levkoff, J., Robertson, A., and Bernasek, S. L.,** *Dynamics of Surfaces (Proc. 17th Jerusalem Symp.)*, North-Holland, Dordrecht, 1984, 117.
260. **Mantell, D. A., Man, Y. F., Ryali, S. B., Haller, G. L., and Fenn, J. B.,** *J. Chem. Phys.*, 78, 4250, 1983.
261. **Assher, M., Guthrie, W. L., Lin, T. N., and Somorjai, G. A.,** *J. Phys. Chem.*, 78, 3233, 1984.
262. **Cross, J. A. and Lurie, J. B.,** *Chem. Lett.*, 100, 174, 1983.
263. **Assher, M., Somorjai, G. A., and Zeir, Y.,** *J. Chem. Phys.*, 81, 1507, 1984.
264. **Rettner, C. T., Kimman, J., Fabre, E., Auerbach, D. J., and Morawitz, A.,** *Surf. Sci.*, 192, 107, 1987.
265. **Gadzuk, J. W. and Holloway, S.,** *Phys. Rev. Lett.*, B33, 4298, 1986.
266. **Houston, P. L. and Merrill, R. P.,** *Chem. Rev.*, 88, 657, 1988.
267. **Vach, H., Hager, J., and Walther, H. J.,** *Chem. Phys.*, 90, 6701, 1989.
268. **Russel, J. N., Chorkendorff, I., Lanzilotto, A. M., Alvey, M. D., and Yates, J. T.,** *J. Chem. Phys.*, 85, 6186, 1986.
269. **Russel, J. N., Gates, S. M., and Yates, J. T.,** *J. Chem. Phys.*, 85, 6742, 1986.
270. **Yates, J. T. and Russel, N. J.,** *Kinetics of Interface Reactions (Proc. Workshop)*, Grunze, M. and Kreuzer, H. J., Eds., Springer-Verlag, Berlin, 1987, 71.
271. **Harris, J., Rahman, T., and Yang, K.,** *Surf. Sci.*, 198, 2312, 1988.
272. **Paz, Y. and Naaman, R.,** *Chem. Phys. Lett.*, 172, 120, 1990.
273. **Kisljuk, M. U. and Bakuleva, T. N.,** *Izv. AN SSSR Ser. Khim.*, N12, 2699, 1990.
274. **Rendulic, K. D., Anger, G., and Winkler, A.,** *Surf. Sci.*, 208, 104, 1989.
275. **Hamza, A. V. and Madix, R. J.,** *J. Phys. Chem.*, 89, 5381, 1985.
276. **Madix, R. J. and Hamza, A. V.,** *J. Vac. Sci. Technol.*, A4, 1506, 1986.
277. **Harten, U. and Toennies, J. P.,** *J. Chem. Phys.*, 85, 2249, 1986.
278. **Berger, H. F. and Rendulic, K. D.,** *Surf. Sci.*, 253, 385, 1991.
279. **Anger, G., Winkler, A., and Rendulic, K. D.,** *Surf. Sci.*, 221, 1, 1989.
280. **Harris, J.,** *Surf. Sci.*, 221, 335, 1989.
281. **Hayden, B. E. and Lamont, C. L. A.,** *Surf. Sci.*, 243, 31, 1991.
282. **Campbell, J. M., Domagala, C. T., and Campbell, C. T.,** *J. Vac. Sci. Technol.*, A9, 1581, 1991.
283. **Kara, A. and De-Pristo, A. F.,** *J. Chem. Phys.*, 88, 5240, 1988.
284. **Lee, J., Madix, R. J., Schlaegel, J. E., and Auerbach, D. J.,** *Surf. Sci.*, 143, 626, 1984.
285. **Rettner, C. T. and Auerbach, D. J.,** in *Kinetics of Interface Reactions (Proc. Workshop)*, Grunze, M. and Kreuzer, H. J., Eds., Springer-Verlag, Berlin, 1987, 300.
286. **Rettner, C. T., DeLouise, L. A., and Auerbach, D. J.,** *J. Chem. Phys.*, 85, 1131, 1986; *J. Vac. Sci. Technol.*, A4, 1491, 1986.
287. **Rettner, C. T. and Mullins, C. R.,** *J. Chem. Phys.*, 94, 1626, 1991.
288. **Brown, J. R. and Luntz, A. R.,** *Chem. Phys. Lett.*, 186, 125, 1991.
289. **Kohrt, C. and Gomer, R.,** *Surf. Sci.*, 40, 71, 1973.
290. **Steinruck, H. P., D'Evelin, M. P., and Madix, R. J.,** *Surf. Sci.*, 172, L561, 1986.
291. **Harris, J. and Luntz, A. R.,** *J. Chem. Phys.*, 91, 6421, 1990.
292. **Steinruck, H. P. and Madix, R. J.,** *Surf. Sci.*, 185, 36, 1987.
293. **Rendulic, K. P., Winkler, A., and Kerner, H.,** *J. Vac. Sci. Technol.*, A5, 462, 1987.
294. **Akazawa, H. and Murata, Y.,** *J. Chem. Phys.*, 88, 3317, 1988.
295. **Kang, H., Shuler, T. R., and Rabalais, J. W.,** *Chem. Phys. Lett.*, 128, 348, 1986.
296. **Kang, H., Srinadan, K., and Rabalais, J. W.,** *J. Chem. Phys.*, 86, 3753, 1987.
297. **Kang, H., Kasi, S. R., and Rabalais, J. W.,** *J. Chem. Phys.*, 88, 5882, 1988.
298. **Rettner, C. T., Kimman, J., and Auerbach, D. J.,** *J. Chem. Phys.*, 94, 734, 1991.

299. **Becker, J. H., Böwering, N., Volkmar, M., Pawlitzky, B., and Heinemann, U.,** *Surf. Sci.*, 250, L169, 1990.
300. **Hodgson, A., Mossyl, J., and Zhao, H.,** *Chem. Phys. Lett.*, 182, 152, 1991.
301. **Michelsen, H. A. and Auerbach, D. J.,** *J. Chem. Phys.*, 94, 7502, 1991.
302. **Norskov, J. K., Stoltze, P., and Nielsen, U.,** *Catal. Lett.*, 9, 178, 1991.
303. **Caesciatore, N., Capitelli, M., and Billing, J. D.,** *Surf. Sci.*, 217, L391, 1989.
304. **Crisa, M., Doyen, G., and van Trentini, E.,** *Surf. Sci.*, 163, 120, 1986.
305. **Lee, M. V., Beckerle, J. D., Tang, S. L., and Ceyer, S. T.,** *J. Chem. Phys.*, 81, 723, 1987.
306. **Misevich, J. and Loy, M. M. T.,** *J. Chem. Phys.*, 84, 1939, 1986.
307. **Vach, H., Häger, J., and Walter, H.,** *J. Chem. Phys.*, 90, 6701, 1989.
308. **Gdowski, E., Hamza, A. V., D'Evelyn, M. P., and Madix, R. P.,** *J. Vac. Sci. Technol.*, A3, 1561, 1985; *Surf. Sci.*, 167, 451, 1985.
309. **Misevich, J., Plum, C. N., Blyholder, C., Houston, R. P., and Merril, R. P.,** *J. Chem. Phys.*, 78, 4245, 1983.
310. **Clary, D. C. and De-Pristo, A. F.,** *J. Chem. Phys.*, 81, 5167, 1984.
311. **Bras, J. E.,** *J. Chem. Phys.*, 87, 170, 1987.
312. **Rettner, C. T. and Stern, H. J.,** *J. Chem. Phys.*, 87, 770, 1987.
313. **Stewart, C. N. and Ehrlich, G.,** *J. Chem. Phys.*, 62, 4672, 1975.
314. **Slater, N. B.,** *Theory of Unimolecular Reactions*, Cornell University Press, Ithaca, 1959, 230.
315. **Yates, J. T., Zinck, J. J., Shears, S., and Weinberg, W. H.,** *J. Chem. Phys.*, 70, 2266, 1979.
316. **Rettner, C. T., Pfnur, H. E., and Auerbach, D. J.,** *J. Chem. Phys.*, 84, 4163, 1986.
317. **Lo, T. S. and Ehrlich, G.,** *Surf. Sci.*, 179, L19, 1987; 205, L813, 1988.
318. **Ceyer, S. T., Beckerle, J. D., Lee, M. B., Tang, S. L., Yang, Q. Y., and Hines, M. A.,** *J. Vac. Sci. Technol*, A5, 501, 1987.
319. **Xi, G., Wang, J., Li, S., and Shao, S.,** *J. Vac. Sci. Technol.*, A9, 1581, 1991.
320. **Arumayanagam, C. R., McMaster, M. C., Schoofs, G. P., and Madix, R. J.,** *Surf. Sci.*, 222, 213, 1989.
321. **Schoofs, G. R., Arumayanagam, C. R., McMaster, M. C., and Madix, R. J.,** *Surf. Sci.*, 215, 1, 1989.
322. **Luntz, A. C. and Bethune, D. S.,** *J. Chem. Phys.*, 90, 1259, 1989.
323. **Beckerle J. D., Yang, Q. Y., Johnson, A. D., and Ceyer, S. T.,** *J. Chem. Phys.*, 86, 7236, 1987.
324. **Coltrin, M. E. and Kay, B.,** *Surf. Sci.*, 204, 2375, 1988; 205, 2805, 1988.
325. **Luntz, A. C. and Harris, J.,** *Surf. Sci.*, 258, 397, 1991.
326. **Mollins, C. B. and Weinberg, W. H.,** *J. Chem. Phys.*, 92, 4508, 1990.
327. **Steinruck, M. P., Hamza, A., and Madix, R. J.,** *Surf. Sci.*, 173, L571, 1986.
328. **Arunamayanagam, C. R., McMaster, M. C., and Madix, R. J.,** *Surf. Sci.*, 237, L424, 1990.
329. **Arunamayanagam, C. R., Schoofs, G. R., McMaster, M. C., and Madix, R. J.,** *J. Phys. Chem.*, 95, 1041, 1991.
330. **Hamza, A. V. and Madix, R. J.,** *Surf. Sci.*, 179, 25, 1987.
331. **Weinberg, W. H.,** *J. Chem. Phys.*, 89, 2497, 1985.
332. **Hamza, A. V., Steinruck, H. P., and Madix, R. J.,** *J. Chem. Phys.*, 86, 6506, 1986.
333. **Halpern, B. and Almitaz, I.,** *Chem. Eng. Sci.*, 41, 899, 1986.
334. **Arakawa, R. and Rabinovitch, B. S.,** *J. Phys. Chem.*, 86, 4774, 1982.
335. **Wei Yuan, Rabinovitch, B. S.,** *J. Chem. Phys.*, 80, 1687, 1984.
336. **Oref, J. and Rabinovitch, B. S.,** *J. Chem. Soc. Faraday Trans.*, 80, 769, 1984.
337. **Gerber, R. B. and Elber, R.,** *Chem. Phys. Lett.*, 102, 466, 1983; 107, 141, 1984.
338. **Botari, F. J. and Greene, E. F.,** *J. Chem. Phys.*, 88, 4238, 1984.
339. **McCarroll, J. J. and Thomson, S. J.,** *J. Catal.*, 19, 144, 1970.

340. **Prada-Silva, G., Kester, K., Löffler, D., Haller, G. L., and Fenn, J. B.,** *Rev. Sci. Instrum.*, 48, 897, 1977.
341. **Löffler, D., Haller, G. L., and Fenn, J. B.,** *J. Catal.*, 57, 960, 1979.
342. **Prada-Silva, G., Löffler, D., Halpern, B. L., Haller, G. L., and Fenn, J. B.,** *Surf. Sci.*, 83, 459, 1979.
343. **Tsou, L., Haller, G. L., and Fenn, J. B.,** *J. Chem. Phys.*, 91, 2654, 1987.
344. **Prada-Silva, G., Haller, G. L., and Fenn, J. B.,** *Proc. 4th N. Am. Meet. Catalysis Soc.*, Preprint 1975, 39.
345. **Tschukov-Roux, E.,** *J. Phys. Chem.*, 73, 3891, 1969.
346. **Smith, J. N.,** *Experimental Methods in Catalytic Research,* Vol. 3, Anderson, R. B., Ed., Academic Press, New York, 1979, 149.
347. **Salmeron, M., Gale, R., and Somorjai, G. A.,** *J. Chem. Phys.*, 70, 2807, 1979.
348. **West, L. A. and Somorjai, G. A.,** *J. Vac. Sci. Technol.*, 9, 71, 1968; 9, 668, 1971; **Bernasek, S. L. and Somorjai, G. A.,** *J. Chem. Phys.*, 65, 3149, 1976.
349. **Goltharp, R. N., Scott, J. T., and Muschlitz, E. E.,** *J. Chem. Phys.*, 51, 5180, 1969.
350. **Smith, J. N. and Palmer, R. L.,** *J. Chem. Phys.*, 49, 5027, 1968.
351. **Palmer, R. L. and Smith, J. N.,** *J. Chem. Phys.*, 60, 1453, 1974.
352. **Chuang, T. J.,** *Surf. Sci. Rep.*, 3, 3, 1983.
353. **George, T. F., Lin, J., Beri, A. C., and Murphy, W. C.,** *Prog. Surf. Sci.*, 16, 139, 1984.
354. **Chuang, T. J.,** *Surf. Sci.*, 178, 763, 1986.
355. **Basov, N. G., Belenov, E. M., Isakov, V. A., Leonov, Ju. S., Markin, E. R., Oraevsky, A. N., Romanenko, V. I., and Ferapontov, N. E.,** *Pisma Zh. Eksp. Teor. Fiz.*, 22, 221, 1975.
356. **Lin, C. L., Atvarz, T. D. Z., and Pessine, F. B. T.,** *J. Appl. Phys.*, 46, 1720, 1977; *J. Chem. Phys.*, 68, 4233, 1978.
357. **Petrov, Ju. N.,** *Izv. Akad. Nauk SSSR Ser. Fiz.*, 50, 671, 1986.
358. **Chuang, T. J.,** *J. Chem. Phys.*, 72, 6303, 1980.
359. **Goodman, F. O.,** *Surf. Sci.*, 109, 3, 1981.
360. **Beri, A. C. and George, T. F.,** *J. Chem. Phys.*, 78, 4288, 1983.
361. **Zare, R. M. and Levine, R. D.,** *J. Chem. Phys. Lett.*, 140, 593, 1987.
362. **Dzhidzhoev, M. S., Osipov, A. I., Panchenko, V. Ja., Planovchenko, V. T., Khokhlov, R. V., and Shaitan, K. V.,** *Zh. Fiz. Khim.*, 47, 684, 1978.
363. **Huang, H.-Y., George, T. F., Yuan, J. M., and Narducci, L. M.,** *J. Phys. Chem.*, 38, 5772, 1984.
364. **Hussla, J. and Chuang, T. J.,** *Ber. Bunsenges Phys. Chim.*, 105, 565, 1985; *Proc. 17th Jerusalem Symp.*, 1984, 310.
365. **Kay, B. D. and Raymond, T. D.,** *Chem. Phys. Lett.*, 130, 79, 1986.
366. **Chuang, T. J.,** *J. Chem. Phys.*, 76, 3828, 1982; *Phys. Rev. Lett.*, 49, 382, 1982.
367. **Macchni, J. and Hass, R.,** *Appl. Phys.*, B28, 224, 1982.
368. **Schäfer, B. and Hass, R.,** *Chem. Phys. Lett.*, 105, 563, 1984.
369. **Creighton, J. R. and White, J. M.,** *Surf. Sci.*, 136, 449, 1984.
370. **Perouns, A., Bacchia, M., Darville, J., and Girles, J. M.,** *J. Electron Spectrosc.*, 54/55, 13, 1990.
371. **Gortel, Z. W., Kreuzer, H. J., Piercy, P., and Teshima, R.,** *Phys. Rev. Lett.*, B28, 2119, 1983.
372. **Hussla, I., Seki, H., Chuang, T. J., Görtel, Z. W., Kreuzer, H. J., and Piercy, P.,** *Phys. Rev. Lett.*, B32, 3482, 1985.
373. **Kreuzer, H. J.,** *Faraday Disc.*, 80, 265, 1985.
374. **Görtel, Z. W., Piercy, P., Teshima, R., and Kreuzer, H. J.,** *Surf. Sci.*, 179, 179, 1987.
375. **Tro, N. J., Arthur, N. J., and George, S. J.,** *J. Chem. Phys.*, 90, 3389, 1990.
376. **Heidberg, J., Stein, H., and Weiss, H.,** *Surf. Sci.*, 184, L431, 1987.

377. **Heidberg, J., Brase, B., Stadmer, K. M., and Suhier, M.,** *Appl. Surf. Sci,* 46, 44, 1990.
378. **Feim, P. M., Kurtz, S. R., Pearl, K. A., and McCleland, G. M.,** *Phys. Rev. Lett.,* B24, 2602, 1987.
379. **Murphy, W. C. and George, T. F.,** *Surf. Sci.,* 144, 183, 1982; *J. Phys. Chem.,* 36, 4481, 1982.
380. **Lin, Y. S., Chiang, S. W., and Bacon, F.,** *Appl. Phys. Lett.,* 38, 1981, 1981; **Flory, C.,** *Phys. Rev. Lett.,* 52, 2076, 1984.
381. **Murphy, W. C. and George, T. F.,** *J. Chem. Phys.,* 80, 5303, 1984.
382. **Lichtman, D. and Shapira, Y.,** *Crit. Rev. Solid State Mater. Sci.,* 8, 93, 1978.
383. **Koel, B. E.,** *Adv. Chem. Ser.,* 184, 27, 1980.
384. **Moiseenko, I. P., Glebovskii, A. A., and Lisachenko, A. A.,** *React. Kinet. Catal. Lett.,* 28, 117, 1985.
385. **Artamonov, P. O., Moiseenko, I. F., and Lisachenko, A. A.,** *Poverchnost,* N 3, 57, 1990.
386. **Van Hieu, N. and Litchtman, D.,** *Surf. Sci.,* 103, 536, 1981; *J. Vac. Sci. Technol.,* 18, 49, 1981.
387. **Baurdon, E. B., Cowin, J. P., Harrison, J., Polanyi, J. C., Segner, J., Stannus, C. D., and Young, P. A.,** *J. Chem. Phys.,* 88, 6100, 1984.
388. **Sesselmann, W., Marinero, E. E., and Chuang, T. J.,** *Surf. Sci.,* 175, 787, 1986.
389. **Lu, P. H., Li, Y. L., and Qin, Q. Z.,** *Surf. Sci.,* 238, 245, 1990.
390. **Gluck, N. S., Ying, Z., Bartosch, C. E., and Ho, W.,** *Surf. Sci.,* 86, 4957, 1987.
391. **Swanson, J. R., Flitsch, F. A., and Friend, C. M.,** *Surf. Sci.,* 226, 147, 1990.
392. **Germer, T. A. and Ho, W.,** *J. Vac. Sci. Technol.,* A7, 1878, 1989.
393. **Weide, D., Anderson, P., and Freund, M. J.,** *Chem. Phys. Lett.,* 176, 106, 1987.
394. **Bunin, S. A., Richter, L. J., King, D. S., and Cavanagh, R. R.,** *J. Vac. Sci. Technol.,* A7, 1880, 1989.
395. **Richter, L. J., Bunin, S. A., and Cavanagh, R. R.,** *J. Electron Spectrosc.,* 54/55, 181, 1990.
396. **Zhu, X. Y., Hatch, S. R., Campion, A., and White, J. M.,** *Surf. Sci.,* 226, 147, 1990.
397. **Wolf, M., Nettesheim, S., White, J. M., Hasselbrink, E., and Ertl, G.,** *J. Chem. Phys.,* 92, 1509, 1990.
398. **Natzle, W. C., Padowitz, D., and Sibener, S. J.,** *J. Chem. Phys.,* 88, 7975, 1988.
399. **Burgess, D., Cavanagh, R. R., and King, D. S.,** *J. Chem. Phys.,* 88, 6556, 1988.
400. **Prybyla, J. A., Heinz, T. F., Misewich, J. A., Loy, M. H. T., and Glownin, J. H.,** *Phys. Rev. Lett.,* 64, 1537, 1990.
401. **Chekali, S. V., Golovlev, V. V., Kozlov, A. A., Matveyets, Yu. A., Yartsev, A. P., and Letokhov, V. S.,** *J. Phys. Chem.,* 92, 6855, 1988.
402. **Sun Yian-Xia and Quin Qi-Zong,** *Ann. Chim. Sinica,* 43, 19, 1985.
403. **Belikov, A. P., Borman, V. D., and Nikolaev, B. I.,** *Khim. Fiz.,* 4, 1030, 1985.
404. **Umstead, M. E. and Lin, M. C.,** *J. Phys. Chem.,* 82, 2047, 1978.
405. **Umstead, M. E., Talley, L. D., Tevault, D. E., and Lin, M. C.,** *Opt. Eng.,* 19, 94, 1980.
406. **Bass, H. F. and Fanchi, J. R.,** *J. Chem. Phys.,* 64, 4417, 1976.
407. **Naik, P. D., Ramarao, K. V. S., and Mittal, J. P.,** *Chem. Phys. Lett.,* 175, 59, 1990.
408. **Hemminger, J. C., Carr, R., Lo, W. J., and Somorjai, G.,** *Adv. Chem. Ser.,* 194, 233, 1980.
409. **Goucher, G. M., Parson, C. A., and Harris, C. B.,** *J. Phys. Chem.,* 88, 4200, 1984.
410. **Wood, B. J. and Wise, H.,** *J. Phys. Chem.,* 65, 1976, 1961.
411. **Melin, G. A. and Madix, R. J.,** *Trans. Faraday Soc.,* 67, 2711, 1971.
412. **Dickens, P. G., Linnett, J. W., and Palczweska, W.,** *J. Catal.,* 4, 141, 1965.
413. **Kisljuk, M. U., Tretakov, I. I., and Korchak, V. N.,** *Kinet. Katal.,* 17, 963, 1976.

414. **Comsa, G., David, R., and Schumacher, B. J.,** *Surf. Sci.,* 95, 135, 1985.
415. **Lin, T. N. and Somorjai, G. A.,** *J. Chem. Phys.,* 87, 135, 1985.
416. **Eenshuistra, P. J., Bonnie, J. H. M., Los, J., and Hopman, H. J.,** *Phys. Rev. Lett.,* 60, 377, 1988.
417. **Kratzer, P. and Brenig, W.,** *Surf. Sci.,* 254, 275, 1991.
418. **Kolasinski, K. W., Shane, S. F., and Zare, R. H.,** *J. Chem. Phys.,* 95, 5482, 1991.
419. **Schröter, L., David, R., and Zacharias, H.,** *J. Vac. Sci. Technol.,* A9, 1712, 1991.
420. **Halpern, B. L. and Rosner, D. E.,** *J. Chem. Soc. Faraday Trans.,* 74, 1883, 1987.
421. **Thorman, B. P., Andersson, D., and Bernasek, L.,** *J. Chem. Phys.,* 18, 6498, 1981.
422. **Ogryzlo, E. A. and Pearson, A. E.,** *J. Phys. Chem.,* 72, 2913, 1968.
423. **Hall, R., Cade, Z. I., Laudran, M., Pichon, F., and Shermann, P.,** *Phys. Rev. Lett.,* 60, 377, 1987.
424. **Mannella, G. G. and Harteck, P.,** *J. Chem. Phys.,* 34, 2177, 1961.
425. **Weinraub, B. P. and Mannella, G. G.,** *J. Chem. Phys.,* 34, 2172, 1961.
426. **Haller, G. L. and Coulston. G. W.,** in *Catalysis, Science and Technology,* Vol. 9, Anderson, J. P. and Boudart, M., Eds., Springer-Verlag, Berlin, 1991, 131.
427. **Smith, J. N. and Palmer, R. L.,** *J. Chem. Soc. Faraday Trans.,* 54, 13, 1972.
428. **Smith, J. N., Palmer, R. L., and Vroom, D. A.,** *J. Vac. Sci. Technol.,* 10, 373, 1973.
429. **Comsa, G. and David, R.,** *Chem. Phys. Lett.,* 49, 512, 1977.
430. **Becker, C. A., Cowin, J. P., Wharton, L., and Auerbach, G. J.,** *J. Chem. Phys.,* 67, 3394, 1977.
431. **Segner, J., Campbell, C. T., Doyen, G., and Ertl, G.,** *Surf. Sci.,* 138, 505, 1984.
432. **Brown, L. S. and Sibener, S. J.,** *J. Chem. Phys.,* 90, 2807, 1989.
433. **Matsushima, T. and Asada, H.,** *Chem. Phys. Lett.,* 120, 412, 1985; *J. Chem. Phys.,* 85, 1658, 1986.
434. **Matsushima, T.,** *J. Phys. Chem.,* 91, 6192, 1987.
435. **Matsushima, T. J.,** *J. Chem. Phys.,* 91, 5722, 1990.
436. **Ohno, Y. and Matsushima, T.,** *Surf. Sci.,* 239, L521, 1990.
437. **Matsushima, T.,** *Proc. 8th Int. Congr. on Catalysis,* Weinheim, Basel, 4, 1984, 121.
438. **Matsushima, T. and Ohno, Y.,** *J. Chem. Phys.,* 93, 1464, 1990.
439. **Matsushima, T.,** *Surf. Sci.,* 242, 482, 1991.
440. **Matsushima, T., Ohno, Y., and Nagai, K.,** *J. Chem. Phys.,* 94, 704, 1991.
441. **Mantell, D. A., Ryali, S. B., Halpern, B. L., Haller, G. L., and Fenn, J. R.,** *Chem. Phys. Lett.,* 81, 181, 1981.
442. **Mantell, D. A., Ryali, S. B., Haller, G. L.,** *Chem. Phys. Lett.,* 102, 37, 1983.
443. **Mantell, D. A., Kunimori, K., Ryaali, S. B., Haller, G. L., and Fenn, J. B.,** *Surf. Sci.,* 172, 281, 1986.
444. **Coulston, G. M. and Haller, G. L.,** *J. Chem. Phys.,* 92, 5661, 1990; 95, 6932, 1991.
445. **Brown, L. S. and Bernasek, S. L.,** *J. Chem. Phys.,* 82, 2110, 1985.
446. **Kori, M. and Halpern, B. L.,** *Chem. Phys. Lett.,* 54, 532, 1986.
447. **Ben-Shaul, A., Haas, Y., Kompa, K. K., and Levine, R. D.,** *Laser and Chemical Change,* Vol. 10, Springer-Verlag, Berlin, 1980, 23.
448. **Kori, M. and Halpern, B. L.,** *Chem. Phys. Lett.,* 98, 32, 1983.
449. **Kisljiuk, M. U., Migulin, V. V., Savkin, V. V., Vlasenko, A. G., Skliarov, A. V., and Tretjakov, I. I.,** *Kinet. Katal.,* 31, 1392, 1990.
450. **Ceyer, S. T., Guthrie, W. L., Lin, T. H., and Somorjai, G. A.,** *J. Chem. Phys.,* 78, 6982, 1983.
451. **Hoffbauer, M. A., Hsu, D. S. Y., and Lin, M. L.,** *J. Chem. Phys.,* 532, 1986; *Surf. Sci.,* 184, 25, 1987.
452. **Washington, T., Ljungstrom, S., Rosen, A., and Kasemo, B.,** *Surf. Sci.,* 234, 419, 1990.
453. **Fridell, F., Westblom, U., Aldon, M., and Rosen, A.,** *J. Catal.,* 128, 92, 1991.
454. **Hsu, D. S. Y., Squire, D. W., and Lin, M. C.,** *J. Chem. Phys.,* 89, 2861, 1988.

455. **Arnold, G. S. and Coleman, D. J.,** *Chem. Phys. Lett.*, 177, 279, 1991.
456. **Ardebild, M. H., Grice, R., Hudges, C. J., and Whitehead, J.,** *J. Chem. Soc. Faraday Trans.*, 88, 399, 1992.
457. **Guillory, J. P. and Shiblom, C. M.,** *J. Catal.*, 54, 24, 1978.
458. **Savkin, V. V., Bakuleva, T. N., Boeva, O. A., Vlasenko, A. G., Kisliuk, M. U., Sizova, N. N., and Skliarov, A. U.,** *Kinet. Katal.*, 30, 1455, 1989.
459. **Mödl, A., Budde, F., Gritsch, T., Chuang, T. J., and Ertl, G.,** *J. Vac. Sci. Technol.*, A5, 522, 1987.
460. **Novakovski, L. V. and Hsu, D. S. Y.,** *J. Chem. Phys.*, 92, 1999, 1990.
461. **Sawin, H. H. and Merrill, R. P.,** *J. Chem. Phys.*, 73, 996, 1980.
462. **Wood, B. J. and Wise, H.,** *J. Catal.*, 39, 471, 1975.
463. **Kiela, J. B., Halpern, B. L., and Rosner, D. E.,** *J. Chem. Phys.*, 88, 4522, 1984.
464. **Foner, S. N. and Hudson, R. L.,** *J. Vac. Sci. Technol.*, A1, 1261, 1983; *J. Chem. Phys.*, 80, 518, 1984.
465. **Krylov, O. V.,** *Dokl. Akad. Nauk SSSR*, 130, 1063, 1960.
466. **Vajo, J. J., Tsai, W., and Weinberg, W. H.,** *J. Phys. Chem.*, 90, 6531, 1986.
467. **Harris, J. and Kasemo, B.,** *Surf. Sci.*, 105, L282, 1981.
468. **Kozhushner, M. A. and Shub, B. R.,** *Probl. Kinet. Katal.*, 17, 11, 1978.
469. **Ovchinnikov, A. A.,** *Zh. Exsp. Tear. Fiz.*, 57, 263, 1969.
470. **Tretjakov, I. I., Shub, B. R., and Skljarov, A. A.,** *Probl. Kinet. Katal.*, 15, 131, 1973.
471. **Nitzan, A., Mukamel, S., and Jortner, J.,** *Mol. Phys.*, 25, 716, 1973.
472. **Nitzan, A., Mukamel, S., and Jortner, J.,** *J. Chem. Phys.*, 60, 3920, 1974.
473. **Nitzan, A., Mukamel, S., and Jortner, J.,** *J. Chem. Phys.*, 63, 200, 1975.
474. **Sune, H. Y. and Rice, S. A.,** *J. Chem. Phys.*, 42, 3826, 1965.
475. **Tully, J. C.,** *J. Electron Spectrosc.*, 54/55, 1, 1991.
476. **Tobin, H.,** *Surf. Sci.*, 183, 226, 1987.
477. **Persson, B. N.,** *J. Phys. Chem.*, 17, 4741, 1984.
478. **Persson, B. N., Schumacher, P., and Otto, A.,** *Chem. Phys. Lett.*, 178, 204, 1991.
479. **Shafranovsky, P. A., Prostnev, A. S., Kuzin, L. A., Zhizhin, G. N., and Shub, B. R.,** *Dokl. Akad. Nauk SSSR*, 311, 87, 1990.
480. **Heilweil, E. J., Casassa, M. P., Cavanagh, R. R., and Stephenson, J. C.,** *J. Chem. Phys.*, 81, 2856, 1984; *J. Vac. Sci. Technol.*, A3, 1655, 1985.
481. **Misevich, J., Houston, R. L., and Merrill, R. P.,** *J. Chem. Phys.*, 82, 2577, 5216, 1985;. 84, 2361, 1986.
482. **Heilweil, E. J., Cavanagh, R. R., Casassa, M. P., and Stephenson, J. C.,** *Chem. Phys. Lett.*, 117, 185, 1985; 129, 48, 1986. *J. Vac. Sci. Technol.*, A3, 1665, 1985.
483. **Cavanagh, R. R., Casassa, M. P., Heilweil, E. J., and Stephenson, J. C.,** *J. Vac. Sci. Technol.*, A5, 469, 1987.
484. **Morin, M., Levinos, N. J., and Harris, A. L.,** *J. Chem. Phys.*, 96, 3950, 1992.
485. **Cavanagh, R. R., Beckerle, J. D., Casassa, M. P., Heilweil, E. J., and Stephenson, J. C.,** *Surf. Sci.*, 269, 113, 1992.
486. **Heilweil, E. J., Stephenson, J. C., and Cavanagh, R. R.,** *J. Phys. Chem.*, 92, 6099, 1988; *J. Chem. Phys.*, 88, 5251, 1988.
487. **Beckerle, J. D., Cavanagh, R. R., Casassa, M. P., Heilweil, E. J., and Stephenson, J. C.,** *J. Chem. Phys.*, 95, 5401, 1991.
488. **Harris, A. L., Rothberg, L., Dhar, L., Levinos, N. J., and Dubois, L. H.,** *J. Chem. Phys.*, 94, 2338, 1991.
489. **Harris, A. L., Levinos, N. J., Rothberg, L., Dubois, L. H., Dhar, L., Shane, S. F., and Morin, M.,** *J. Electron Spectrosc.*, 54/55, 9, 1990.
490. **Guyot-Sionnest, P., Dumas, P., Chabal, Y. L., and Higashi, G. S.,** *Phys. Rev. Lett.*, 64, 2156, 1990.
491. **Chang, H. C. and Ewing, G. E.,** *Phys. Rev. Lett.*, 65, 2125, 1990; *J. Phys. Chem.*, 94, 7635, 1990.

492. **Vach, H.,** *Chem. Phys. Lett.*, 166, 1, 1990.
493. **Avouris, P. and Perssun, B. N.,** *J. Phys. Chem.*, 88, 837, 1984.
494. **Dreshage, K. N.,** *Ber. Bunsenges. Phys. Chem.*, 72, 329, 1968; 73, 1179, 1969.
495. **Campion, A., Gallo, A. R., Harris, C. B., Robota, H. J., and Whitmore, P. N.,** *Chem. Phys. Lett.*, 73, 447, 1980.
496. **Wokaun, A., Luntz, H. P., King, A. P., Wild, U. P., and Ernst, R. R.,** *J. Chem. Phys.* 79, 509, 1983.
497. **Anfiniad, P. A., Cansgrove, T. P., and Strave, W. V.,** *J. Phys. Chem.*, 90, 5882, 1986.
498. **Persson, B. N. J. and Avouris, P.,** *J. Chem. Phys.*, 79, 5157, 1986.
499. **Waldeck, D. M., Alivisatos, A. P., and Harris, C. B.,** *Surf. Sci.*, 159, 109, 1985.
500. **Thomas, J. K.,** *J. Phys. Chem.*, 91, 267, 1987.
501. **McClellan, M. R., McFeely, F. R., and Gland, J. L.,** *Surf. Sci.*, 129, 188, 1983.
502. **Schmidt, L. D.,** *Phys. Basis Heterogeneous Catalysis Mater. Sci. Colloq.*, Batelle Inst., New York, 1975, 451.
503. **Chambers, R. S. and Ehrlich, G.,** *Surf. Sci.*, 126, L325, 1987.
504. **Norton, P. R., Tapping, R. L., and Goodale, J. W.,** *Surf. Sci.*, 72, 33, 1977.
505. **Schmeisser, D., Jacobi, K., and Kolb, P. M.,** *Appl. Surf. Sci.*, 11/12, 364, 1982.
506. **Grunze, M.,** *Surf. Sci.*, 129, 109, 1984.
507. **Tully, J. C.,** *Faraday Discuss.*, 80, 291, 1985.
508. **Steinbruchel, C. S. and Schmidt, L. D.,** *Phys. Rev. Lett.*, 32, 594, 1974; *Phys. Rev.*, B10, 4209, 1974.
509. **Rettner, C. T., Schweizer, E. K., and Stein, H.,** *J. Chem. Phys.*, 93, 1442, 1990.
510. **Harris, J., Kasemo, B., and Tornquist, E.,** *Surf. Sci.*, 105, L281, 1981.
511. **Cassuto, A. and King, D. A.,** *Surf. Sci.*, 112, 325, 1981.
512. **Boheim, J., Breig, W., and Leuthauser, H.,** *Surf. Sci.*, 131, 258, 1983.
513. **Doren, D. J. and Tully, J. C.,** *J. Chem. Phys.*, 94, 8428, 1990.
514. **Ehrlich, G.,** *J. Phys. Chem. Solids*, 5, 47, 1958.
515. **Kisliuk, P.,** *J. Phys. Chem. Solids*, 3, 95, 1957; 5, 78, 1958.
516. **Adams, J. E. and Doll, J. D.,** *Surf. Sci.*, 103, 472, 1981.
517. **Alnot, M. and Cassuto, A.,** *Surf. Sci.*, 112, 325, 1981.
518. **Monroe, D. R. and Merrill, R. P.,** *J. Catal.*, 65, 461, 1980.
519. **Gland, J. L. and Korchak, V. N.,** *J. Catal.*, 53, 9, 1978.
520. **Kozhushner, M. A., Prostnev, A. S., and Shub, B. R.,** *Dokl. Akad. Nauk SSSR*, 279, 1401, 1984.
521. **Prostnev, A. S., Kozhushner, M. A., and Shub, B. R.,** *Khim. Fiz.*, 5, 85, 1986.
522. **Prostnev, A. S.,** *Theoretical Aspects of Surface Migration and Chemical Reactions of Adsorbed Atoms at Small Coverages*, Ph.D. thesis, Institute of Chemical Physics, 1986 (in Russian).
523. **Müller, E. M.,** *Z. Elektrochem.*, 61, 43, 1957.
524. **Gomer, R., Wortman, R., and Lundy, R.,** *J. Chem. Phys.*, 26, 1147, 1957.
525. **Wortman, R., Gomer, R., and Lundy, R.,** *J. Chem. Phys.*, 27, 1099, 1957.
526. **Lewis, R. and Gomer, R.,** *Surf. Sci.*, 17, 333, 9, 1969.
527. **Gomer, R. and Lewis, R.,** *Surf. Sci.*, 12, 157, 1968.
528. **Ehrlich, G.,** *Crit. Rev. Solid State Mater. Sci.*, 10, 391, 1982.
529. **Gomer, R.,** *Surf. Sci.*, 38, 373, 1973.
530. **Mazenko, G., Banavar, J. R., and Gomer, R.,** *Surf. Sci.*, 107, 459, 1981.
531. **Gomer, R.,** *Aspects and Dynamics Surfaces Reaction Workshop*, New York, 1986, 207.
532. **Chen, J.-R. and Gomer, R.,** *Surf. Sci.*, 79, 413, 1979.
533. **Chen, J.-R. and Gomer, R.,** *Surf. Sci.*, 81, 589, 1979.
534. **Di Foggoli, R. and Gomer, R.,** *Phys. Rev.*, B25, 3490, 1982.
535. **Gonchar, V. V., Kagan, Ju. M., and Kanash, O. V.,** *Zh. Eksp. Tekh. Fiz.*, 84, 249, 1983.
536. **Braun, O. M. and Pashatsky, E. A.,** *Poverchnost*, N7, 49, 1984.

537. **Ehrlich, G. and Hudda, F. G.,** *J. Chem. Phys.*, 44, 1039, 1966.
538. **Kisljuk, M. U., Tretjakov, I. I., and Nartikoev, R. K.,** *Kinet. Katal.*, 23, 1191, 1982.
539. **Bassett, D. W.,** *NATO ASI Ser. B,* 86, 63, 1983.
540. **Tsong, T. T.,** *NATO ASI Ser. B,* 86, 109, 1983.
541. **Metiu, H., Kitahara, K., Silbey, R., and Ross, J.,** *Chem. Phys. Lett.*, 43, 189, 1976.
542. **Kitahara, K.,** *Surf. Sci.*, 75, 383, 1978.
543. **Efrima, S. and Metiu, H.,** *J. Chem. Phys.*, 65, 2871, 1976.
544. **Efrima, S. and Metiu, H.,** *J. Chem. Phys.*, 69, 2286, 1978.
545. **Banavar, J. R., Cohen, M. H., and Gomer, R.,** *Surf. Sci.*, 107, 113, 1981.
546. **Zhdanov, V. P.,** *Elementary Physical and Chemical Processes at Surface,* Edition of Siberian Division of Acad. Sci. USSR, Novosibirsk, 1988, 319 (in Russian).
547. **Kramers, H. A.,** *Physica,* 7, 284, 1940.
548. **Tully, J. C., Gilmer, C. H., and Shugart, M.,** *J. Chem. Phys.*, 71, 1630, 1979.
549. **Mrusik, M. R. and Pound, G. M.,** *J. Phys. Chem.*, 11, 1403, 1981.
550. **Doll, J. D. and McDowell, H. K.,** *J. Chem. Phys.*, 77, 479, 1982.
551. **Doll, J. D., McDowell, H. K., and Valou, S. M.,** *J. Chem. Phys.*, 78, 5276, 1983.
552. **Kozhushner, M. A., Prostnev, A. A., Rozovsky, M. O., and Schub, B. R.,** *Phys. Status Solidi,* 136, 557, 1986.
553. **Prostnev, A. S. and Shub, B. R.,** *Kinet. Katal.*, 28, 1402, 1987.
554. **Zhdanov, V. P.,** *Dokl. Akad. Nauk SSSR,* 254, 392, 1980.
555. **Voyevodsky, V. V., in** *Chemical Kinetics and Chain Reactions,* Kondratiev, V. N., Ed., Nauka, Moscow, 1966, 214.
556. **Spiridonov, K. N., Krylov, O. V., and Gati, G.,** *Probl. Kinet. Katal.*, 17, 149, 1978.
557. **Golovina, O. A., Roginsky, S. Z., Sakharov, M. M., and Eidus, Ya. T.,** *Probl. Kinet. Katal.*, 9, 76, 1957 (in Russian).
558. **Bell, A. T.,** *Catal. Rev.*, 23, 203, 1981.
559. **Rozovsky, A. Ja.,** *Kinet. Katal.*, 8, 1143, 1967.
560. **Mole, T., Bett, G., and Seddon, D.,** *J. Catal.*, 84, 43, 1983.
561. **Nijs, H. H. and Jacobs, P. A.,** *J. Catal.*, 66, 401, 1980.
562. **Volodin, Ju. E., Barelko, V. V. and Khal'zov, V. I.,** *Dokl. Akad. Nauk SSSR,* 221, 1114, 1975.
563. **Barelko, V. V. and Merzhanov, A. G.,** *Probl. Kinet. Katal.*, 17, 182, 1978.
564. **Krupka, R. P., Kaplan, H., and Laidler, K. J.,** *Trans. Faraday Soc.*, 62, 2754, 1966.
565. **Boudart, M.,** *Ind. Eng. Chem. Fundam.*, 25, 70, 1986.
566. **Menzel, D.,** in *Kinetics of Interface Reactions,* Grunze, M. and Kreuzer, H. J., Eds., Springer-Verlag, Berlin, 1987, 2.
567. **De Donder, Th.,** *L'Affinite,* Gauthier-Villars, Paris, 1936.
568. **Boudart, M. and Djega-Mariadassou, G.,** *Kinetics of Heterogeneous Catalysis Reactions,* Princeton University Press, Princeton, 1984.
569. **Happel, J.,** *Catal. Rev.*, 6, 221, 1972.
570. **Boudart, M.,** *J. Phys. Chem.*, 87, 2786, 1983.
571. **Temkin, M. I.,** in *Advances in Catalysis,* Ed., Vol. 28, Academic Press, New York, 1979, 173.
572. **Prigogine, I., Guter, P., and Herbo, C.,** *J. Phys. Colloid Chem.*, 52, 321, 1948.
573. **Kreuzer, H. J.,** *J. Chem. Soc. Faraday Trans.*, 86, 1299, 1990.
574. **Kreuzer, H. J. and Payne, S. H.,** *Surf. Sci.*, 231, 213, 1990.
575. **Temkin, M. I. and Pyzhov, V. N.,** *Acta Physicochim. USSR,* 12, 327, 1940.
576. **Kemball, C.,** *Discuss. Faraday Soc.*, N 41, 190, 1966.
577. **Boudart, M.,** *Catal. Lett.*, 3, 111, 1989.
578. **Sinfelt, J. H.,** *J. Phys. Chem.*, 64, 892, 1969.
579. **Hunt, P. M., Hunt, K. L. C., and Ross, J.,** *J. Chem. Phys.*, 92, 2572, 1990.
580. **Hunt, K. L. C., Hunt, P. M., and Ross, J.,** *Annu. Rev. Phys. Chem.*, 41, 409, 1990.
581. **Glansdorf, P. and Prigogine, I.,** *Thermodynamic Theory of Structure, Stability, and Fluctuations,* John Wiley & Sons, New York, 1971.

582. **Shchukarev, S. A.,** *Zh. Russ. Fiz. Khim. Ova.*, 47, 1646, 1915.
583. **Tsitovskaja, I. L., Al'tshuller, O. V., and Krylov, O. V.,** *Dokl. Akad. Nauk SSSR*, 112, 1400, 1973.
584. **Ukharsky, A. A., Slin'ko, M. M., Berman, A. D., and Krylov, O. V.,** *Kinet. Katal.*, 22, 1353, 1981.
585. **Slin'ko, M. M. and Slin'ko, M. G.,** *Usp. Khim.*, 49, 561, 1980; *Kinet. Katal.*, 23, 1421, 1982.
586. **Ertl, G.,** *Adv. Catal.*, 37, 1, 1991.
587. **Ertl, G., Norton., P. R., and Rustig, J.,** *Phys. Rev. Lett.*, 49, 171, 1982.
588. **Thiel, P. A., Behm, R. J., Norton, P. R., and Ertl, G.,** *Surf. Sci.*, 121, L533, 1982.
589. **Hösler, W., Ritter, E., and Behm, R. J.,** *Ber. Bunsenges. Phys. Chem.*, 90, 205, 1986.
590. **Moller, R., Wetzl, L., Eiswirth, M., and Ertl, G.,** *J. Chem. Phys.*, 85, 5328, 1986.
591. **Eiswirth, M. and Ertl, G.,** *Surf. Sci.*, 177, 90, 1986.
592. **Eiswirth, M., Moller, R., Wetzl, R., Imbihl, R., and Ertl, G.,** *J. Chem. Phys.*, 90, 510, 1989.
593. **Krischer, K., Eiswirth, M., and Ertl, G.,** *Surface Sci.*, 201/202, 900, 1991.
594. **Vishnevski, A. L. and Elokhin, V. I.,** in *Unsteady State Processes in Catalysis (Proc. Int. Symp.)*, Matros, Y. Sh., Ed., USPC, Utrecht, 1990, 203.
595. **Ehsasl, M., Block, J., Christman, K., and Hirschwald, W.,** *J. Vac. Sci. Technol.*, A5, 801, 1987.
596. **Ehsasi, M., Rezaie-Serej, S., Block, J. H., and Christman, K.,** *J. Chem. Phys.*, 92, 1596, 1990.
597. **Sander, M., Imbihl, R., and Ertl, G.,** *J. Chem. Phys.*, 95, 6162, 1991.
598. **Basset, M. R., Aris, R., and Imbihl, R.,** *J. Chem. Phys.*, 93, 811, 1990.
599. **Schmidt, L. D. and Aris, R.,** in *Unsteady State Processes in Catalysis (Proc. Int. Symp.)*, Matros Yu. Sh., Ed., USPC, Utrecht, 1990, 200.
600. **Hwang, S. J. and Schmidt, L. D.,** *J. Catal.*, 114, 230, 1988.
601. **Fink, Th., Dath, J.-P., Imbihl, R., and Ertl, G.,** *Surf. Sci.*, 251/252, 983, 1991.
602. **Vayenas, C. G., Georgakis, C., Michaels, J., and Thormo, J.,** *J. Catal.*, 67, 348, 1981.
603. **Vayenas, C. G. and Michaels, J. N.,** *Surf. Sci.*, 120, L40, 1982.
604. **Berry, R. J.,** *Surf. Sci.*, 120, L405, 1975.
605. **Amariglio, A., Benali, O., and Amariglio, H.,** *J. Catal.*, 118, 164, 1989.
606. **Kadushin, A. A., Matyshak, V. A., and Krylov, O. V.,** *Izv. Akad. Nauk SSSR Ser. Khim.*, N 5, 911, 1979.
607. **Jaeger, N. J., Moller, K., and Plath, P. J.,** *J. Chem. Soc. Faraday Trans.*, 82, 3315, 1986.
608. **Slin'ko, M. M., Jaeger, N. I., and Svensson, P.,** *J. Catal.*, 118, 164, 1989.
609. **Scheintuch, M.,** *J. Catal.*, 96, 326, 1985.
610. **Eng, D., Stockides, M., and McNally, T.,** *J. Catal.*, 106, 343, 1987.
611. **Ehsasi, M., Frank, C., Block, J. H., and Christmann, K.,** *Chem. Phys. Lett.*, 165, 111, 1990.
612. **Einswirth, M., Krischer, K., and Ertl, G.,** *Surf. Sci.*, 202, 565, 1981.
613. **Barelko, V. V. and Merzhanov, A. G.,** *Probl. Kinet. Katal.*, 17, 182, 1978.
614. **Barelko, V. V., Kurochka, I. I., and Merzhanov, I. G.,** *Dokl. Akad. Nauk SSSR*, 229, 898, 1976.
615. **Cox, M. P., Ertl, G., and Imbihl, R.,** *Phys. Rev. Lett.*, 54, 1725, 1985.
616. **Rotermund, H. H., Jakubith, S., von Gertzen, A., and Ertl, G.,** *J. Chem. Phys.*, 91, 4942, 1990.
617. **Falta, J., Imbihl, R., and Henzler, M.,** *Phys. Rev. Lett.*, 69, 1409, 1990.
618. **Davis, W.,** *Philos. Mag.*, 17, 233, 1934; 19, 309, 1935.
619. **Boreskov, G. K., Slin'ko, M. G., and Filippova, A. G.,** *Dokl. Akad. Nauk SSSR*, 92, 353, 1953.

620. **Slin'ko, M. G., Beskov, V. S., and Dubjaga, N. A.,** *Dokl. Akad. Nauk SSSR,* 204, 1174, 1972.
621. **Slin'ko, M. G. and Jablonsky, G. S.,** *Probl. Kinet. Katal.,* 17, 154, 1978.
622. **Jablonsky, G. S., Bykov, V. I., and Elokhin, V. I.,** *Kinetics of Selected Reactions in Heterogeneous Catalysis,* Nauka, Novosibirsk, 1984 (in Russian).
623. **Labastida-Bardalas, J. R., Hodgins, R. R., and Silveston, P. L.,** *Can. J. Chem. Eng.,* 67, 418, 1989.
624. **Chuang, G. K., Jaenicke, S., and Lee, J. Y.,** *Appl. Catal.,* 72, 51, 1991.
625. **Madix, R. J., Falconer, J., and McCarty, J.,** *J. Catal.,* 31, 320, 1974.
626. **Falconer, J., McCarty, J., and Madix, R. J.,** *Surf. Sci.,* 42, 329, 1974.
627. **Mars, P. and Van Krevelen, D. V.,** *Chem. Eng. Sci. Suppl.,* 3, 41, 1954.
628. **Langmuir, I.,** *J. Am. Chem. Soc.,* 38, 2221, 1916.
629. **Wagner, C. and Hauffe, K.,** *Z. Elektrochem.,* 44, 172, 1938.
630. **Boreskov, G. K.,** *Zh. Fiz. Khim.,* 14, 1337, 1940.
631. **Bruns, B. P.,** *Zh. Fiz. Khim.,* 21, 1011, 1947.
632. **Sobyanin, V. A.,** *React. Kinet. Catal. Lett.,* 25, 153, 1984.
633. **Wagner, C.,** *Ber. Bunsenges. Phys. Chem.,* 74, 401, 1970.
634. **Greger, M., Ihme, B., Kotter, M., and Rieckert, L.,** *Ber. Bunsenges. Phys. Chem.,* 88, 427, 1984.
635. **Greger, M. and Rieckert, L.,** *Ber. Bunsenges. Phys. Chem.,* 91, 1007, 1987.
636. **Rozentuller, B. V., Spiridonov, K. N., and Krylov, O. V.,** *Kinet. Katal.,* 22, 797, 1981; *Dokl. Akad. Nauk SSSR,* 259, 893, 1981.
637. **Makarova, M. A., Rozentuller, B. V., and Krylov, O. V.,** *Kinet. Katal.,* 28, 1143, 1395, 1987; 29, 872, 876, 1988.
638. **Krylov, O. V.,** *Kinet. Katal.,* 28, 61, 1987.
639. **Hodnett, B. K. and Delmon, B.,** *Appl. Catal.,* 15, 141, 1985.
640. **Najbar, B.,** *Abstr. 10th Int. Symp. Reactivity of Solids, Dijon,* 1984, 70.
641. **Cunningham, S. L., Ho, W., Weinberg, W. H., and Dobrzinski, L.,** *Appl. Surf. Sci.,* 1, 33, 1978.
642. **Volkenstein, F. F.,** *Adv. Catal.,* 12, 189, 1960.
643. **Jacobi, K., Zwieker, G., and Gutmann, A.,** *Surf. Sci.,* 141, 189, 1984.
644. **Stone, F. S.,** *An. R. Soc. Esp. Fis. Quim.,* 61, 109, 1965.
645. **Henrich, V. E., Kurtz, R. L., and Saleghi, H. R.,** *J. Vac. Sci. Technol.,* A1, 1074, 1983; *Rep. Prog. Phys.,* 48, 1481, 1989.
646. **Lad, R. J. and Henrich, V. E.,** *Phys. Rev.,* B 38, 10860, 1988; B39, 13478, 1989.
647. **Boreskov, G. K.,** *Catalysis. The Problems of Theory and Practice,* Nauka, Novosibirsk, 1987 (in Russian).
648. **Forzatti, P., Villa, P. L., Ferlazzo, N., and Jones, D.,** *J. Catal.,* 76, 188, 1982.
649. **Somorjai, G. A. and Van Hove, M. A.,** *Adsorbed Monolayers on Solid Surface,* Springer-Verlag, Berlin, 1979.
650. **Ertl, G. and Küppers, J.,** in *Low Energy Electrons and Surface Chemistry,* VCH Verlag, Weinheim, Germany, 1985, 374.
651. **Conrad, H., Ertl, G., and Küppers, J.,** *Surf. Sci.,* 76, 323, 1978.
652. **Engel, M. and Ertl, G.,** *Adv. Catal.,* 28, 1, 1979.
653. **Ritter, F., Behm, H., Potschke, G., and Winterlin, J.,** *Surf. Sci.,* 181, 403, 1987.
654. **Chang, S. L. and Thiel, P. A.,** *Phys. Rev. Lett.,* 59, 296, 1987.
655. **Berman, A. D. and Krylov, O. V.,** *Probl. Kinet. Katal.,* 17, 102, 1978.
656. **Berman, A. D. and Krylov, O. V.,** *Dokl. Akad. Nauk SSSR,* 227, 122, 1976.
657. **Delmon, B.,** *Introduction a la Cinetique Heterogene,* Technip., Paris, 1969.
658. **Heicklen, J. and Luria, M.,** *Int. J. Chem. Kinet.,* 7, 567, 641, 1975.
659. **Homer, J. B. and Prothero, A.,** *J. Chem. Soc. Faraday Trans.,* 69, 673, 1973.
660. **Berman, A. D. and Krylov, O. V.,** *Kinet. Katal.,* 27, 1019, 1986.
661. **Lowe, A. T. and Eyring, H.,** *J. Solid State Chem.,* 14, 383, 1975.

662. **Porter, S. K.,** *J. Chem. Soc. Faraday Trans.*, 79, 2043, 1983.
663. **Martin, M. and Schmalzried, H.,** *Ber. Bunsenqes. Phys. Chem.*, 89, 124, 1985.
664. **Vedula, Yn. S., Loburets, A. T., Lyuksyutov, I. F., Naumovets, A. G., and Poplavsky, V. V.,** *Kinet. Katal.*, 2, 31, 315, 1990.
665. **Ziff, R. M., Gulari, E., and Barshad, J.,** *Phys. Rev. Lett.*, 56, 2553, 1986.
666. **Meakin, P. and Scalapino, D. J.,** *J. Chem. Phys.*, 87, 731, 1988.
667. **Fischer, P. and Titulaer, U. M.,** *Surf. Sci.*, 221, 409, 1989.
668. **Dumont, M., Dufour, P., Sento, B., and Dagonnier, R.,** *J. Catal.*, 125, 55, 1990.
669. **Mai, J., von Niessen, W., and Blumen, A.,** *J. Chem. Phys.*, 93, 3685, 1991.
670. **Bagnoli, F., Sento, B., Dumont, M., and Dagonnier, R.,** *J. Chem. Phys.*, 94, 777, 1991.
671. **Ehsasi, M., Matloch, M., Frank, O., Block, J. H., Christmann, K., Rys, F., and Hirschwald, W.,** *J. Chem. Phys.*, 91, 4949, 1990.
672. **Heinz, K.,** in *Kinetics of Interface Reaction (Proc. Workshop)* Grunze, M. and Kreuzer, H. J., Eds., Springer-Verlag, Berlin, 1986, 202.
673. **Suhl, R.,** *Phys. Rev.*, B11, 2011, 1975.
674. **Delmon, B. and Roman, A.,** *Trans. Faraday Soc.*, 67, 971, 1971.
675. **Kiselev, V. F., Levshin, N. L., and Poroikov, S. Yn.,** *Dokl. Akad. Nauk SSSR*, 317, 1408, 1991.
676. **Boreskov, G. K., Savchenko, V. I., Dadayan, K. A., Ivanov, V. P., and Bulgakov, N. M.,** *Probl. Kinet. Katal.*, 17, 115, 1978.
677. **Ilchenko, N. I., Juza, V. A., and Roiter, V. A.,** *Dokl. Akad. Nauk SSSR*, 172, 133, 1967.
678. **Bychkov, V. Ju., Sinev, M. Ju., Kuznetsov, B. N., Aptekar, E. L., Korchak, V. N., and Krylov, O. V.,** *Kinet. Katal.*, 28, 665, 1987.
679. **Vinetsky, V. A. and Kholodar, G. A.,** *Static Interaction of Electrons with Semiconductor Defects*, Naukova Dumka, Kiev, 1969 (in Russian).
680. **Kiselev, V. F., Kozlov, S. N., and Levschin, N. L.,** *Phys. Status Solidi*, 66, 93, 1981.
681. **Golovanova, G. F., Kiselev, V. F., Silaev, E. L., and Stepanova, T. S.,** *Kinet. Katal.*, 24, 1177, 1983.
682. **Il'ichev, A. N. and Rufov, Ju. N.,** *Kinet. Katal.*, 20, 437, 1979.
683. **Sakun, V. P., Rufov, Ju. N., Aleksandrov, I. V., Vladimirova, V. I., and Il'ichev, A. N.,** *Kinet. Katal.*, 20, 441, 1979.
684. **Roginsky, S. Z. and Rufov, Ju. N.,** *Kinet. Katal.*, 11, 383, 1970.
685. **Eremenko, A. M.,** *Zh. Fiz. Khim.*, 71, 541, 1981.
686. **Spenser, M. S.,** *J. Catal.*, 67, 259, 1981.
687. **Weisz, P. B.,** *Adv. Catal.*, 13, 137, 1962.
688. **Roginsky, S. Z., Janovsky, M. I., and Gaziev, G. A.,** *Dokl. Akad. Nauk SSSR*, 140, 1125, 1961; *Kinet. Katal.*, 3, 529, 1962.
689. **Roginsky, S. Z., Janovsky, N. I., and Berman, A. D.,** *The Background of Catalytic Applications of Chromatography*, Nauka, Moscow, 1972 (in Russian).
690. **Ponec, V.,** *Faraday Discuss.*, N72, 93, 1981.
691. **Schmalzried, H.,** *Reactivity Solid*, 1, 117, 1986.
692. **Lagun, W. and Schmalzried, H.,** *Oxid. Mater.*, 15, 379, 1981.
693. **Krylov, O. V.,** *Vestn. Akad. Nauk SSSR*, N1, 26, 1983.
694. **Krylov, O. V., Ed.,** *Probl. Kinet. Katal.*, 19, *Partial Oxidation of Organic Compounds*, Nauka, Moscow, 1985 (in Russian).
695. **Krylov, O. V.,** *Physical Chemistry, Modern Problems*, Khimija, Moscow, 1986, 41 (in Russian).
696. **Maksimov, Ju. V., Zurmukhtashvili, M. Sh., Suzdalev, I. P., Margolis, L. Ya., and Krylov, O. V.,** *Kinet. Katal.*, 25, 948, 1984.
697. **Krylov, O. V., Maksimov, Yu. V., and Margolis, L. Yu.,** *J. Catal.*, 25, 948, 1984.
698. **Isaev, O. V., Udalova, O. V., and Krylov, O. V.,** *Kinet. Katal.*, 25, 1016, 1984.

699. **Sleigt, A. W.,** *Science,* 208, 895, 1980.
700. **Forzatti, P., Trifiro, F., and Villa, P. L.,** *J. Catal.,* 55, 52, 1978.
701. **Bart, J. C. J., Giordano, N., and Forzatti, P.,** *Am. Chem. Soc. Symp. Ser. Washington,* 279, 89, 1985.
702. **Cotter, M., Reikert, L., and Weyland, F.,** *Preparation of Catalysis, Proc. 3rd Int. Symp. on Scientific Bases for the Preparation of Heterogeneous Catalysts,* Delmon, B., Ed., Elsevier, Louvain-la-Neuve, 1982.
703. **Cullis, C. F. and Hucknalt, D. J.,** in *Catalysis,* Vol. 5, Kemball, C., Ed., Roy Soc. Chem., London, 1982, 273.
704. **Vejux, A. and Courtine, P.,** *J. Solid State Chem.,* 23, 93, 1978.
705. **Courtine, P.,** *Am. Chem. Symp. Ser. Washington,* N 279, 37, 1985.
706. **Yabrov, A. A., Ismailov, E. G., Boreskov, G. K., Ivanov, A. A., and Anufrienko, V. F.,** *React. Kinet. Catal. Lett.,* 3, 237, 1975.
707. **Cole, D. J., Cullis, C. F., and Hucknail, D. J.,** *J. Chem. Soc. Faraday Trans.,* 71, 2185, 1976.
708. **Bordes, E. and Courtine, P.,** *J. Catal.,* 88, 232, 1984.
709. **Batis, N. H., Batis, H., Chorbel, A., Vedrine, J. C., Volta, J. C.,** *J. Catal.,* 128, 248, 1991.
710. **Ioffe, I. I., Ezhkova, Z. I., and Ljubarsky, G. K.,** *Kinet. Katal.,* 3, 184, 1962.
711. **Berry, F. L. and Brett, V. E.,** *J. Catal.,* 88, 232, 1984.
712. **Krylov, O. V. and Kiselev, V. F.,** *Adsorption and Catalysis on Transition Metals and Their Oxides,* Springer-Verlag, Berlin, 1989.
713. **Wagner, J. B.,** *in Transport in Nonstoichiometric Compounds,* Simkovich, G. and Stubican, V. S., Eds., NATO ASI Ser., Plenum Press, New York, N 129, 1985, 3.
714. **Ruz, P., Zhow, M., Remy, T., Machej, T., Aoun, F., Doumain, B., and Delmon, B.,** *Catal. Today,* 1, 181, 1967.
715. **Delmon, B. and Matralis, M.,** in *Unsteady State Process in Catalysis (Proc. Int. Conf.),* Matros, Yu. Sh., Ed., VSTP, Utrecht, 1990, 29.
716. **DiCosimo, P., Burrington, J. D., and Grasselli, P.,** *J. Catal.,* 102, 1, 1986.
717. **Gryaznov, V. M., Ermilova, M. M., Orekhova, N. V., and Makhota, N. A.,** *Proc. 5th Int. Symp. on Heterogeneous Catalysis, I,* Bulgarian Acad. Sci., Sofia, 1983, 225.
718. **Gryasnov, V. M. and Slinko, M. G.,** *Faraday Discuss.,* N72, 35, 1981.
719. **Neikam, W. C. and Vannice, M. A.,** *J. Catal.,* 20, 210, 1971.
720. **Matyshak, V. A., Slin'ko, M. M., and Kadushin, A. A.,** *Kinet. Katal.,* 26, 1123, 1985.
721. **Wang, C. J. and Eckerdt, J. C.,** *J. Catal.,* 80, 172, 1983.
722. **Bianchi, D., Borcar, S., Teule-Gay, F., and Bennett, C. O.,** *J. Catal.,* 82, 442, 1983.
723. **Rozovsky, A. Ja.,** *Kinet. Katal.,* 21, 97, 1980.
724. **Berry, F. J. and Smith, M. R.,** *J. Chem. Soc. Faraday Trans. I,* 85, 467, 1989.
725. **Butt, J. B.,** *Catal. Lett.,* 7, 61, 83, 1990.
726. **Jung, H. and Thomson, W. J.,** *Catalysis,* 134, 654, 1992.
727. **Morozova, O. S., Maksimov, Yn. V., Shaskin, D. P., Shiryaev, P. A., Matveyev, V. V., Zhorin, V. A., and Krylov, O. V.,** *Appl. Catal.,* 78, 227, 1991.
728. **Colley, S. E., Copperthwaite, R. G., and Hutchings, G. E.,** *Catal. Today,* 9, 203, 1991.
729. **Sachtler, W. M. A.,** *Proc. 8th Int. Congr. on Catalysis,* Vol. 1, Dechema, Weinheim, Germany, 1984, 151.

INDEX

A

N

O

P

Q

R

T